27th International Conference on CADCAM, Robotics and Factories of the Future 2014

IOP Conference Series: Materials Science and Engineering Volume 65

London, United Kingdom
22 – 24 July 2014

ISBN: 978-1-63439-876-3
ISSN: 1757-8981

Printed from e-media with permission by:

Curran Associates, Inc.
57 Morehouse Lane
Red Hook, NY 12571

Some format issues inherent in the e-media version may also appear in this print version.

Printed by Curran Associates, Inc. (2015)

For permission requests, please contact the Institute of Physics
at the address below.

Institute of Physics
Dirac House, Temple Back
Bristol BS1 6BE UK

Phone: 44 1 17 929 7481
Fax: 44 1 17 920 0979

techtracking@iop.org

Additional copies of this publication are available from:

Curran Associates, Inc.
57 Morehouse Lane
Red Hook, NY 12571 USA
Phone: 845-758-0400
Fax: 845-758-2634
Email: curran@proceedings.com
Web: www.proceedings.com

TABLE OF CONTENTS

PRODUCT DEVELOPMENT AND SUSTAINABILITY

MODELLING AND SIMULATION

AUTOMATION, ROBOTICS AND HANDLING SYSTEMS

ADVANCED QUALITY SYSTEMS TOOLS AND QUALITY MANAGEMENT

HUMAN ASPECTS IN ENGINEERING ACTIVITIES

EMERGING SCENARIOS IN ENGINEERING EDUCATION AND TRAINING

EMERGING TECHNOLOGIES IN FACTORIES OF THE FUTURE

POSTERS

IOP Conf. Series: Materials Science and Engineering 65 (2014) 011001 doi:10.1088/1757-899X/65/1/011001

Preface

It is a great pleasure to welcome you to the 27th International Conference on CADCAM, Robotics and Factories of the Future, sponsored by the International Society for Productivity Enhancement, Middlesex University, Festo Limited GB, National Instruments UK & Ireland, the Sector Skills Council for Science, Engineering and Manufacturing Technologies and our proceedings publisher Institute of Physics Publications.

This is the second time Middlesex University has played host to this longstanding international conference, last time being the 12th edition in 1996. The subject content of the conference remains current, focusing on cutting edge developments in research. The conference themes this year are divided into seven themes, Product Development and Sustainability, Modelling and Simulation, Automation, Robotics and Handling Systems, Advanced Quality Systems Tools and Quality Management, Human Aspects in Engineering Activities, Emerging Scenarios in Engineering Education and Training, and Emerging Technologies in Factories of the Future. The conference is organised into seven sessions running in parallel over three days, providing a platform to speakers from 16 different countries. The programme also features four eminent keynote speakers and a hands-on workshop organised by National Instruments.

Organising an event such as this would not be possible without the help of many colleagues. I am grateful to the members of the Organising Committee, the International Scientific Committee, our sponsors and all those colleagues who helped in the review of many abstracts and consequently full papers. This required meticulous attention to detail and strict adherence to very tight deadlines. However large or small a conference is, the effort required to make the local arrangements work for all is not insignificant. The conference organisers acknowledge the particular efforts of Miss Mita Vaghi in providing her expertise in event management and her diligent support and Anete Ashton of IoP Publications for her guidance and help in producing the conference proceedings and online listing. The organisers also recognise the support provided by our sponsors and in particular to Richard Roberts and David Baker from National Instruments, Babak Jahanbani and Phil Holmes from Festo Ltd. Their continued support over the course of the planning period and also during the event itself is very much appreciated.

We are also indebted to all the contributors to the conference, particularly the researchers, and practitioners.

Professor Mehmet Karamanoglu
Conference Chair

IOP Conf. Series: Materials Science and Engineering 65 (2014) 011001 doi:10.1088/1757-899X/65/1/011001

Our sponsors

We are grateful for the generosity of our sponsors:

Middlesex University, London

International Society for Productivity Enhancement

Festo GB and Festo AG & Co. KG

National Instruments UK & Ireland

Sector Skills Council for Science, Engineering and Manufacturing Technologies

IOP Conf. Series: Materials Science and Engineering 65 (2014) 011001 doi:10.1088/1757-899X/65/1/011001

Keynote speakers

Brian Prasad, Ph.D.
Knowledge Architect and LPD
Thought Leader
Control Systems Division
Parker Aerospace
Irvine, Californi
USA

On Knowledge-based Enterprising Strategy for Lean Product Development

Keynote Summary:

Today, more advanced and rich PLM technologies exist to capture valuable product development processes, gather the right knowledge required for the process, sort through all those processes to determine the right process/knowledge to employ, make the right decisions for engineering the product, lay the right product design details, perform a right set of (appropriate) analyses, develop right sizing algorithms, and perform right checks to validate results, etc. If all of these 'right' process knowledge be embedded into a framework of dynamic and reconfigurable Product models (called SMART PRODUCT CONFIGURATOR), supported by a set of reconfigurable dynamic SmartPart templates (in each categories say Specs, Sizing and Geometry), one could achieve remarkable results. Companies could be able to engineer-to-order most products in half the time or in one third of the time, what it takes today. The group of experts in each domain could become the knowledge-keeper for those domain templates. Today, indeed, several leading edge aerospace and automotive companies are moving toward supporting this vision. Today, reusable (and plug and play) knowledge templates are being defined upfront, based either on standardized rules and constraints, or on a standardized components library that adapts to the particulars of a product family, or project, with minimum efforts. This talk will discuss the process used for building those Product Configurators. You will also hear how engineers could use them for 'engineering-to-order' new 'production-ready' products – never done before.

About Keynote Speaker:

Brian Prasad is an award-winning, internationally recognized thought leader in the area of knowledge capture & reuse, knowledge management (KM), knowledge-based engineering (KBE) and KM architecture. As a practicing KBE manager, consultant, author and researcher, he has been actively performing and studying these areas for more than twenty-five years. His latest book is titled Concurrent Engineering Fundamentals – 2 volume set. He has held leadership positions with the NASA Langley, UCI, Caltech, Unigraphics Solutions, Ford Motor, EDS, General Motors. He has consulted with more than 50 organizations in US and abroad.

Dr Prasad's achievements have resulted in recognition as a Fellow of ASME and bibliographic entries in Who's Who of Aerospace, Who's Who in Science and Engineering, and other COE recognitions and OCEC distinguished engineering Merit Award. His entertaining but clear and concise insights make him a sought after speaker, lecturer and consultant. He is the founding Editor-in-Chief of Concurrent Engineering – An International Journal (published by SAGE publishing since 1991: http://cer.sagepub.com/).

IOP Conf. Series: Materials Science and Engineering **65** (2014) 011001 doi:10.1088/1757-899X/65/1/011001

Professor William Wong
Professor of Human-Computer Interaction
School of Science and Technology
Middlesex University
London
UK

What is Visual Analytics and how different disciplines make use of it?

About Keynote Speaker:

Dr B L William Wong is Professor of Human-Computer Interaction and Head, Interaction Design Centre, at Middlesex University's School of Science and Technology, London, UK. Prior to academia, Professor Wong worked in the Republic of Singapore Air Force. In his last appointment he was Head, Systems and Communications Operations Branch, HQ RSAF.

His research interest is in the representation design of information to support decision making in naturalistic environments, from a Cognitive Work Analysis perspective. His research includes air traffic control, hydro-electricity dispatch control, emergency ambulance command and control, intelligence analysis, and visual analytics, emphasising information uptake, decision making and situation awareness in real-time dynamic environments.

He is a recipient of over US$25.3 million in grants, and project coordinator for several US-UK and European Union multi-institution R&D project consortiums. Together with his students and colleagues, he has published over 100 scientific peer reviewed articles. He currently leads an 18-organisation UK–Europe–US R&D consortium funded by the European Commission FP7 Project VALCRI, Visual Analytics for Sense-making in Criminal Intelligence Analysis.

Dr Nina Gaissert
Corporate Bionic Research
Festo AG & Co. KG
Germany
www.festo.com/bionics

The Bionic Learning Network of Festo – Production of the future inspired by nature

Keynote summary:
Whether it's energy efficiency or light-weight construction, function integration or the ability to learn and to communicate, throughout evolution, nature has developed a wealth of optimisation strategies for adapting to its environment, and these strategies can be applied to the world of engineering. That is why Festo set up the Bionic Learning Network in 2006, a research network linking Festo to well-known universities, institutes, development companies and private inventors. Within this network the engineers, designers and researchers are aiming to learn from nature's vast resources of intelligent solutions to provide fresh impulses for technology and industrial applications. Since then more than 40 projects have been perfected. Here we present the most stunning like the AquaJellies, the Bionic Kangaroo or the Bionic Handling Assistant. We will explain what we learned from swarms of jellyfish, kangaroos or elephants about human-machine-interaction and machine-machine-communication and how this knowledge will shape the production of the future.

About Keynote Speaker:
Nina Gaissert is responsible for corporate bionic research at Festo. The Bionic Learning Network of Festo is a cooperation between renowned universities, research institutes, technology companies and Festo itself, a leading world-wide supplier of automation technology and the performance leader in industrial training and education programs.

Within the Bionic Learning Network the engineers, designers and researchers are aiming to get inspired by nature and to learn from nature's vast resources of intelligent solutions to provide fresh impulses for technology and industrial applications. Within this network Nina Gaissert is responsible for corporate bionic research, i.e. providing the biological background for ongoing and future projects and accompanying the projects scientifically.

Prior to joining Festo in 2011 Nina Gaissert studied technical Biology at the University of Stuttgart and the MIT, MA, USA. In 2011 she obtained a PhD of Natural Sciences from the Max Planck Institute for Biological Cybernetics, studying the neural integration of the visual and the haptic senses.

IOP Conf. Series: Materials Science and Engineering **65** (2014) 011001 doi:10.1088/1757-899X/65/1/011001

Professor Janusz Szpytko, C.Prof., D.Sc., Ph.D., M.Sc., B.Sc., C.Eng.
Safety and Reliability, Transport Technology and Telematics
AGH University of Science and Technology
Krakow
Poland

From Industrialization to Re-industrialization and Factory of the Future

Keynote Summary:

Characteristics of the transformation process from traditional societies into industrial, results of urbanization. The impact of industrialization on the environment. Industrialization, its causes and effects: reallocations, outside-resource-using, service society. Information based society. Neo–industrialization based on intelligent networking, circular economy, shared value and promoting competitiveness and sustainable development. Factory of the future: today dream and future new visions. New manufacturing concepts, advanced materials, communication system, human aspects in engineering activities. Whether the industry is needed?

About Keynote Speaker:

Professor dr hab. inz. Janusz Szpytko, AGH University of Science and Technology, Poland. Professor, educator, researcher, manager in education and research. Type of sector: university, R&D and education, mechanical engineering, automation and metrology, transportation and logistics, automation and robotics, manufacturing, quality and control, sensing and diagnostics/decision making, mechatronics. Specialty in Science and Technology: transport systems and devices, safety and reliability, automation and robotics, engineering logistics. Industrial replacements and training in: Poland, Sweden, Germany; professional training and visiting professor: UK, Italy, France, Greece, Germany, Norway, Laos, and others.

Author or co-author of app. 500 publications in science and education, promoter more then 140 students on B.Sc., M.Sc., Ph.D. levels. Member of STSTKT PAN, IFAC, PTD, ISPE, PTB, PSRA, ISA, SITPH, SEFI and others. Co-coordinator and member of several R&D projects both on national and internationals level. President of the SITPH, and member of management boards of various organizations national and international. Organiser and member of several professional scientific and programme committees of international and national conferences and symposiums, member of editorial boards projects and both more than few hundred scientific papers and project reviewer, editor of scientific journals; biographical are published also in: Who is who in America, Who is Who in in Science and Engineering and others.

Professor Szpytko was the organiser of the 21st International Conference on CAD/CAM, Robotics and Factories of the Future, 17–20 July, 2005, Krakow, Poland.

IOP Conf. Series: Materials Science and Engineering **65** (2014) 011001 doi:10.1088/1757-899X/65/1/011001

Conference organisation

PATRON
Professor Michael Driscoll
Vice Chancellor Middlesex University, UK

ADVISORS
Professor Martin Loomes
Dean of School of Science and Technology
Middlesex University, UK

Professor Raj Gill
President of ISPE
(Pro Vice Chancellor, Middlesex University, UK)

Professor Richard Comley
Director of Research
Middlesex University, UK

Professor Chanan Syan
Vice President Conferences (ISPE)
University of the West Indies

IOP Conf. Series: Materials Science and Engineering **65** (2014) 011001 doi:10.1088/1757-899X/65/1/011001

International Scientific Committee 2014

Conference Chair
Professor Mehmet Karamanoglu

Conference Hon Secretary
Dr Rui Loureiro

Conference Hon Treasurer / Finance
Jim McClemont Louise Story

Local Arrangements
Miss Mita Vaghji

Publicity / Website
Miss Shazia Mowlabaccus
Tom Lowenstein
Louis Slabbert

Local Technical Committee
Professor William Wong
Professor Balbir Barn
Professor Anthony White
Professor Martin Smith
Dr Aleksander Zivanovic
Dr Vaibhav Gandhi
Dr Andrew Tizzard
Dr Xin-She Yang
Miss Helena Ambrosio
Dr Zhijun Yang

Peer review statement

All papers published in this volume of *IOP Conference Series: Materials Science and Engineering* have been peer reviewed through processes administered by the proceedings Editors. Reviews were conducted by expert referees to the professional and scientific standards expected of a proceedings journal published by IOP Publishing.

IOP Conf. Series: Materials Science and Engineering **65** (2014) 012001 doi:10.1088/1757-899X/65/1/012001

Design Approach for Spiral Milling Parts Using Knowledge Based Engineering

M Humpa[1] and P Köhler[1]
[1]Institute of Product Engineering, Computer Aided Engineering,
University of Duisburg-Essen, Lotharstr. 1, 47057 Duisburg, Germany,

E-mail: marcin.humpa@uni-due.de

Abstract. From the engineering point of view there are still problems regarding geometry modelling that cannot be satisfactorily solved despite efficient CAD systems. This applies in particular to virtual models when they are meant to match with the resulting manufacturing geometry of components as closely as possible. In this paper, relationships between modelling and manufacturing strategies are discussed. The focus is directed to the relationship between the modelling strategy and the manufacturing process by spiral milling with a ball cutter, whereby especially the influence of the tool geometry and the tool paths for the modelling strategy will be investigated. By means of an appropriate example of a design of spiral mandrel die from the field of plastics processing an approach will show how functional design and manufacturing-oriented design can be coupled and automated.

1. Introduction

In the field of mechanical engineering problems concerning the design of the 3D model occur in relation to manufacturing. The development process of flow-conducting components, which focuses on the dimensioning of the geometry in the field of fluid mechanics, thermal, and structural mechanics goals, is the relevant example. To reach an appropriate dimensioning, analysis and simulation applications are used. The usage of 3D CAD systems, calculation- and simulation-tools has become the common practice in today's enterprises. The Computer Aided Manufacturing (CAM) technologies are also applied. With this technology manufacturing data can be transferred directly from the virtual product model (3D CAD model). To secure the required quality by the virtual model up to the real part not only a consistent usage of a common data basis (3D CAD model) is essential, but also the implementation of manufacturing information and knowledge in the 3D CAD model. For efficient manufacturing the manufacturing strategy has to be implemented as early as possible during the product development process. For this reason it is necessary to find a reasonable compromise between the optimal, functional design and geometry resulting from the specified manufacturing processes.
How can this be realised?
By the use of methods and tools of Knowledge Based Engineering (KBE) both manufacturing process information and manufacturing knowledge can be integrated into the 3D product model (see figure 1).

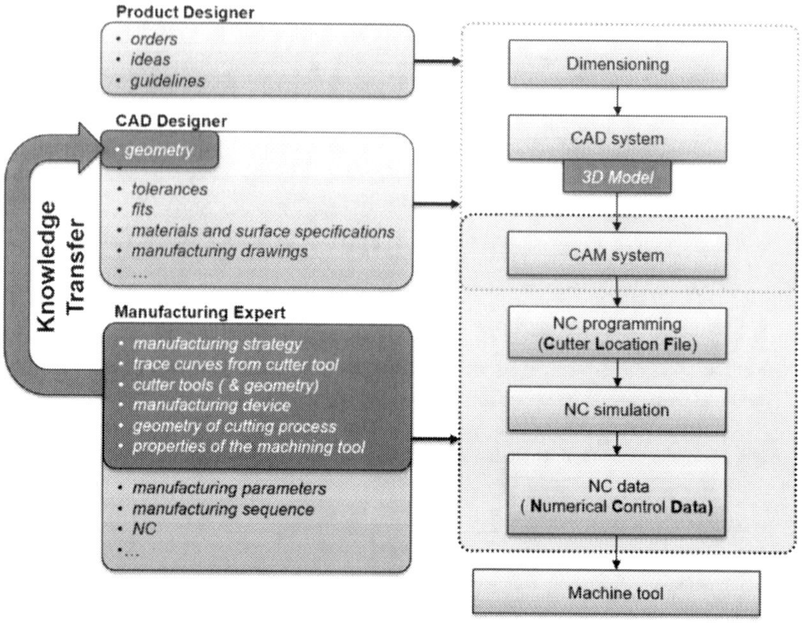

Figure 1. Transfer of the manufacturing knowledge into the design development process

One useful approach in this case is the Feature Technology. This strategy offers the possibility to find a meaningful compromise between design and manufacturing geometry. This will ensure that the CAD geometry is created according to the required manufacturing steps. Proceeding this way can ensure that the simulation results, which are based on the virtual product model, can be transferred to the manufactured part. Regarding this, the present paper propounds a possible strategy which allows the consideration of manufacturing constraints during the development process of a spiral mandrel die for plastics extrusion.

2. Today's development process of spiral mandrel die

Extrusion is a technology for the continuous manufacturing of semi-finished plastic products, such as pipes or tubes. A spiral mandrel die is a shaping extrusion die which transforms the plastic melt (provided by an extruder) into the product cross section. The quality of a spiral mandrel die has a main influence on the product quality indicators, such as wall thickness distribution, surface characteristics, or mechanical properties. Today, the development process of spiral mandrel dies (see figure 2) is mostly based on the use of empirical knowledge. This manual development approach is usually carried out iteratively, which means that the pre-dimensioned geometry proceeding from the first calculations is iteratively changed until the goals (e. g. uniform flow rate at the die outlet) are reached [1]. The quality of the resulting product is analysed according to the properties as criteria for the wall thickness distribution or maximal and minimal shear stress by conducting Computational Fluid Dynamic simulations (CFD) [2, 3]. Current research activities in this field focus on a complete automation of the development process in order to minimise today´s time-consuming, repeatable and manually performed steps [4, 5, 6]. Those available approaches have in common that designed CAD models of the tool have to be transformed into a geometry suitable for manufacturing in each separate step (e. g. milling, turning). Based thereupon it is possible to generate the essential manufacturing documents (drawings, CNC data, etc.). As technical manufacturing restrictions are not usually taken into account during the prior development process further adjustments of the design are usually necessary before the manufacturing data can be generated.

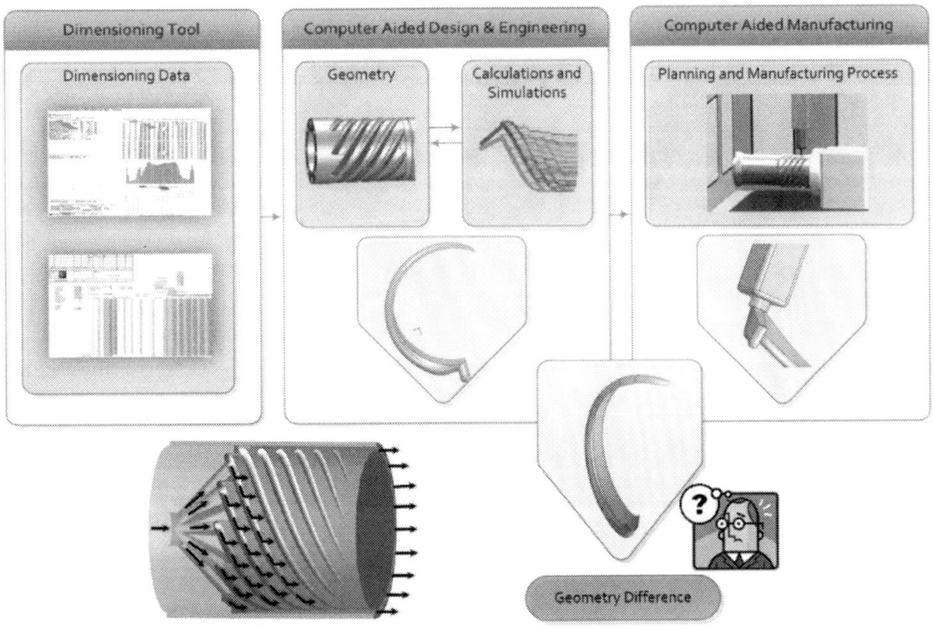

Figure 2. Today's development process of cylindrical spiral mandrel die

The spiral milling on the 4-axis machine can be applied to the manufacturing of helical flow channels of the spiral mandrel die. The spiral milling, which is the result of the helical feed motion, is also used for worm milling, thread milling and thread whirling [7]. The advantages of this manufacturing method mainly lie mainly in the high productivity, low set-up and tool change times and the easy chips disposal [8]. It is easy to imagine that this strategy is characterised by the presence of an interdisciplinary know-how and great wealth of experience on the part of the design engineer, but also by high costs and time consumption.

3. CAD-based Approach

To generate the accurate geometry of the real spiral milling process in the 3D CAD model using the Feature Technique, a feature is required which corresponds to the spatial movement of solid objects and creates the three-dimensional swept volume in the CAD system. However, present-day CAD systems do not deliver feasible solutions for this problem. Hence Multi Body Simulations (MBS) deliver swept envelope as triangulated volumes, which is not satisfying due to data size and data quality. The focus of the developed approach is based on generating swept tool envelope surfaces, which encloses the swept volume of moving tool. This issue has been investigated before from a mathematical point of view [9, 10, 11, 12]. These approaches mostly define the opportunities for analytical or differential geometric surface descriptions of different surface types. However, these methodologies cannot be applied by the CAD system users because the CAD-internal geometry description is barely adaptable. For this reason a CAD-based approach has been developed which allows the automated generation of swept tool envelope surfaces for the component design by application of KBE methods (see figure 3).

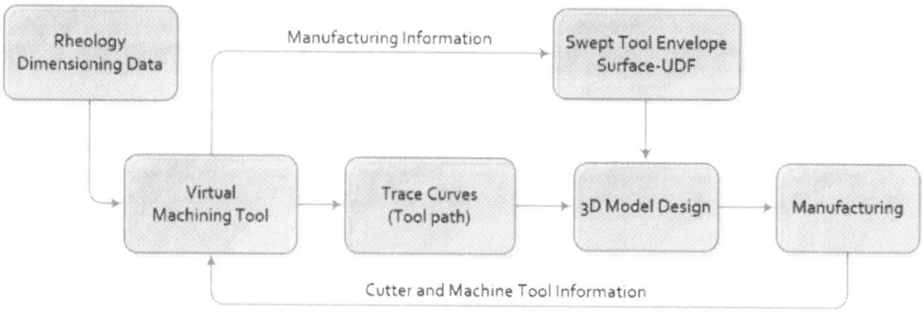

Figure 3. CAD-based coupling approach of the dimensioning, design and manufacturing data

3.1. Virtual Machining Tool and Trace Curves (tool path)

The development process of the spiral mandrel die commences upon dimensioning. Based on the relevant process parameters, the geometry data such as the number and pitch of the spirals, depth profile and width of the spiral channel, the width of the overflow gap and other design details such as radii, chamfers and sealing surfaces can be calculated for the 3D model design of a spiral mandrel die [1]. Here the most important dimensioning design data are the values of the depth profile of the spiral channel. They are represented by the 2D curve, which can be interpreted as a unfold trace curve of milling cutter (tool path). For the design of the spiral channels in the CAD system the 3D curve is necessary, therefore the 2D curve should be transformed by appropriate methods into 3D representation. Furthermore, its suitability and feasibility for acting as the trace curve of a milling cutter (tool path) on the 4-axis machine has to be verified. For the proper analysis of this concern, the knowledge of the spatial arrangement of the milling cutter and the workpiece as well as of the kinematic chain of the machine tool is primarily important. These conditions and restrictions describe the specific form of the milling cutter motion envelope surface which arises from the tool path during the spiral milling process as well as the primary design data of the extrusion-die. This transfer of the "real" tool motions and cutting geometries into the CAD development process leads to improvements of the initial die design. The description of the tool path can be realised through various approaches, e.g. by analytical geometry description methods or by using MBS.

From the mathematical perspective it is a helical curve, which is often referred to as a helix curve. The general equation is shown in figure 4 (a). On condition that the dimensioning data for the depth profile of the spiral channel are provided or described in the form of coordinate tuples or a mathematical or graphical function, the linking of the data to the analytical description of a 3D curve in the CAD system is easy to realise (see figure 4 (b) and (c)). To secure the spatial definiteness for the modelling of the swept tool envelope surface, a second 3D curve is required to be followed by the cutter tool. In case that during the manufacturing process the cutter tool axis is aligned perpendicularly to the workpiece axis, the analytic description of the second cutter tool trace curve has to be supplemented by the distance "*a*". In case that additional rotation by the angle "*α*" is considered, the analytical description has to be completed by component "*b*" (see figure 4 (c)).

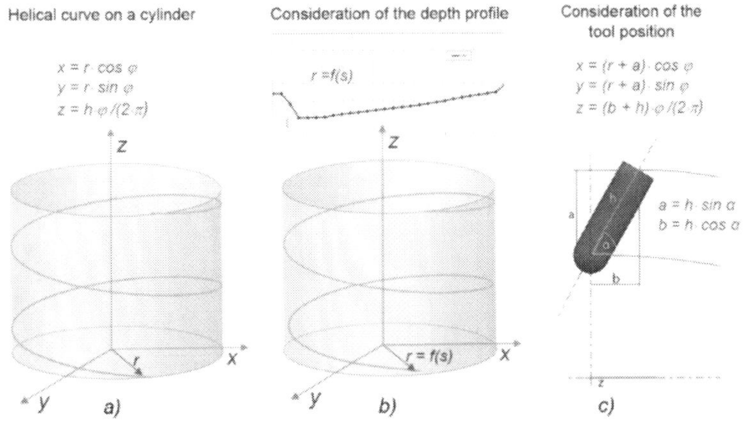

Figure 4. Analytical description of trace curves (tool path)

Another way to generate the "real" and "feasible" tool path directly without mathematical calculations is the use of trace curves of the MBS. These are 3D curves which can be automatically generated by the CAD systems as simulation results by defining characteristic points on the cutter tool faces and axis. These curves are used for suitable surface modelling and the automated generation of the swept milling cutter envelope surfaces in the subsequent steps of the presented approach (see figure 5). In order to unambiguously integrate the dimensioning data into the simulation the position of the tool's characteristic points is significant. For the case presented here the first point is defined in the centre of the ball cutter and the second on the ball cutter axis. By setting a course profile in the form of a 2D curve for height control of the cutter tool or for the machine table, which corresponds to the helix depth profile from dimensioning data, the real tool paths can be automatically generated pursuant to the simulation process.

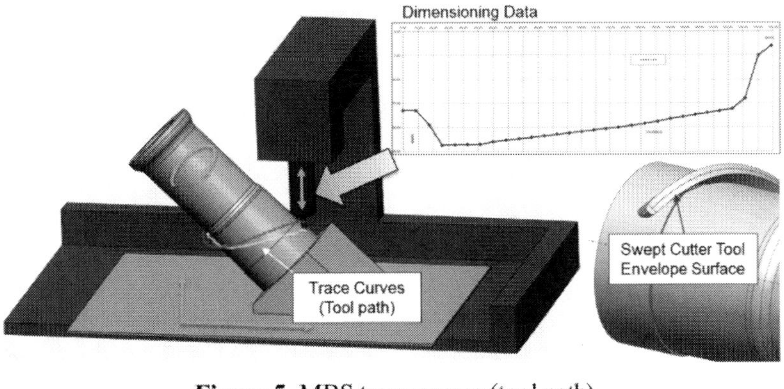

Figure 5. MBS trace curves (tool path)

The difficulty in the creation of the presented *"Virtual Machining Tool"* based on the MBS is the determination and mapping of the real machine kinematics with the help of virtual joints. They do not necessarily stay in 1:1 relation with the real joint connections. However, the use of the kinematic method is a distinct advantage, because the "accuracy" and "feasibility" of the analytically calculated trace curves from cutter tool can be verified in a simple way.

3.2. Swept Tool Envelope Surface-UDF

Generally, trajectory features are used for the geometry design of such spatial characteristics of manufacturing processes in the CAD systems. They allow the sweeping of a 2D section along one or more trajectory curves (also known as guide curves). In contrast to manufacturing operations, such as drilling or pocket milling, the spiral milled channels can be depicted for simple exceptions by one single feature in CAD systems (see figure 6). The presented case refers to the manufacturing process on a 4-axis milling machine with helical feed and ball mill as cutter tool. The rotation axis of the cutter can be arranged perpendicularly or obliquely to the workpiece axis. In both cases geometric deviations were obtained compared to the real manufacturing geometry, which is shown by the control sections.

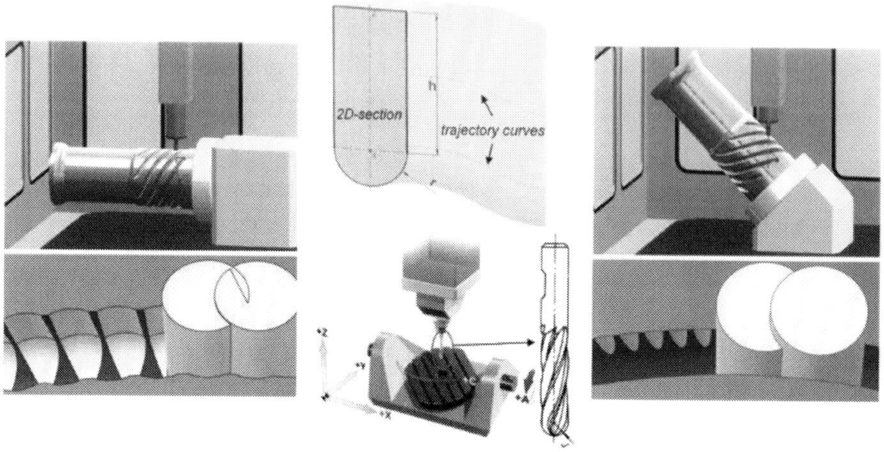

Figure 6. Geometry problems by the design of spiral channels in CAD systems

The cause lies in the fact that depending on how the spatial arrangement of the tool is in relation to the workpiece, a differently complex geometry of the cutting process may result. In order to map a precise manufacturing geometry using a geometry feature in a 3D CAD model it is necessary to provide the possibility of spatial movement of solid objects and creates the tree-dimensional swept volume. This is, however, still very problematic in today's CAD systems or hardly possible. Therefore, another solution approache is needed. Since we are basically dealing with ruled surfaces respectively screw surfaces and their degenerations, a surface-based solution strategy can be applied. For the spiral mandrel die in conjunction with a ball cutter and the spiral milling on a 4-axis machine the ball part of the cutter can be depicted as a swept tube surface. The cylindrical part is formed by a surface offset from the centre area, which is defined as ruled surface along the kinematically or analytically determinated tool guide curves (see figure 7).

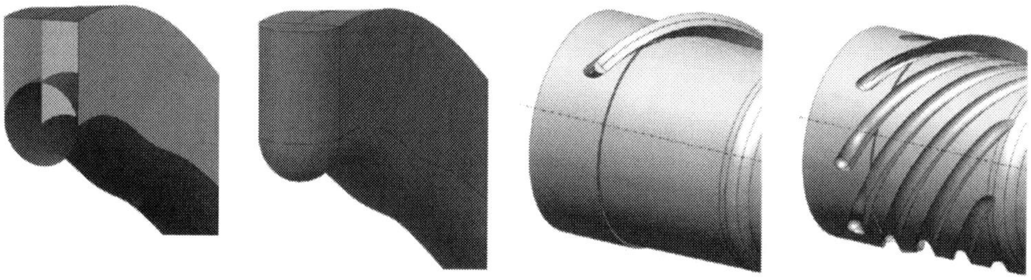

Figure 7. Modelling of a swept cutter tool envelope surface

By further surface design operations the surfaces can be merged into a single associated collecting surface and subsequently be subtracted from the solid (blank). Through the use of UDF not only the single modelling steps can thus be combined to generate tool motion envelopes proper to the specific manufacturing and be provided in a database for future use; it also allows the integration of manufacturing-relevant information on to-be-used tools and machine parameters for the subsequent CAM processes.

4. CAD-application for the design of a spiral milling parts

The presented CAD-based approach to design spiral milled 3D CAD models has been exemplary implemented in an application. For this purpose, the Application Programming Interfaces (API) of 3D CAD systems CATIA V5 were used, as well as SolidWorks 2010 and Creo 2.0, in order to develop an application program in Visual Studio 2010 (see figure 8).

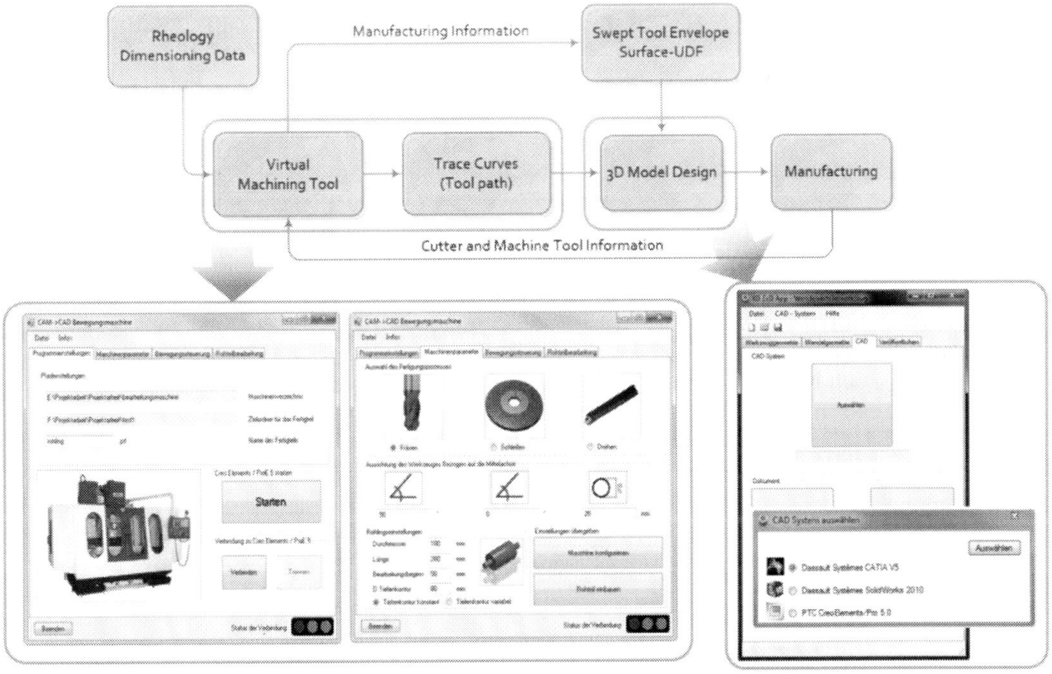

Figure 8. User design application for a spiral milling parts

With this application it is possible to design a spiral milled channel with a ball cutter in mandrel as well as housing suite to manufacture in the above mentioned 3D CAD systems. Furthermore, it allows to create the 3D CAD models of cylindrical mandrel dies under consideration of dimensioning and simulation aspects. The generation of the tool paths by an analytical or kinematic approach is feasible. For the kinematic approach a sub-program has been developed in conjunction with the CAD system Creo 2.0, which allows the user to configure a virtual manufacturing machine with all relevant parameters. In this application three machining methods (turning, spiral milling, and cutting) have been implemented, so that, motion analysis (MBS) can be performed automatically, depending on the choice of the method, the tool parameters, the position of the tool to the workpiece axis and other process-specific parameters. Thereby it is possible to distinguish between the generation mode as well as verification mode of the tooling process for the spiral milling with a ball cutter. For all the mentioned manufacturing methods the tool- and machine-dependent motion envelopes can be generated as a triangulated volume for further geometry testing.

5. Summary

In this paper a practicable CAD-based geometry design approach for spiral milling parts is presented as well as concrete subjects of the development, design and manufacturing process of a spiral mandrel die for plastics extrusion are discussed. The analysis pointed out a potential problem in transforming the geometry design with the CAD-Systems to the manufacturing geometry using a ball cutter and spiral milling process on a 4-axis machine. On this basis a data coupling approach of the dimensioning, designing and manufacturing has been proposed and an appropriate surface-based modelling strategy has been developed. The use of KBE methods, such as UDF, in combination with hybrid modelling techniques and rule based model design leads to an improved process. Using this approach, the modelling steps of swept cutter envelope surfaces for spiral milling can be stored in a database for future designs. In addition, the integration of tool and machine parameters for the subsequent CAM processes shortens the subsequent steps and allows a positive influence on the products' costs, quality and on the development- and production time. The developed approach ensures that the virtual geometry of the part corresponds to the "real" geometry generated by the spiral milling manufacturing process. The suitability of this approach is proved by optical measurements of a spiral mandrel die.

6. References

[1] K. Saul, PhD Thesis, Automatic design of extrusion dies, University of Duisburg-Essen Germany, 2011

[2] Wortberg, J., Saul, K., Köhler, P., Klemme, U.: Die Quality Criteria for fully Automized Optimization of Melt Distribution Systems, Proceedings of Polymer Processing Society 27th, PPS, Marrakesch, Marokko, 2011

[3] Wortberg, J., te Heesen, O.: CFD-Based optimization of extrusion dies by analyzing the dwell time distribution and flow history, Proceedings of Polymer Processing Society 28th, PPS, Pattaya, Thailand, 2012

[4] Cretu, Ionel Michael: Analyse, Auslegung und Optimierung von Wendelverteilsystemen, Dissertation an der Universität Stuttgart, 2008, ISBN 978-3-8322-775-5

[5] Hopmann, C., Eilbracht, S.: Towards automatic optimization of flow channel geometries in complex multi-level dies for film extrusion, International Conference on Engineering Optimization, EngOpt 2012 – 3rd, ISBN 978-85-7650-344-6

[6] Metwally, H., M., London, M.: Die Balancing: Coupling Flow Simulation, Shape Deformation & Optimization, Society of Plastics Engineers, SPE, Antec 2006, S. 835-839

[7] Paucksch, E.; Holsten, S.; Linß, M.; Tikal, F.: Zerspantechnik. 12 Aufl. Wiesbaden: Vieweg + Teubner 2008. (Studium).

[8] Witt, G., u.a.: Taschenbuch der Fertigungstechnik. Fachbuchverlag Leipzig im Carl Hanser Verlag, 2006. – ISBN 3-446-22540-4

[9] Aras, E.: Generating cutter swept envelopes in five-axis milling by two-parameter families of spheres. Computer-Aided Design (41/2009), S. 95–105.

[10] Lee, N. J.; Huang, C. B.; Chen, T. C.: Toolpath generation method for four-axis NC machining of helical rotor. Journal of Achievements in Materials and Manufacturing Engineering (2008) Volume 31, Issue 2.

[11] Lee, S. W.: Beitrag zur Generierung der Hüllvolumina für die NC-Fertigung mittels Gaußabbildung. Dresden. 2006.

[12] Lee, S. Y.; Chiou, J. C.: Swept Tool Envelope and Machining Potential Field for 5-Axis Sculptured Surface Machining. Computer-Aided Design and Applications (2006) Vol. 3, No. 6, S. 751–760.

IOP Conf. Series: Materials Science and Engineering **65** (2014) 012002 doi:10.1088/1757-899X/65/1/012002

Optimization of rotating equipment in offshore wind farm

O A Okunade

37, Street A7, Mount Pleasant Estate (Citec) Mbora Abuja Nigeria.
Email: femmyokunade@yahoo.com

Abstract. The paper considered the improvement of rotating equipment in a wind farm, and how these could maximise the farm power capacity. It aimed to increase capacity of electricity generation through a renewable source in UK and contribute to 15 per cent energy-consumption target, set by EU on electricity through renewable sources by 2020. With reference to a case study in UK offshore wind farm, the paper analysed the critique of the farm, as a design basis for its optimization. It considered power production as design situation, load cases and constraints, in order to reflect characteristics and behaviour of a standard design. The scope, which considered parts that were directly involved in power generation, covered rotor blades and the impacts of gearbox and generator to power generation. The scope did not however cover support structures like tower design. The approaches of detail data analysis of the blade at typical wind load conditions, were supported by data from acceptable design standards, relevant authorities and professional bodies. The findings in proposed model design showed at least over 3 per cent improvement on the existing electricity generation. It also indicated overall effects on climate change.

Keywords: wind farm power capacity, load cases, load conditions, wind turbine.

1. Introduction

According to conservation law of energy (first law of thermodynamics), energy cannot be created nor destroyed, but can be transferred from one form to another. While the energy in a system may take various forms (like heat, kinetic etc), depending on its source, the energy can get diminished. Thus, the source of energy generation can be said to be either renewable or non-renewable. Unlike non-renewable (fossil fuels), renewable energy source (solar, wind, geothermal etc) is not exhaustible and emits lower levels of carbon dioxide emissions.

In order to reduce the effects of emissions being caused by non-renewable source and avoid bad global warming, then there is a need to integrate the use of renewable source of energy in the national consumption. This need was echoed by European Union (EU) by setting 15 per cent (15 %) target of all energy consumption in UK to come from renewable source by 2020 (Carbon Trust, 2008). This target could cut across the energy consumption categories of transport, heat and electricity. The percentage is three-quarter of the 20 per cent (20 %) target that was set for the entire Europe by the Union in March 2007, according to the Carbon Trust. This target set for UK means that at least 29 GW (91 TWh) of offshore wind power could be needed to meet the 15 % EU renewable energy, even as the global leader for offshore wind energy (DECC, 2011), and wind power, as the primary renewable energy source (IMechE, 2013).

An offshore wind farm is a group of wind turbines located in the seabed or base of the ocean, to produce electric power. A wind turbine is device that converts the kinetic energy in fast-moving flow of wind to produce mechanical power, via wind medium to rotate a bladed rotor. The device consists of a foundation, tower, nacelle and a rotor. The tower holds up the rotor and a nacelle (box). While the nacelle contains components like main axle, gearbox, generator, transformer and the control system, the rotor is made of blades and hubs. The rotary equipment converts the kinetic energy into mechanical power, and therefore, the rotary equipment can be said to influence the mechanical power output, among other factors.

DECC indicated that between 33-58 TWh offshore wind technology would make up for 234 TWh, which is equivalent to the estimated 15% target for year 2020. By projection, offshore wind power could cost-effectively deliver 20-50% of total electricity generation by 2050 (LCICG, 2012).

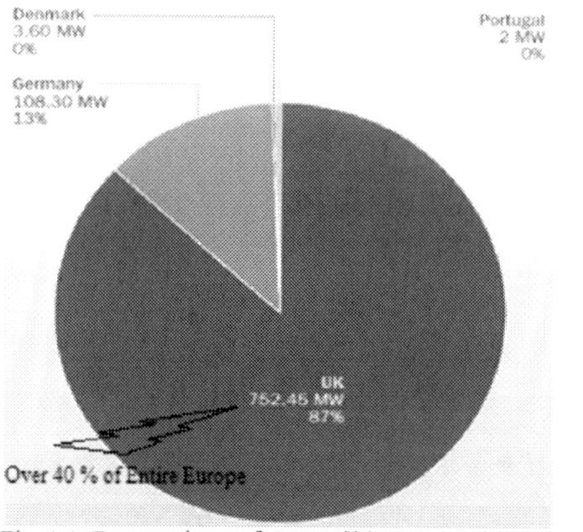

Fig. 1.1: Europe share of 2011 offshore wind farm installations in MW. (Source: EWEA, 2012)

Table 1.1: Technology breakdown (TWh) for central view of deployment in 2020. (Source: DECC, 2011).

TECHNOLOGIES	CENTRAL RANGE FOR 2020 (TWH)
Onshore wind	24 - 32
Offshore wind	**33 - 58**
Biomass electricity	32 - 50
Marine	1
Biomass heat (Non-domestic	36 - 50
Air- and ground-source heat pumps	16 - 22
Renewable transport	Up to 48
Others (Hydro, geothermal, solar)	14
ESTIMATED 15 % TARGET	234

Prominent Source

2. Methodology

As the largest and operational offshore wind farm in UK (as at the time of conceiving this as case study), it was believed that Greater Gabbard, being commissioned in August 2012, was among the latest commissioned wind farm that could still be improved. The improvements could be instrumental to meeting the 15 per cent renewable target by the year 2020. Its latest commissioning and being the largest in term of its capacity therefore, provided motivation for choosing it as case study.

Using turbines within the designed conditions is vital for design layout. The minimum acceptable turbine spacing is dependent on the nature of the terrain and the wind rose on a particular site. While a tight spacing turbines are affected by turbulence, positioning turbines spaced closer than five rotor diameters (5D) is likely to result in high wake losses, according to Renewable Energy World Magazine (REWM). Hence, minimum space required in positioning two turbines apart can be said to be sum of blade radius for turbine 1, minimum acceptable turbine spacing, and blade radius for turbine 2, assuming uniform blade diameter in a farm location. This is spacing efficiency.

i.e. $r_1 + 5D + r_2$ => where $r_1 = r_2 = r$ for uniform blade diameter $= D + 5D$

Table 2.1: Classes of Wind Turbine (Source: IEC 61400-1, 2005).

The design lifetime for turbine classes shall be at least 20 years.

Wind Turbine Class	I	II	III	S
V_{ref} (m/s)	50	42.5	37.5	Values specified by designer
A I_{ref} (-)	0.16			
B I_{ref} (-)	0.14			
C I_{ref} (-)	0.12			

A = Higher Turbulence characteristics of the site
B = Medium Turbulence
C = Lower Turbulence
I_{ref} = Expected Mean Value of Hub-height Turbulence at 10 mins average wind speed of 15 m/s.
S = Special Class Design (when special wind or external conditons are required by designers or customers).

With respect to operation, loading and durability of the wind turbine, regulation standard (IEC 61400-1, 2005) indicated that turbines can be subjected to external condition. For the purpose of design analysis, this external condition can be categorised into normal and extreme conditions, based on parameters wind speed and turbulence intensities.

Normal Wind Profile Model (NWP) : A normal external condition occurs when there is a continual structural loading, i.e. when the loading (wind) occurs frequently during the normal operation of a wind turbine. Models under this condition are called normal wind profile models (NWP). From standard, in order to model for this profile model (NWP) with ultimate strength, then

$$v_z = v_{hub}\,(z / z_{hub})^{\alpha} \quad (1)$$

where v_z =average wind speed as a function of height z above the sea from standard

v_{hub} = standard 15m/s at standard 10m/s z_{hub}; α = Power law exponent = 0.2

Extreme Wind Speed Model (EWP) : An extreme condition arise when there is an unusual or rare external design conditions .i.e. when having a 1-year or 50-year recurrence period. Models under this conditions are called extreme wind speed models (EWP). The design of load cases therefore consists of external conditions and other design situations. From standard,(IEC 61400-1, 2005), in order to model for this profile model (EWP) with ultimate strength, then

$$v_{e50\,(z)} = 1.4\ v_{ref}\,(z / z_{hub})^{0.11} \quad (2)$$

$$v_{e1\,(z)} = 0.8\ v_{e50\,(z)} \quad (3)$$

where v_{e50} = Turbulent extreme wind speed with recurrence period of 50 years.

v_{e1} = Steady extreme wind model with a recurrence period of 1 year.

2.1 Mathematical Model

The usable produced power is only some fraction of the available power in the wind, according to BETZ LIMIT. Hence, the need to calculate the power available in the wind. The algorithms to model output power are by calculating the power in the wind, extract it and then average output over a year.

2.1.1 Power in the wind. This is the amount of power present in the wind at a particular wind speed to rotate the rotor blade. The kinetic energy (KE) of an object with a mass (m) at a velocity (v) equals the work done in displacing the object. i.e. KE = Work done. From the Newton law of motion, F = ma (where m =mass, and a = acceleration), and by combining with third equation of motion ($v^2 = u^2 + 2as$), then power is produced.

The power available in the wind (P) is proportional to the square of the blade length, cube of the wind speed and the air density.

This is given by Power (P) = Density of Air (ρ) x Swept Area (A) x Cube of Wind Speed (V)

$$P = ½*\rho*A*V^3 \quad (4)$$

where, Power is in watts (W); ρ = (kg/m^3) A = (m^2) V = (m/s)

However, not all these power available in the wind are usable to generate electricity.

2.1.2 Extract Actual Power. This is amount of usable power that generates electricity. According to Betz Limit, in theory, wind mill can possibly extract a maximum of 59.3% of power from the wind only, while in reality, it is 45% maximum. This proportion of power extraction is termed coefficient of performance or efficiency (C_p). Primarily, this energy are extracted through the physical principle of lift and drag forces.

So, power extracted (P_e) from the wind is modified to:

$$P_e = P * C_p = ½ *\rho*A*V^3 * C_p \quad (5)$$

where C_p = Coefficient of performance of the wind turbine.

C_p is the average power coefficient defined by Betz Law with a limit of 0.59 for marine turbine.

Amount of power produced $P_p = P_e * \eta_g * \eta_b$ (6)

where η_g & η_b are efficiencies for generator and gearbox. η_g = 80%; * η_b = 90 to 95%.

2.1.3 Energy in a year. From BSI, average percentage of running turbine in a year is 23 %. This is given as load factor of 40 % from the Case Study data. The load factor, which varies depending on resource, technology and purpose, is determined by evaluating amount of load and amount of time the system operates at that load. For analysis purpose, given data from case study was used.
1MW of electric power = 1GWh in a year (GigaWatts hour)
where 1 year of turbine = 8760 hours; 1MW of load = 0.4 * 8760 hrs
Therefore, energy produced in a year 1 MW = 3.5 GWh (7)

2.2 Case Study Data Analysis
From table 2.2, Turbine Unit Power = 3.6MW, Total Power = 504MW, Rotor Diameter = 130m; Assumed density $\rho = 1.225$ kg/m^3; where Area $A = \pi r^2$.

2.2.1 Calculating for wind speed that produced this unit power. It is assumed that this unit power 3.6 MW is the actual power produced. From equations 5 and 6, and section 2.1.2, then
3.6 = P * 0.59 * 0.8 * 0.925 => Power available in the wind is 8.246 MW
Substituting this in eq. (4) to calculate for wind speed that produced available power in the wind
$8.246 * 10^6 = 0.5 * 1.225 * (3.142*65^2) * v^3$
v = 10.05 m/s
This is equivalent to average speed of 10 m/s of generic class 1 offshore for environmental condition (Camp, 2003).

Table 2.2: Case Study Data - Greater Gabbard Offshore Wind Farm.
(Source: http://www.power-technology.com/projects/greatergabbardoffsho/, 2012)

Location Area	147 km^2	Turbine Height	170 m
Unit Numbers	140	Rotor Diameter	130 m
Installed Capacity	504 MW	Yearly Production	1900 kWh to 530,000 homes
Unit Power	3.6 MW	Cost (Excluding Grid Connection)	£650 m ($1.05 bn)
Load Factor	40 %		

2.2.2 Calculating for energy produced in a year. From eq. (7), 3.6 MW load of unit turbine runs in a year is 3.5 * 3.6 = 12.6 GWh. Hence, 140 units will produce 1764 GWh.

2.2.3 Calculating for spacing efficiency. For Case Study with 130 m rotor diameter, the spacing efficiency (section 2.) equals 130 m + 5(130 m) = 780 m i.e. 780 m is the required spacing in positioning the two turbines apart. But given that total turbine units in the farm are 140, therefore, the amount of space consumption for whole farm = 780 * 140 = 109200 m^2 = 109.2 km^2. Hence 109.2 km^2 does not justify effective use of the available location area, as there still exists 37.8 km^2 of unproductive space.

2.3 Proposed Optimized Analysis
In order to model for the maximum energy output in a year, this is analyzed using three approaches. Each approach followed the algorithms of each of mathematical model (section 2.1). The approaches were achieved by calculating energy produced with proposed 100 m rotor diameter using existing case study wind speed; and according to standard IEC 61400-1, normal wind profile model (NWP) and extreme wind speed model (EWP).

From section 2, spacing efficiency with proposed 100 m rotor diameter, will be 600 m in positioning two turbines apart. Assuming for same numbers of unit of 140 turbines as case study, these require 84000 m^2 = 84 km^2. However, since available case study site area is 147 km^2, then there remains a large space that can still be maximized. In order to achieve this, the number of turbine units were therefore increased.
100 per cent use of the site area will require 245 units of turbines.
i.e. 245 units turbine * 600 m = 147000 m^2 = 147 km^2 (same as case study site area).

2.3.1 Using Calculated Case Study Wind Speed. This is 10.05 m/s from section 2.2.1. Applying this with proposed 100 m diameter, mathematical model evaluation was as follows.
i). Calculate for Power Available in the wind.
From eq. (4), $P = 0.5 * 1.225 * (3.142 * 50^2) * 10.05^3$ $P = 4.88$ MW.
ii). From eqs. 5 & 6, Power produced $P_p = (4.88 * 10^6) * 0.59 * 0.8 * 0.925 = 2.13$ MW.
Therefore total power produced by maximum units of 245 turbines, at 100 % efficient use of location area = 2.13 MW * 245 = 521.85 MW.
iii). From eq. (7), one unit turbine of 2.13 MW will run in a year for 3.5 * 2.13 = 7.46 GWh.
Hence 1827.7 GWh of energy will be produced by 245 units.

2.3.2 Normal Wind Profile Model (NWP). This was modelled according to standard in section 2.0.
i) From eq. (1) average wind speed at given height of case study (170 m)
$v_z = 15\ (170 / 10)^{0.2} = 26.4$ m/s.
ii) From eqs. (5 & 6), $P_p = (0.5 * 1.225 * 3.142 * 50^2 * 26.4^3) * 0.59 * 0.8 * 0.925 = 38.65$ MW.
Hence, a normal wind profile model, at 100 m rotor diameter produces 38.65 MW for a unit.
Actual power produced for 245 units equals 9469.25 MW
iii) Energy produced in a year by a unit turbine with actual power of 38.65 MW would run for 3.5 * 38.65 = 135.28 GWh (eq. 7). Equivalent of 33143.6 GWh energy produced by 245 units.

2.3.3 Extreme Wind Speed Model (EWP). According to standard 61400-1 (section 2.0), this was considered in two cases: as turbulent and steady EWP with recurrence periods of 50 years and 1 year respectively. From table 2.1, assume v_{ref} for turbulent EWP is 50 m/s, then from eq. 2,
i). average wind speed at height of case study, 170 m
$v_{e50\,(z)} = 1.4 * 50 * (170/10)^{0.11} = 95.59$ m/s.
Hence, for extreme wind speed model with recurrence in 50 years period, the wind speed shall be 95.59 m/s. This means that with this speed, the turbine must be able to withstand the extreme events possible with a return period of 50 years.
For steady EWP, then eq. 3 becomes $v_{e1\,(z)} = 0.8 * 95.59$ m/s = 76.5 m/s.
Now was considered model for analysis of 1 year reoccurrence.

ii). From eq. 5 & 6, actual power produced $P_p = (½ * \rho * A * v_{e1\,(z)}^3) * C_p * \eta_g * \eta_b$
$P_p = (0.5 * 1.225 * 3.142 * 50^2 * 76.5^3) * 0.59 * 0.8 * 0.925 = 940.42$ MW
Therefore, a steady EWP with 1 year reoccurrence produces 940.42 MW for a unit.
Total Output for 245 units equals 230402.9 MW

iii). Energy produced in a year - If actual power produced by a turbine due to steady EWP was 940.42 MW, then according to eq. 7, then 3291.47 GWh amount of energy would be produced in a year. Hence, 806410.15 GWh of energy will be produced by 245 units.

3. Results
The tabulated results show effects of analytical model and redesign of blade to the proposed diameter, and efficient location areas, while maximising the energy productivity.

Table 3.1: Proposed Model Results as against Case Study Data

ATTRIBUTES	CASE STUDY WIND FARM	PROPOSED		
		Approach 1 (Calculated Wind speed)	Approach 2 - NWP	Approach 2 - EWP for 1year reoccurrence
Rotor diameter (m)	130	100	100	100
Calculated Wind speed (m/s)	10.05	10.05	26.4	76.5
Total Area (km²)	147	147	147	147
Unit Numbers	140	245	245	245
Unit capacity (MW)	3.6	2.13	38.65	940.42
Total capacity (MW)	504	521.85	9469.25	230402.9
Calculated energy produced by a unit turbine in a year (GWh)	12.6	7.46	135.28	3291.47
Calculated energy produced by total units of turbines in a year (GWh)	1764	1827.7	33143.6	806410.15

4. Discussions

While the wind farm capacity in first approach showed an increase in total units capacity of actual power produced and energy in a year, occurrence of frequent wind loading in the second approach resulted in the wind speed almost triple the initial case capacity. Consequently, the unit power capacity an energy produced in a year also got increased. In the third approach, the unusual event of design with recurrence 1 year period resulted in unusual increase of wind speed, and consequently the unit power and energy produced in a year. The variations in these wind speeds according to different approaches limits the capacity of power generation, and also are adequate position of the rotor blade in the wind speed direction, altitude of the tower among others. With estimated 91 TWh of offshore wind power predicted to be needed to meet the 15 % target (DECC, 2011), this would mean an approximate numbers of 50 wind farms will be needed, should each farm consists of 245 numbers of turbines, and each generating 1827.7 GWh like that in the first approach.

5. Conclusions

The first approach showed 3.5 per cent increase of actual power produced by total units of turbines, and 3.6 percent of energy produced in a year. The unit power and energy produced were almost 10 per cent increase each as against the given case study data in the second approach, while there was a rare very high percentage increase resulted in the third approach. With this least percent increase of 3.5 on a wind farm, this would mean a greater addition of percentage increase on all operational wind farms in the country. The percentage increments on all operational wind farm would mean serving greater number of homes and hence beneficial to everyone, as it makes environments more friendly, due to its source. As the most significant gases in the atmosphere, the effects of carbon dioxide concentration are eliminated, hence, there is less severity of changes, both physical and ecological, which could affect national population, productivities etc. Also, as predicted by the Department of Energy and Climate Change that offshore wind technology would produce most amount of energy (table 1.1), and couple with the percentage increase obtained via the analysis, it could be said that government supports would be of great importance at realizing this prediction. This could contribute immensely to meeting set target of the need for energy consumption in the country to come from renewable source, and increase chances of meeting it earlier than predicted. The redesign optimization of existing farms could save money, time and other resources instead of construction new ones. Hence, the economy thrives, while maximizing the operational qualities and productions of the existing farms.

References

[1]. Camp, T.R., Morris, M.J. *et al* (2003) *Design methods for offshore wind turbine at exposed sites*. Bristol: Garrad Hassan & Partners Ltd.
[2]. Carbon Trust (2008) *Offshore wind power: big challenge, big opportunity*. London: Carbon Trust Enterprise.
[3]. Department of Energy and Climate Change (2011) *UK renewable energy roadmap*. London: The Stationery Office.
[4]. European Wind Energy Association EWEA (2012) *The European offshore wind industry key trends and statistics 2011 by Justin Wilkes, Jacopo Moccia et al*. Brussels: EWEA.
[5]. IMechE (2013) ETI backs iow turbine project, *PE Magazine* 26 (1), pp. 6.
[6]. Low Carbon Innovation Co-ordination Group LCICG (2012) *Technology Innovation Needs Assessment (TINA): Offshore Wind Power Summary Report*. London: The Stationery Office.
[7]. IEC (2005) *International standard IEC 61400-1*. (3rd Ed.). Switzerland: International Electrotechnical Commission (IEC) Publication.

Solar-hydrogen energy as an alternative energy source for mobile robots and the new-age car

A Sulaiman[1*], F Inambao[2] and G Bright[3]

[1,2,3]Mechanical Engineering, University of KwaZulu-Natal,
Howard College Campus, Durban, South Africa

200300000@stu.ukzn.ac.za

Abstract. The disastrous effects of climate change as witnessed in recent violent storms, and the stark reality that fossil fuels are not going to last forever, is certain to create renewed demands for alternative energy sources. One such alternative source, namely solar energy, although unreliable because of its dependence on available sunlight, can nevertheless be utilised to generate a secondary source of energy, namely hydrogen, which can be stored and thereby provide a constant and reliable source of energy. The only draw-back with hydrogen, though, is finding efficient means for its storage. This study demonstrates how this problem can be overcome by the use of metal hydrides which offers a very compact and safe way of storing hydrogen. It also provides a case study of how solar and hydrogen energy can be combined in an energy system to provide an efficient source of energy that can be applied for modern technologies such as a mobile robot. Hydrogen energy holds out the most promise amongst the various alternative energy sources, so much so that it is proving to be the energy source of choice for automobile manufacturers in their quest for alternative fuels to power their cars of the future.

1. Introduction

The most powerful storm ever recorded in world history has just struck our planet[4], prompting the leader of the world body[5] to issue a warning on climate change[6]. The utter devastation left in its wake is bound to give impetus to renewed public attention to climate change and mobilise world opinion against the continued reliance on fossil-based energy which is derived from the combustion of fossil fuels. Burning of fossil fuels to generate energy, causes carbon emissions which is considered to be the largest contributing factor to the release of green-house gases into the atmosphere, in turn leading to global warming and consequently climate change [1]. The result is catastrophic disasters such as the type we have witnessed with super typhoon Haiyan that struck the Phillipines.

[1] The author is a candidate for a PhD degree in Engineering in the Mechanical Department, University of KwaZulu-Natal, South Africa.
* Corresponding author

[4] Typhoon Haiyan, which struck the Phillipines on 7 November 2013.
[5] UN Secretary General, Ban Ki Moon.
[6] UN Framework Convention on Climate Change (UNFCC) in Warsaw, Poland, 19 November 2013.

To simply go on relying on fossil fuels for energy, although cheap, is just not sustainable. We are using up fossil fuels, such as coal, petroleum oil and natural gas many times faster than they are being formed in the earth's layers, with the result that these resources are expected to be depleted in a matter of a few decades, at the least [2]. The quest to develop alternative sources of energy has never been more pressing. This study, in some small measure, seeks to address just that. It takes a look at solar and hydrogen energy as an alternative energy source, and demonstrates in an actual case study how it can be applied in modern technologies such as a mobile robot.

Solar energy does not have any of the limitations or draw-backs of the type mentioned above for fossil-based fuels. There are no limitations as to the amount of sunlight that the sun can still continue to generate, and neither does the generation of solar power produce any harmful effects to the environment. It has all the advantages of being both a renewable source of energy, as well as being environmentally clean.

The problem though, is that solar energy is dependent on the availability of sunlight, and the reality is that sunlight is not always present, as for instance at night or even during daylight when cloudy conditions block out sunlight. This makes it necessary to seek an additional source of energy that would be available when no solar energy can be produced, so as to ensure that the supply of energy is constant.

For this, we can turn to Hydrogen, the simplest and most plentiful element in the universe. Although it is the lightest element, yet it has the highest energy content per unit weight of all fuels [1] Hydrogen's energy density is 52,000 Btu/lb, which is three times greater than that of gasoline. NASA has used liquid hydrogen since the 1970s to propel the space shuttle and other rockets into orbit. An atom of hydrogen consists of only one proton and one electron. Yet, despite its simplicity and abundance, it doesn't occur naturally on the earth like coal, oil or natural gas. It is always combined with other elements, as for example water (H_2O), which is a combination of hydrogen and oxygen [1,3,4].

To obtain hydrogen, one needs to separate it from the other element to which it is bound. Water is the most convenient source to obtain hydrogen, and this is achieved by separating it from the other element to which it is bound, namely oxygen, through the process of electrolysis [1-3]. This process requires electricity. However, it would be quite a waste to use up one form of energy, namely electricity, just to create another source of energy i.e. hydrogen, unless of course, the electricity that will be used for that process has been generated from a source that is free, renewable and does not result in any harmful waste products such as carbon emissions and toxic gases, which is the case when producing electricity generated from fossil fuels such as coal. This is where solar energy can provide the answer. Hence, a primary source of energy (i.e. solar energy derived directly from the sun), and a secondary source of energy (hydrogen obtained indirectly through the process of electrolysis) can be combined in an energy system which can serve as an efficient source of energy [5, 6].

2. The Solar-Hydrogen Energy System and its various Components

The process involved in the Solar-Hydrogen Energy System can best be illustrated by the Block Diagram in Figure 1. An explanation of each step in the Solar-Hydrogen system follows thereafter, with particular reference to the components that were used in this case study for the power generation of a mobile robot.

2.1. Photo-voltaic Cells

As a first step, sunlight which is generated from the sun is converted to electricity by the use of Photo-voltaic (PV) cells. The term "photo-voltaic" gets its name from the process whereby light (photons) is converted to electricity (voltage). Hence the term "photo-voltaic". This process requires quite some detail to explain, but briefly put, what happens is that when sunlight strikes the PV cells, it knocks electrons loose from their atoms, allowing them to flow freely through the cell [7]. This flow of electrons is a current, and by placing metal contacts on the top and bottom of the PV cell, we can draw off that current for external use. The electricity that is produced by the PV panels is direct current

(DC). The PV system used in this study consists of 6 PV modules. The total installed power is 1.2kWp in standard conditions. The specifications of the solar modules are determined as follows: maximum power is 200 W; Open circuit voltage (V_{oc}) is 59.5V; Optimum operating voltage (V_{mp}) is 46.1V; Optimum operating current (I_{mp}) is 4.37A.

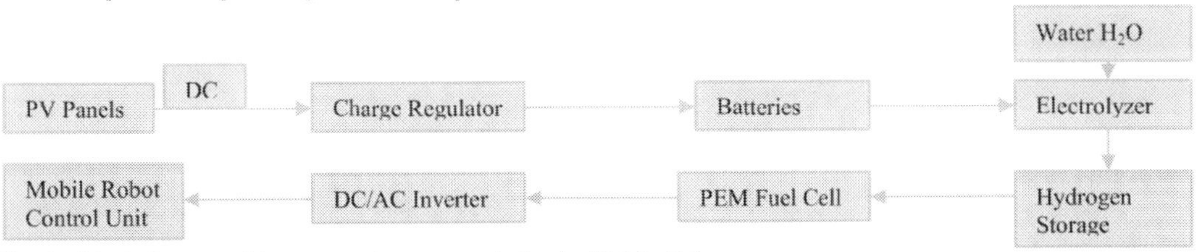

Fig. 1 Block diagram of the proposed power supply for the Mobile Robot

2.2. Charge Regulator and Solar Batteries
The current that is produced by the PV panels, which is in DC voltage, will pass through a Charge Regulator in order to maintain stable current flow, and regulated at 12V. From the Charge Regulator, the DC current will pass to the solar batteries where it will be stored so as to provide a constant supply of power even when the supply of current from the PV system is interrupted during periods when there is no sunlight. Each solar battery will have 12 V to 350 Ah, and there will be 4 solar batteries in total.

2.3. Electrolyzer
From the Solar Batteries, the DC current will pass to the Electrolyzer. Electrolyzers use electricity to split water into its constituent elements, namely hydrogen and oxygen, through the process of water electrolysis. A water molecule is formed by two elements: two positive Hydrogen ions and one negative Oxygen ion, and is held together by the electromagnetic attraction between these ions. When electricity is introduced to water through two electrodes, a cathode (negative) and an anode (positive), these ions are attracted to the oppositely charged electrode. Therefore the positively charged hydrogen ions will collect on the cathode, and the negatively charged oxygen will collect on the anode [1,4]. One now needs to find a suitable means of storing the hydrogen thus collected from the electrolysis process.

2.4. Hydrogen Storage Tank containing Metal Hydrides
The storage of hydrogen poses certain real challenges. That is because it is so light that it occupies so much of space. It is the lightest gas in the atmosphere, even lighter than air. It needs to be somehow squeezed to fit into a smaller space so that it is convenient to store and can be packaged in a form that is more compact and potable enough to be carried on-board a robot or motor vehicle that needs to utilise it as a power source. For this, it has to be pressurized in a container that can handle the extreme pressures of the hydrogen inside. One can achieve higher densities by storing the hydrogen in liquid form. The problem though, is that it will have to be kept at extremely low temperatures of -253°C (referred to as cryogenic temperatures) in order to maintain it in that liquid state. Whether as a compressed gas or cryogenic liquid, hydrogen is quite volatile and requires special handling [9]. It is this very aspect of storage that poses challenges in the use of hydrogen.

A promising alternative would be to store hydrogen in a solid state, using a metal alloy. A metal alloy is like a lattice structure of individual atoms with small spaces in between. When exposed to hydrogen at certain pressures and temperatures, the metal alloy is able to absorb large quantities of the hydrogen gas. The hydrogen gas splits to form two hydride ions. The hydride ion is very small, consisting of only a proton and an electron, and can fit into the small spaces in the metal lattice structure. In other words, the metal alloy acts like a porous sponge that can soak up large quantities of the hydrogen. When this happens, the hydrogen is distributed compactly throughout the metal lattice. When metal alloy combines with hydrogen in this way it forms simple chemical compounds called metal hydrides [10].

Metal hydrides represent an exciting method for storing hydrogen. They have a higher volumetric hydrogen storage capacity. Some hydrides can actually store twice the amount of hydrogen than can be stored in the same volume of liquid hydrogen. When compared with the volume of space that hydrogen occupies in its original form as gas, as a metal hydride the hydrogen is compacted into a solid form which is one thousand times smaller than the original hydrogen gas.

Thus, the component to be used in this experiment for the storage of hydrogen, will be a Hydrogen Tank containing a metal alloy in granular form. Metal alloys have the advantage of not only being able to absorb the hydrogen for the purposes of storage, but they are also able to release (de-sorb) the hydrogen so that it can be used when needed to produce energy, as will be demonstrated in the next step [9,10].

2.5. Fuel Cell and DC/AC Inverter

Just as how electricity is applied to water (H_2O) to produce Hydrogen (H_2) and Oxygen (O) in electrolysis, this whole process can be reversed. By combining Hydrogen with Oxygen one can produce electricity [11]. It is like "reverse electrolysis". This process is achieved by means of a Fuel Cell. The fuel cell is connected to the Hydrogen Storage Tank from which it gets a constant supply of hydrogen. For as long as there is a flow of hydrogen to the Fuel Cell, it never goes dead. Electricity will just simply continue to flow out of the Fuel Cell [12, 13]. The electricity produced by the Fuel Cell is in DC voltage and will be converted to AC current by means of DC/AC inverter so that the current is in the required form to power the mobile robot.

3. The Experiment Set-up

For practical considerations, it may not be possible to incorporate all the components of the Solar-Hydrogen Energy system on-board the mobile robot, particularly with regard to the solar panels which would make a robot unwieldy. In this model it is envisaged that the solar-energy generation process, as well as the electrolysis and production of the hydrogen will be done at a station. Those components that would be incorporated on-board the mobile robot for the power supply, is illustrated in Figure 2.

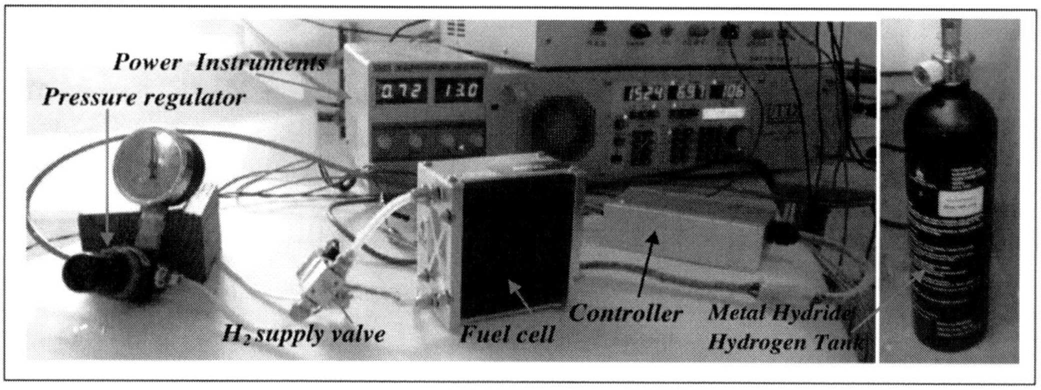

Fig. 2 Experiment set-up showing the Hydrogen Storage System and Power Supply.

4. Mechanical Design and Hardware Architecture

The robot used for this experiment is designed for an industrial and manufacturing setting. It is an Industrial Mobile Robot Platform (IMRP) in the form of a unit load carrier that is capable of carrying a pay-load of up to 40 kg on its deck. The deck is a moving platform, which is powered by a 12V DC Motor. The frame-work for the IMRP has been made from square tubing (38mm x 38mm) (see Figure 3). The overall gross weight of the IMRP will be 120kg.

For improved manoeuvrability, the IMRP has been given omni-directional capabilities by means of mecanum wheels, as illustrated in Figure 3. A mecanum wheel consists of a hub with a number of individual rollers arranged around its circumference. Each roller is mounted on the contact surface of the wheel at a 45° angle to the wheel axis, and is able to rotate about its own axis. Using this type of wheel, a vehicle can be propelled directly in a sideway motion without having to make the vehicle move forward at the same time, as is the case for a conventional vehicle. A vehicle with mecanum wheels can be made to move in various directions by choosing different combinations of wheel rotation direction for the four wheels [14]. This requires that each wheel be driven independently with a separate motor. For the IMRP, four 12V DC motors are used.

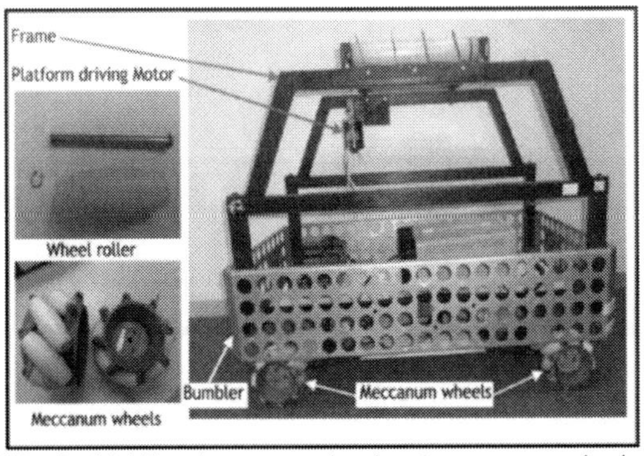

Fig. 3 Mobile robot system showing the mecanum wheels

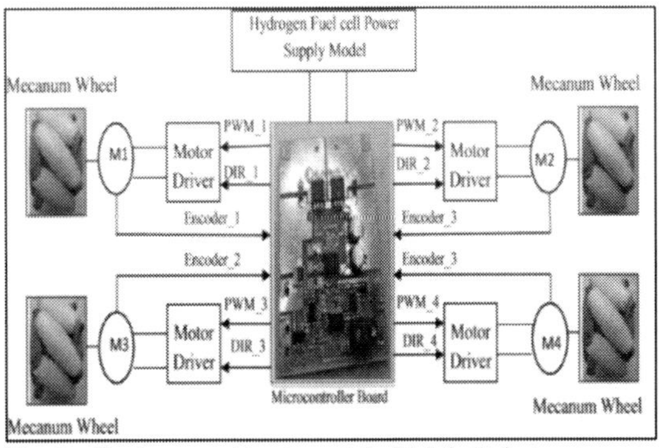

Fig. 4 Hardware architecture of the Mobile Robot system

The control of these motors would require a Motor Driver Circuit Board, which in turn will interface with an on-board processor with intelligent functionality, namely a Micro-controller Board that will have the ability to generate four independent PWM signals for each of the four wheels [15].

The control of movement of each wheel will have to be calculated independently of the other wheels, and the various permutations involved will require higher level computation and this will be carried out by a Main Computer, supported by software specifically written for this purpose [16]. The computer will be controlled by an operator from a remote location, which will transmit digital signals by means of Radio Frequency to the Micro-controller Board, using a Telemetry System.

Figure 4 illustrates that part of the Hardware Architecture that is to be installed on-board the IMRP.

5. The Simulation Model

To analyse the efficiency of the energy system in this study, a software simulation tool is used. The system chosen for this study is TRNSYS (Transient Energy System Simulation) which has a wide range of use in renewable energy and emerging technologies. The TRNSYS model simulates the performance of the entire energy-system.

6. Vibrant interest in Hydrogen for other Technologies : The Automobile Industry

The biggest challenge in the automobile industry in producing a commercially available hydrogen car, has been the high cost associated with the fuel cell technology. For motor cars, one needs a catalyst in the fuel cell that can speed up the chemical reaction so as to yield energy fast enough. Only platinum can achieve the desired effect. But the trouble is that platinum is rarer than gold and more costly. And it is this factor that has stood in the way of making hydrogen powered cars a reality. But new research by Associate Professor Matthias Arenz, of the Department of Chemistry, Copenhagen University, Denmark, has now brought this technology a step closer. "By placing the platinum differently inside the catalyst, we can drastically reduce the platinum requirement. This means that fuel cells will be much cheaper to produce in the future," according to a new study co-authored by him[7]. Already, the

[7] "Hydrogen cars now one step closer" ScienceNordic,, Aug 11,2013 sciencenordic.com

world's biggest car producer (Toyota) is combining forces with the world's biggest luxury car-producer (BMW) on hydrogen fuel cell research, as also Ford, Daimler and Nissan in a separate deal[8].

7. Conclusion

Studies such as this on alternative energy sources, are not merely of academic importance only. The perceived difficulties in alternative energy sources such as solar and hydrogen can be overcome to make these technologies a reality. As this study demonstrates, the seeming unreliability of solar energy because of its intermittent supply, which is due the nature of sunlight which is not always available, can be overcome by combining it in a system with another alternative energy source such as hydrogen. Likewise the challenges posed by hydrogen energy relating to its storage and safe handling can also be overcome by the use of metal hydrides. And finally, the prohibitive costs associated with fuel cell technology which made the concept of hydrogen cars a distant dream, is now one step closer to reality with new emerging research.

References

[1] Pagliaro M, and Konstandopoulo A G 2012 *Solar Hydrogen: Fuel of the Future* (Society of Chemistry UK)

[2] Fang Lin Luo and Hong Ye 2013 *Renewable energy systems: Advanced Conversion Technologies and Applications* (Taylor & Francis Group LLC)

[3] Zuttel A, Borgschulte A and Schlapbach L 2008 *Hydrogen as a Future Energy Carrier* (WILEY-VCH Verlag GmbH & Co. KGaA)

[4] Hordeski M F 2008 *Alternative Fuels: The Future of Hydrogen* (CRC Press)

[5] Yilanci A, Dincer I and Ozturk H K 2009 *A review on solar-hydrogen/fuel cell hybrid energy systems for stationary applications: Progress in Energy and Combustion Science* **35** pp 231

[6] Dincer I and Rosen M A 2011 *Sustainability aspects of hydrogen and fuel cell systems: Energy for Sustainable Development* **15**, pp 137–146

[7] Rand D A J and Dell R M 2008 *Hydrogen Energy Challenges and Prospects* (RSC)

[8] Corbo P, Migliardini F and Veneri O 2011 *Hydrogen Fuel Cells for Road Vehicles - Green Energy and Technology* (Springer-Verlag London Limited)

[9] Broom D P 2011 *Hydrogen Storage Materials the Characterization of Their Storage Properties* (Springer-Verlag London Limited)

[10] Hirscher M 2010 *Handbook of Hydrogen Storage New Materials for Future Energy Storage* (WILEY-VCH Verlag GmbH & Co. KGaA, Weinheim)

[11] Larminie J and Dicks A 2003 *Fuel Cell Systems Explained* (John Wiley & Sons Ltd)

[12] Ali S M and Andrews J 2005 *Low-cost storage options for solar hydrogen systems for remote area power supply. In: 16th world hydrogen energy conference (Lyon, France)*

[13] Rubio M A, Urquia A and Dormida S 2009 Diagnosis of Performance Degradation Phenomena in PEM Fuel Cells *Int. J. Hydrogen Energy*, pp. 1-5

[14] Xu W L, Bright G and Cooney J A 2003 Visual dead-reckoning for motion control of a Mecanum-wheeled mobile robot *Mechatronics International Journal* vol 14 (Elsevier Ltd) pp 623–37

[15] Lee S 2010 Development of a 600 W Proton Exchange Membrane Fuel Cell Power System for the Hazardous Mission Robot *J. Fuel Cell Sc. and Tech.*

[16] McComb G 2003 *Robot Builder's Sourcebook* (McGraw-Hill)

[8] "Ford, Daimler and Nissan commit to Fuel Cells" MIT Technology Review, Jan. 28, 2013 www.technologyreview.com

IOP Conf. Series: Materials Science and Engineering **65** (2014) 012004 doi:10.1088/1757-899X/65/1/012004

Development of a structured approach for decomposition of complex systems on a functional basis

Unal Yildirim[1] and I. Felician Campean[1]

[1]Bradford Engineering Quality Improvement Centre (BEQIC),
University of Bradford, United Kingdom, BD7 1DP

E-mail: u.yildirim@student.bradford.ac.uk

Abstract. The purpose of this paper is to present the System State Flow Diagram (SSFD) as a structured and coherent methodology to decompose a complex system on a solution-independent functional basis. The paper starts by reviewing common function modelling frameworks in literature and discusses practical requirements of the SSFD in the context of the current literature and current approaches in industry. The proposed methodology is illustrated through the analysis of a case study: design analysis of a generic Bread Toasting System (BTS).

1. Introduction

Product development (PD) organizations are facing the increasing challenge of engineering of complex systems driven by the pressure for delivery of customer requirements and the need to address environmental and safety concerns (e.g. emission standards for an automotive manufacturer) within a highly competitive market.

Within a customer focused engineering approach, systems engineering design must focus on robust and reliable delivery of customer required functions. Therefore, function analysis plays a central role in the systems engineering design process. It has been recognized that the increasing complexity of systems places a significant challenge for function analysis [1,2], in particular due to the complicated nature of systems interdependencies. Robust systems engineering design must ensure that all systems interfaces are identified, characterized and managed through functional requirements both at the system level and subsequent levels of systems decomposition [3].

Structured approaches to design, based on the axiomatic design framework [4], make the clear argument that a separation between functional domain and the design domain must be maintained, and design development should take place through zig-zagging between the functional domain and the design domain. This ensures that functions are articulated in a solution-neutral manner, and that decomposition is pursued in the functional domain first. However, most engineering practice focused design approaches deviate from this fundamental principle, in that the decomposition of systems is pursued on structural basis, commonly carried out by clustering the elements into architectural and team chunks on a design structure matrix (DSM) [5,6]. The authors' observation of current systems engineering design practice in the automotive industry has highlighted the prevalence of this approach in the decomposition of systems; a clear element for this is that design responsibilities are typically allocated to engineering teams on either a disciplinary basis (mechanical, electrical, software) or structural basis relating to design units or chunks (e.g. taking an automotive example - body, powertrain, chassis), and not functions, which does not guarantee that a flawless systems integration is achieved. There is overwhelming evidence from industry [7] that most field problems occur at systems interfaces which have not been appropriately managed at the design stage.

Another weakness of current practice in function analysis of complex systems stems from the common reliance on brainstorming as the basis for identification of functional requirements [3]. While brainstorming can be a powerful tool, it cannot guarantee completeness of functional requirements. The introduction of structured approaches for systems decomposition based on either the DSM [5] or reverse engineering [8] improve the effectiveness of brainstorming, however, this deviates from the axiomatic design principle that analysis in the function domain should take place first, and often compromises the solution neutral articulation of the function. The effect of this is that opportunities for driving innovation in the design are missed.

The System State Flow Diagram (SSFD) has been recently introduced [9] as a framework for functional representation and decomposition of a complex system on a solution-independent functional basis. This paper presents research to enhance structure of the SSFD framework, focusing in particular on the integrity of the analysis of flows through the system, as well as functional representation and articulation. The organization of the paper is as follows: next section reviews current function modelling frameworks in literature; section 3 introduces the proposed SSFD framework, discussed along with a case study based on the design analysis of a generic Bread Toasting System (BTS); section 4 reflects on the merits of the proposed approach, and outlines further research work.

2. An Overview of Function Modelling Frameworks

An engineered system is generally represented in terms of an input-output relationship and the function of the system is described with respect to this relationship [10]. The system decomposition is developed on the basis of this input and output. The identification of the inputs and the outputs in terms of the flows of energy (E), material (M) and information (I) is common in literature [8,11-12]. The system decomposition is managed based on mapping the flows of energy, material and information through the system; sub-functions are defined as successive operations on flows, with output from one sub-function providing the input to the next [11].

The Functional Basis (FB) model of Otto and Wood [8] uses the flows of energy, material and information for design decomposition of a system. The FB model represents the overall function of a system in terms of interconnected sub-functions which are defined as operations on the flows of E, M and I. The model provides a consistent way for the description of the flows and the functions by introducing a taxonomy for both. The function is described in verb-object form [8]. Design development is carried out through successive decompositions of functions and sub-functions into lower level sub-functions which can be referred to an iterative decomposition in the functional and the design solution domains in the context of axiomatic design [4]. While the FB model is limited to the decomposition of a system based on the flows at successive levels, system integration can be supported at any level by utilizing a DSM based approach [5], which can be used in the identification of the relationships between design solutions [13] or grouping design solutions into structure-function units [6].

Several methods have also addressed visualization of functional structure of a system on the basis of flows, but in different manners, e.g. Function Flow Block Diagram and Integrated Definition for Function Modeling (IDEF0) [14].

The Contact and Channel Approach (C&C^2-A) [15] describe the input and the output of a system as a Working Surface Pair (WSP) and specifies a Channel and Support Structure (CSS) to connect two WPSs. The C&C^2-A suggests that the description of a function should include at least two WSPs, the connecting CSS and at least two Connectors which embed the model into the environment. A WSP is described as a state characterized by measurable attributes and the function of a system is described as a transfer of one state into another [16]. Design decomposition is carried out through describing the basic elements WSP, CSS and Connectors at different levels of the system.

The Object-Attribute-Function (OAF) framework [17] and TRIZ [18] propose similar approaches. The Object-Attribute-Function (OAF) framework defines the input and the output as generic objects which are described by their measurable attributes (e.g. mass), with a clear taxonomy developed to describe both. TRIZ describes a function with respect to interaction between two components of a system and identifies an element which provides the interaction.

Object-Process Methodology (OPM) [19] represents a system both graphically (Object-Process Diagram (OPD)) and textually (Object-Process Language (OPL)). The method describes a function as a function sentence which begins with the list of processes followed by objects and represents the function as a box named function box that encloses at least one object and one process. The method can represent a system at any level through its complexity management mechanisms (e.g. in-zooming/out-zooming).

Description of the functional structure of a system with respect to its intended behavior and structure at different phases of the design process is proposed by [20-22]. The FBS framework [20] suggests that functions are defined in relation to states of the structures (design objects), which are represented by entities, their attributes, and relations between entities. Within FBS the functions are defined by the combination of verb-object which relate to the designer intentions, and behaviors expressed through adjectives, which can instantiate the function, thus embedding the time dimension into the function definition.

Several methods possess similar characteristics, e.g. description of a function in verb-object format. Implementation of some methods in engineering practice lends itself to useful software implementation and automation. The FBS framework is an example of this. Decomposition of a complex system in a solution independent manner is a generic issue which is not addressed by any framework. System decomposition tends to be based on brainstorming at any level, i.e. by asking the question "how is this function achieved?"

Based on a review of current methods and practices in industry, which emphasized the need for a structured and coherent methodology for the decomposition of a complex system on a solution-independent functional basis, the requirements for a functional framework have been summarized as follows [23]:

- To be integrated with other tools commonly used in industry to encourage broad take-up of the framework;
- To have a graphical representation to provide shared understanding of the analysis of a system;
- To promote the primacy of function and solution-neutral thinking in systems engineering design analysis;
- To be based on tools and methodology which can be applicable across disciplines (i.e. electro-mechanical, control and software).

3. System State Flow Diagram

3.1. The Basis of the System State Flow Diagram

An engineered system is commonly represented as a system block diagram, which is a black-box representation of the system [11]. The overall function of the system is shown in the box, with the inputs and outputs to the system also indicated on the box. Figure 1 illustrates a system block diagram for a Bread Toasting System (BTS), which will be used as a reference case study.

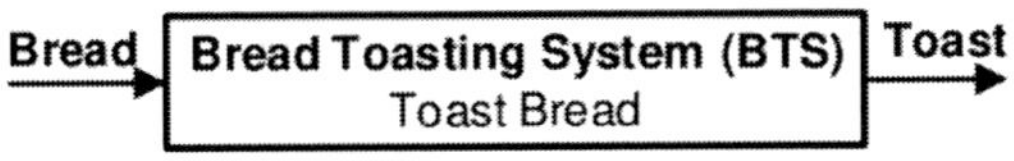

Figure 1. Representation of a Bread Toasting System as a system block diagram

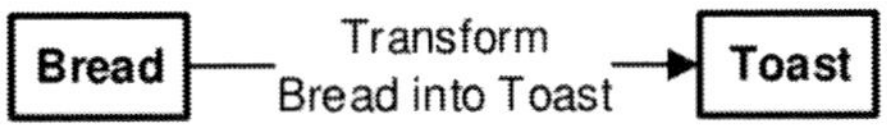

Figure 2. A system state diagram of a BTS

Most function modelling frameworks define the inputs and the outputs of the system in terms of the flows of energy, material, or information [10]. The function of the system is commonly articulated in verb-noun format in relation to flows.

The SSFD, illustrated in Figure 2, follows the general principles of state diagrams [24]: a box denotes a state and an arrow denotes the function required to achieve the state transition. Coherent with [17,19-20], a state can be thought of as a generic object described by a set of measurable attributes. The function can be described in verb-object structure which is common practice in literature (see [10] for a review). Coherent with [17], this characterization is related by the rule that the verbs correspond to the operations on the attributes and the nouns to the objects. The function is articulated in respect of relevant object attribute change required to transition between states.

3.2. *Functional Decomposition of a System Based on the System State Flow Diagram*

The SSFD starts by generation of a high level system state flow diagram which represents input and output states of the system. This follows the identification of the main flow through the system. This requires the identification of intermediate states between the input state and the output state described in the high level system state flow diagram. An intermediate state is thought of as an object and its measurable attributes like for the input state and the output state. The flow of the states is mapped through the system and then functions required to achieve these state transitions are articulated. Next, the auxiliary flows of the system are identified based on the main flow. The structure of the auxiliary flows follows the same logic and structure as the one used for the main flow. Once all the flows of the system are identified they are aggregated into a single model. Lastly, design elements are identified for the fulfilment of the functions. In this paper, all design decisions will be taken based on the structure of a common slotted toaster as an example.

Step 1: Generate a high level system state flow diagram

The SSFD provides a high level representation of a system as a state diagram, showing the input state and the output state to the system. From an engineering point of view, the analysis should start by defining the fundamental working principle of the system to describe its input and output states. In fundamental terms bread toasting is achieved by heating the bread surface to 310°F [25], when chemical transformations (known as the Maillard reaction) in the bread are triggered generating the characteristic flavours of toast. Further analysis of physics shows that increasing the temperature of a slice of bread changes its size (i.e. length and thickness), temperature, weight, composition (i.e. sugars and starches start to caramelize and turn brown) and moisture capacity (i.e. vaporization of the moisture of the bread which directly reduces the weight of the bread). Location of the bread/toast (i.e. transport, orientation, position, etc.) is also considered during transition between states. The system should address these attributes for the generation of a slice of toast. Using this information the high level SSFD for the BTS can be updated, as shown in Figure 3.

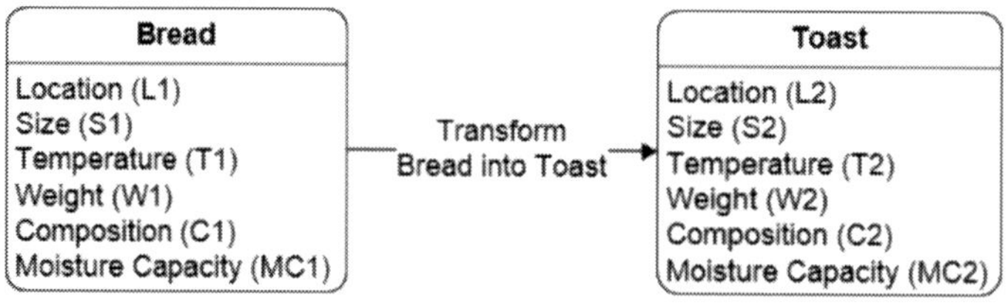

Figure 3. BTS high level system state flow diagram

Step 2: Create the main flow through the system for the input-output state

The main flow through the system represents the system's required functionality by identifying required state transitions with reference to the input and the output states of the system shown in

Figure 3. As discussed, the fundamental working principle of the system is associated with the increase of the bread temperature. A design decision should be made to heat the bread. There are three basic ways of heat transfer: conduction, convection, and radiation and there are a variety of ways of increasing the temperature of the bread on the basis of these modes, e.g. over an open fire, focused sunlight and radiant heat. A common slotted toaster uses radiant heat to heat the bread. In this case, the system must apply radiant heat directly to the bread slice [25]. The usage of radiant heat in the system points out the requirement for the orientation of the bread relative to the heat source. According to this requirement, the flow should start by positioning the bread relative to the heat source (i.e. load bread). Once the bread is toasted (i.e. heat bread) it comes out of the system (i.e. eject toast). Figure 4 shows the main flow through the system. The box around the system defines the "system boundary" which shows the limits of the scope for responsibility for the design team.

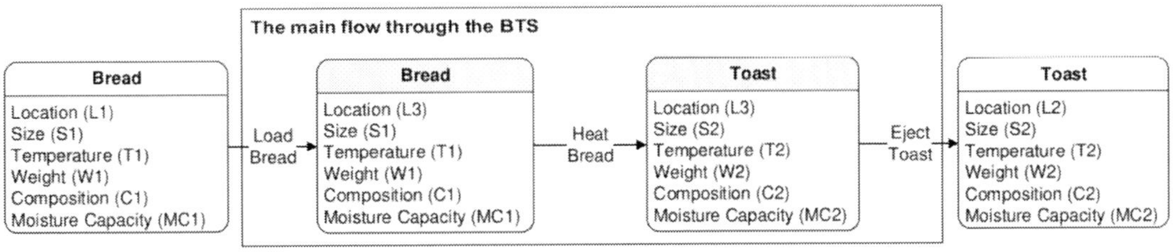

Figure 4. The main flow through the BTS

Step 3: Identify the auxiliary flows based on the main flow of the system
As mentioned earlier, the BTS engineering principle is associated with the increase of the bread temperature through applying radiant heat directly to the bread slice. This step requires a way of delivering the function `heat bread` through generating radiant heat from a given energy source. A variety of energy sources (e.g. mains electricity, sunlight) can be considered to achieve the function `heat bread`. A common slotted toaster uses mains electricity as energy source, which should be converted into radiant heat. Figure 5 represents the high level SSFD for the auxiliary flow.

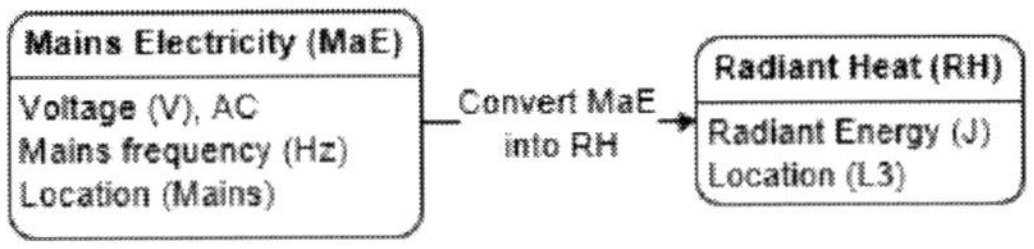

Figure 5. The high level system state flow diagram for the auxiliary flow

Reflecting the energy source choice made (i.e. mains electricity supply), the auxiliary flow should provide "controlled supply of mains electricity" (i.e. conversion of AC into DC) and then its conversion into Radiant Heat (i.e. convert EE into RH) which is used (i.e. channel RH) to achieve the function `heat bread`. Figure 6 shows the auxiliary flow through the BTS.

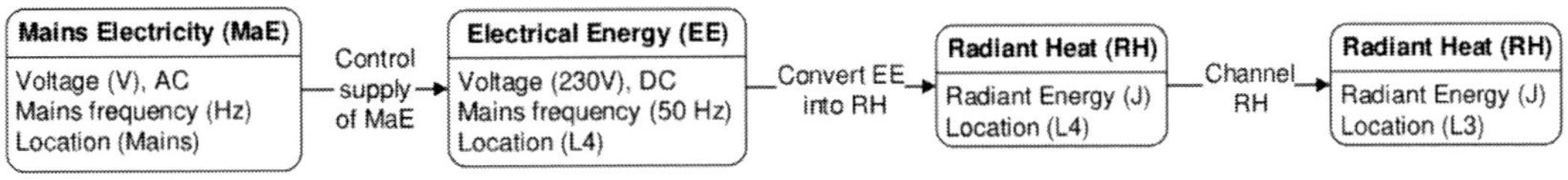

Figure 6. The auxiliary flow through the BTS

Step 4: Aggregate function chains into a functional model
In this step, all of the function chains from Step 2 and Step 3 are aggregated into a single model. Figure 7 shows the SSFD for the BTS. The dashed state is used to connect the distinct chains together.

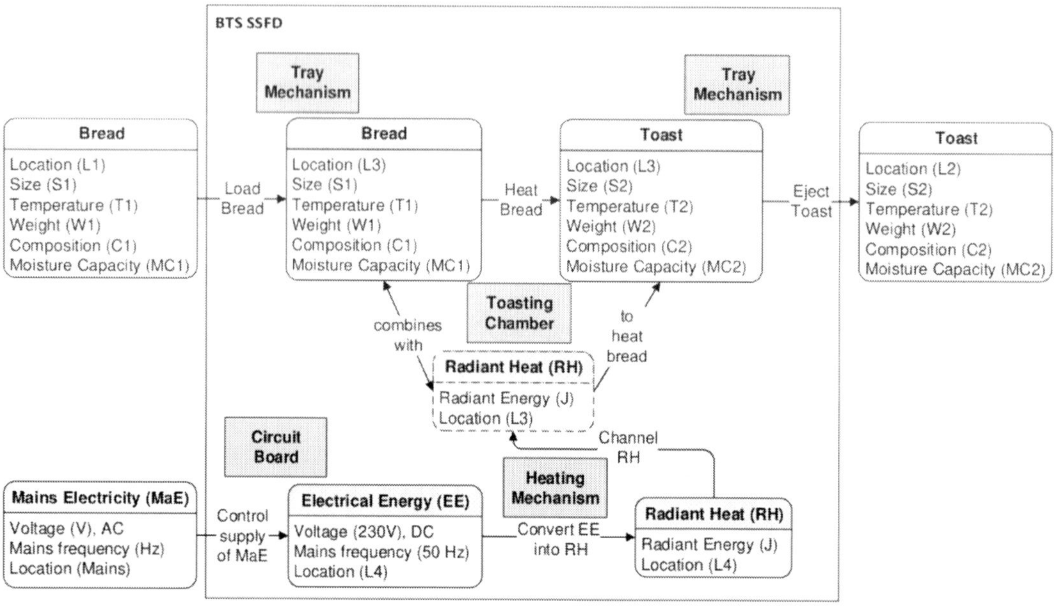

Figure 7. BTS SSFD

An engineered function can completely be defined in terms of the triad of an input state, an output state and a design element required to transition between states coherent with [16-18]. Figure 7 includes design elements for the fulfillment of the functions.

From this SSFD representation a high level BTS Function Tree and a BTS boundary diagram can be directly extracted as shown in Figure 8 and in Figure 9 respectively.

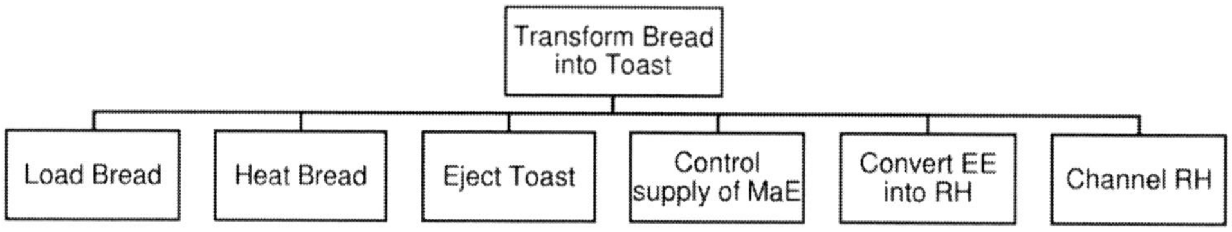

Figure 8. BTS Function Tree

A System Boundary Diagram (SBD) is a graphical representation of the system and it includes the design elements identified on the SSFD along with the flows of energy, material and information required to transform system inputs into system outputs [11].

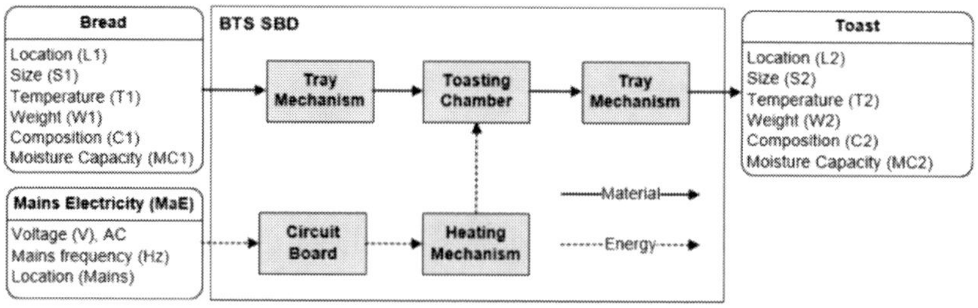

Figure 9. BTS SBD

The diagram is extracted from the BTS SSFD by deleting all information other than Input/Output states and Design elements boxes. Then, the design elements boxes are joined with arrows denoting the flows of Energy, Material and Information between them.

4. Discussion and Conclusion

The aim of this paper was to present the System State Flow Diagram (SSFD) as a structured framework for the decomposition of a complex system on a solution-independent functional basis. The SSFD draws from the methodology of Paul and Beitz [11] and Otto and Wood [8] for system decomposition, and from the C&C^2-A [16], the OAF framework [17] and TRIZ [18] for function modelling. The SSFD offers a more structured way for system decomposition based on its function representation model, addressing the weaknesses of the current practice, discussed in the first section of the paper.

Through the implementation of the SSFD on the Bread Toasting System case study it is demonstrated that the requirements outlined in section 2 are met, as discussed below:

- The SSFD can be integrated with other tools currently used in the systems engineering design. A function tree can be extracted from a SSFD in a solution neutral way rather than conduct brainstorming (How-Why) which is the common approach in practice [8]. A SSFD also facilitates the development of a system boundary diagram which enables interface functions to be identified through interface analysis [7] (see [23] for an example).
- It improves communication within the engineering team in the sense that its system representation is easy to understand and portable across multiple engineering disciplines. Thus it supports the achievement within an engineering team of a common understanding of the functional decomposition of the system in a solution neutral way.
- Thinking of the input and the output in terms of an object and its measurable attributes is an important feature of the SSFD which divorces the consideration of function from the consideration of the design solution, that is, it promotes solution independent thinking in the design of a system, for example, a slice of bread can be toasted in a variety of ways (e.g. Electromagnetic waves, over an open fire) as long as the bread's surface temperature reaches about 310°F.

Further work will concentrate on the development of the SSFD to promote identification of further functional requirements through its integration with the interface analysis method [7] to provide a comprehensive interface management to ensure all system interfaces are identified, characterized and managed through function requirements both at the system level and the subsystem level.

References

[1] Lu SC-Y and Suh N-P 2009 Complexity in design of technical systems *CIRP Annals* **58** pp 157-160
[2] Tomiyama T, D'Amelio V, Urbanic J and ElMaraghy W 2007 Complexity of multi-disciplinary design *CIRP Annals* **56**(1) pp 185-188
[3] Campean I F and Henshall E J 2012 Failure mode avoidance in automotive systems engineering design *Proceedings of the 2nd International Workshop on Modelling and Management of Engineering Processes (MMEP)* P Heisig, P J Clarkson eds. University of Cambridge Cambridge UK pp 15-29
[4] Suh N P 2001 *Axiomatic design: advances and applications* (New York: Oxford University Press)
[5] Browning T R 2001 Applying the design structure matrix to system decomposition and integration problems: a review and new directions *IEEE Transactions on Engineering Management* **48**(3) pp 292-306

[6] Pimmler T U and Eppinger S D 1994 Integration analysis of product decompositions *Proceedings of the ASME 6th International Conference on Design Theory and Methodology* Minneapolis
[7] Webb R D 2002 *Investigation into the application of robustness and reliability tools to the design process* MSc thesis University of Bradford
[8] Otto K and Wood K 2001 *Product design: techniques in reverse engineering and new product development* (New Jersey: Prentice Hall)
[9] Campean I F, Henshall E, Yildirim U, Uddin A and Williams H 2013 A Structured Approach for Function Based Decomposition of Complex Multi-disciplinary Systems *Proceedings of the 23th CIRP design conference* M Abramovici, R Stark eds. Bochum Germany pp 113-123
[10] Srinivasan V, Chakrabarti A and Lindemann U 2012 A framework for describing functions in design *Design methods: proceedings of the 12th international design conference* Dubrovnik 21-24 May the Faculty of Mechanical Engineering and Naval Architecture University of Zagreb pp 1111-1122
[11] Pahl G, Beitz W, Feldhusen J and Grote K H 2007 *Engineering design: A systematic approach* 3rd edn. (London: Springer)
[12] Ullman D G 2010 *The Mechanical design process* 4th edn. (Singapore: McGraw-Hill)
[13] Jarratt T A W 2004 *A Model-based approach to support the management of engineering change* Ph.D. thesis University of Cambridge Cambridge
[14] Department of Defence System Management College 2001 *Systems Engineering Fundamentals* (Virginia: Defense Acquisition University Press)
[15] Matthiesen S and Ruckpaul A 2012 New Insights on the Contact&Channel-Approach – Modelling of Systems with Several Logical States *Proceedings of the International Design Conference (DESIGN)* Dubrovnik Croatia pp 1019-1028
[16] Albers A, Oerding J and Alink T 2011 Abstract Objectives can Become More Tangible with the Contact and Channel Model (C&CM) *Proceedings of the 20th CIRP Design Conference* A Bernard ed. Springer Berlin pp 203-213
[17] Sickafus E N 1997 *Unified Structured Inventive Thinking: How to Invent* (Michigan: Ntelleck)
[18] Fey V and Rivin E 2005 *Innovation on Demand* (New York: Cambridge University Press)
[19] Dori D 2002 *Object-Process Methodology: a Holistic Systems Paradigm* (Germany: Springer)
[20] Umeda Y, Ishii M, Yoshioka M, Shimomura Y and Tomiyama T 1996 Supporting Conceptual Design based on the Function-Behaviour-State Modeller *AI EDAM* **10**(4) pp 275-288
[21] Gero, J S 1990 Design prototypes: A knowledge representation schema for design *AI Magazine* **11**(4) pp 26-36
[22] Goel A, Rugaber S and Vattam S 2009 Structure, behaviour, and function of complex systems: the structure, behaviour, and function modelling language *AI EDAM* **23**(1) pp 23-35
[23] Campean I F, Henshall E, Brunson D, Day A, McLellan R, and Hartley J 2011 A Structured Approach for Function Analysis of Complex Automotive Systems *SAE Int. J. Mater. Manuf.* **4**(1) pp 1255-1267
[24] Harel D 1987 Statecharts: A Visual Formalism for Complex Systems, *Science of Computer Programming* **8** pp 231-274
[25] Mital A, Desai A, Subramania A and Mital A 2008 *Product Development: a Structured Approach to Consumer Product Development, Design and Manufacture* (Oxford: Butterworth-Heinemann)

27th International Conference on CADCAM, Robotics and Factories of the Future 2014 IOP Publishing
IOP Conf. Series: Materials Science and Engineering **65** (2014) 012005 doi:10.1088/1757-899X/65/1/012005

Complex Product Architecture Analysis using an Integrated Approach

Amad Uddin[1, a], Ioan Felician Campean[1,b] and Mohammed Khurshid Khan[1, c]

[1] Bradford Engineering Quality Improvement Centre (BEQIC),
University of Bradford, United Kingdom, BD7 1DP.

[a]a.ud-din1@bradford.ac.uk, [b]f.campean@bradford.ac.uk, [c]m.k.khan@bradford.ac.uk,

Abstract. Product design decomposition and synthesis is a constant challenge with its continuously increasing complexity at each level of abstraction. Currently, design decomposition and synthesis analytical tasks are mostly accomplished via functional and structural methods. These methods are useful in different phases of design process for product definition and architecture but limited in a way that they tend to focus more on 'what' and less on 'how' and vice versa. This paper combines a functional representation tool known as System State Flow Diagram (a solution independent approach), a solution search tool referred as Morphology Table, and Design Structure Matrix (mainly a solution dependent tool). The proposed approach incorporates Multiple Domain Matrix (MDM) to integrate the knowledge of both solution independent and dependent analyses. The approach is illustrated with a case study of solar robot toy, followed by its limitations, future work and discussion.

1. Introduction

Complexity in product development is very common and emerges from many domains of design world. The product can be complex in its requirement, function, and form domains. Researchers often consider many other domains necessary and useful for the product design and architecture analysis [1]. Engineers tend to deal with product complexity by breaking down the complex problem into smaller problems [2, 3]. Many tools and methods are available to the engineers to manage, organize, and integrate the information of product complexity. There are two major aspects of product modelling in engineering design: *functional modelling* (also referred to as solution independent analysis) and *structural basis* (solution dependent analysis) [4].

In the *functional modelling*, the function of the product is decomposed into various sub-functions so that engineers can search solutions for each of the decomposed sub-function [3]. The function modelling technique has been discussed in many texts with various methods employment at the early stages in the design process. For example, Pahl et al. [5] describe structured methods to establish function structures in the conceptual design stage, where the product is modelled as a series of sub-functions via three distinct types of input and output: energy, material, and information; and thereafter to search for working principles and structures that fulfil sub-functions.

The product is also modelled through various methods by using *structural basis*. Among various methods, Design Structure Matrix (DSM) is readily available in literature and deals with the design of a product through decomposing and integrating it on structural basis and handling interactions or dependencies between the decomposed parts, mainly via qualitative and quantitative schemes. A comprehensive review of this method is provided by [6] and [7].

The two key modelling aspects, functional and structural basis, have their pros and cons. In practice and in literature [3, 4, 8], it is observed that both aspects are applied in product development and systems engineering organizations. On the one hand, the product architects take interest in solution dependent analysis, where they look at the linkages, attributes and interface relations between the decomposed parts. The functionalists define and analyse the functionalities of the product on the other hand. This paper proposes a concept that aims to capture the information of both solution independent and dependent analysis in a single framework in a coherent and structured manner using a multiple-domain approach.

2. Literature review of product modelling frameworks

A number of product modelling frameworks have been developed over the past two decades in engineering design. For readers, the comprehensive reviews and applications of such modelling frameworks in academic and industry are provided by [9] and [10]. In this paper, a literature of existing modelling frameworks is reviewed to highlight the different information domains. A set of requirements for integrating the solution independent and dependent analysis is thus synthesized, forming the basis of the conceptual approach in this paper.

2.1. Modelling context on functional aspect

A distinct step in function modelling approaches is the establishment of solution neutral *function structure*, which is used as basis for the subsequent design tasks to define the physical parts and structure of the product in the early stage of design [5, 11]. The basic purpose of a function structure is to organize the *functions* of a product in a coherent manner. The functional models have a variety of such function structures ranging from hierarchical tree [12] to flow oriented [5, 13–15] shown in Figure 1(a-c). In flow orientation, one function structure approach which is quite famous, shown in Figure 1(c), is the organization of functions in terms of *flows* between them: material (M), energy (E) and information (I) [5]. Another type of function structure, recently developed by [13, 14, 16] is shown in Figure 1(b), which describes the function between the two states. As a conclusion, functional models allow the possibility to identify the functional requirements and on the basis of which different concepts can be generated and evaluated.

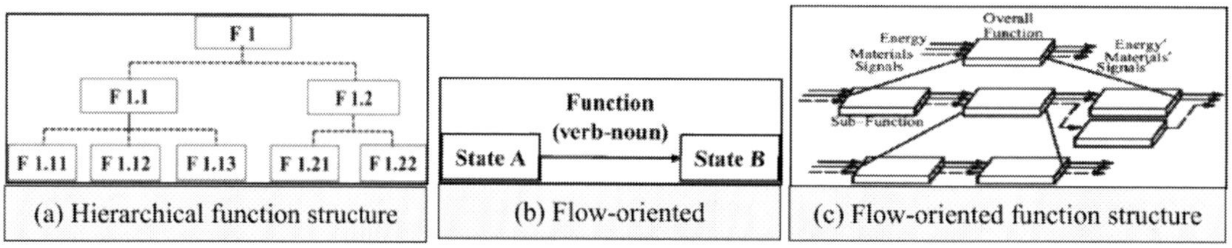

Figure 1. Classification of functional models

2.2. Modelling context on function to structure aspect

Function domain is the intermediary between requirement and *structural* domains. In the Axiomatic Design (AD) [17], a product is modelled hierarchically via a zigzag procedure between functional requirements and design parameters of functional and physical domains respectively. According to [17], design parameters may be *physical parts*, parameters or assemblies. However, concrete decomposition operations on the product description have not been explained [18]. Many *building blocks/elements* such as physical features in Function-Behaviour-State (FBS) model [19], object and process elements in Object-Process Methodology (OPM) [20] and working surface pairs and channel support elements in Contact & Channel Model (C&CM) [15] have been introduced for function to structure *mapping* as shown in Figure 2(a-c). The concept of working principles [5] has also been established to search for conceptual solution via morphology matrix that could satisfy the developed function structure, a network of functions and flows. The key reason behind such diverse building blocks is that a group of researchers believe, particularly related to artificial intelligence (AI) [19, 21, 22], that functions and parts alone are not sufficient in conceptualizing the product design and thus introduced another notions such as *behaviour* and *state* that exist between functions to parts mapping. Many cross-domain matrix approaches, such as Domain Mapping Matrices (DMM) and Multiple-Domain-Matrix (MDM) in [23–25], shown in Figure 2(d), have been developed to model the interactions between functions and parts of a product. As a conclusion, the above approaches strongly define and represent product with different building blocks, domains, *and* support product analysis in the context of function to structure mapping.

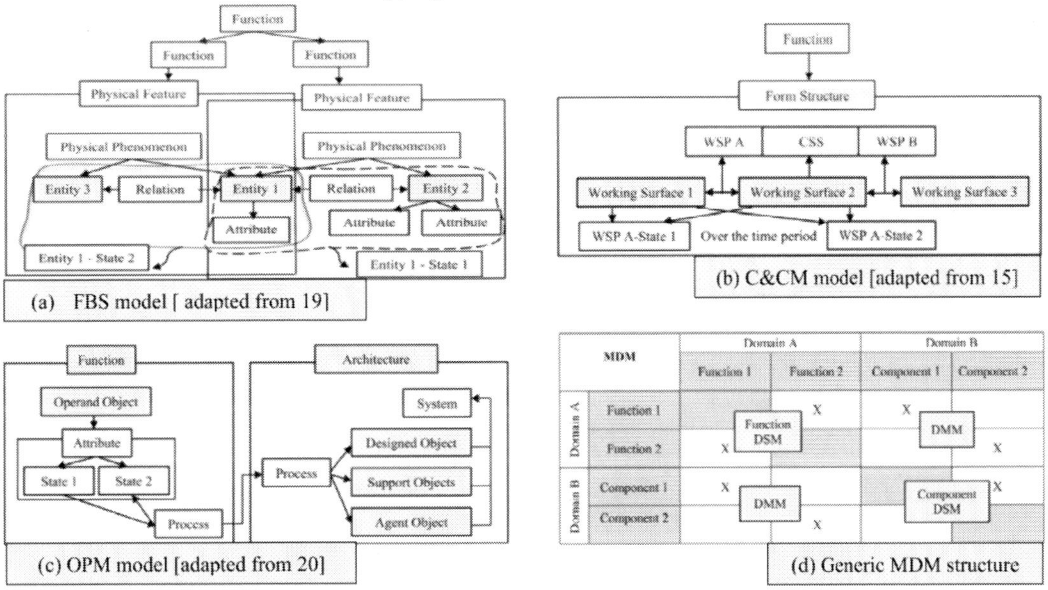

Figure 2. Classification of function to structure modelling approaches

2.3. Modelling context on structural aspect

In literature, a DSM approach is widely recognized to model the product in a single domain. Looking at the physical domain of a product, a DSM approach decomposes the product into its parts/components and represents its architecture by capturing *interactions/relationships* between the components [6]. In a traditional DSM, shown in Figure 3(a), a relationship between two components is described via a qualitative *scheme* such as 'x' or quantitative number e.g. '1'. However, this sort of representation does not describe multiple interactions as realized by researchers in [3, 26, 27]. According to [28], the relationships or exchanges between two components can be of *diverse* nature. The importance of DSM based decomposition of a product and its integration analysis, underpinned by clustering technique in an industry practice, has been highlighted by Pimmler and Eppinger [3]. They used static DSMs to identify and examine alternative product architectures. Pimmler and Eppinger [3] describe four types of interactions referred as spatial (S), material (M), energy (E), and information (I) between the two interacting components along with the quantification scheme that facilitates weighing interactions among them, shown in Figure 3(b). Jarratt [4], looking from engineering change management perspective, also used the concept of S/E/M/I interactions between two interacting components but with addition to multiple-type *linkages* definitions with steady and dynamic states, shown in Figure 3(c). Campean et al [28], looking from product function analysis perspective in physical domain, characterized the exchange nature between two subsystems/components into four: physical (P), and M/E/I. They also considered internal and external interfaces in the interface matrix (IM), shown in Figure 3(d). They argued that there may be several functional requirements associated with just one type of exchange in an interface and such requirements can be documented in tabular format. Martin and Ishii [27] used the concept of specification flows in the product platform architecture. They introduced two indices to develop the decoupled architecture: a generational variety index for analyzing generational change and the coupling index to analyse impact of changes in components. In their component-based DSM, an interface is characterized by specification flows that seem to represent a combination of flowing attributes and parameters, shown in Figure 3(e). As a conclusion, a product is modelled via DSM on structural basis with diverse exchanges and representations in its components.

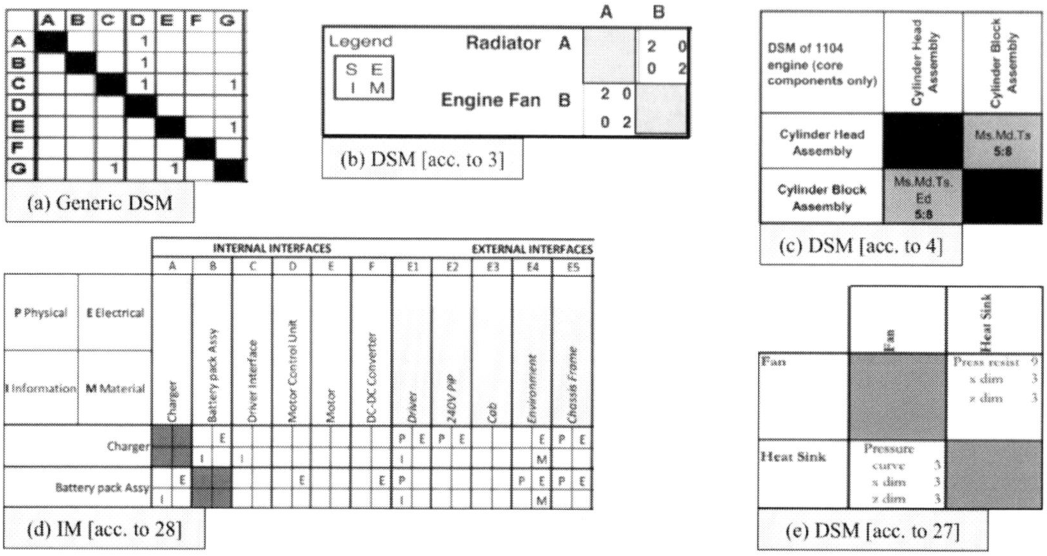

Figure 3. Classification of structural models

3. Analysis and critique of state-of-the-art

A number of product modelling approaches have been discussed in previous section in the context of functional, function to structure mapping, and structural domains. There are a number of commonalities in the modelling approaches. For instance, the concept of multiple flows or exchanges i.e. E/M/I between functional and structural elements is quite distinct as discussed in Section 2. Furthermore, it is also observed that many building blocks and tools have been established and introduced for mapping and describing the relations from function elements to structure elements for conceptualising the product. Theories like AD, C&CM, and OPM, on the one hand, support the product evolvement between different levels, whilst rule-based approaches do not, e.g. FBS. On the other hand, cross domain matrix approaches support the product analysis process but do not support design solutions search and synthesis activities based on functional domain.

The application of product modelling approaches have been widened over the past decades. However, such modelling approaches involve many complex notions, representational elements, and tools for product modelling that practitioners and industrial engineers still often find hard to implement and grasp. One of the critical reasons is the management and tracing of a product's complex information in various tools from solution independent to solution dependent analysis and vice versa. One visual approach that is considered very powerful and helpful in representing the product modelling information of many domains is the Multiple-Domain-Matrix (MDM), shown in Figure 2(d). The applications of MDM approach has continuously increased over the last few years in product design [25], process optimisation and modelling [29] and management of product development project [30] and is also considered in this paper for product architecture analysis.

As an overall conclusion, the discussed methods are valuable in different phases of product design. However, the existing methods still lack to deliver a straightforward analysis and comprehensive integrated solution independent-dependent analysis. To summarise, no current approach exists that provides the information of both solution independent and dependent analysis in a *single tool, and that also takes into consideration* the following issues;

- To *balance* the design effort both across solution independent and dependent analysis
- To deliver a *process* based on existing but minimum tools that are utilised in current practice for product decomposition and synthesis
- To establish a structured, coherent and *common linkage* between solution independent and dependent analysis for the *integration* purpose
- To support the consistent knowledge transferability between functionalists and architects

This paper proposes an integrated approach for complex product architecture analysis starting from solution-neutral to solution dependent analysis based on tools existing in literature.

4. Proposed Approach

The process for the proposed integrated approach is presented, on left side of Figure 4. The approach is based on two key phases: solution independent and dependent phases. The tools' sequence and information flow for the integration of both phase analysis is shown, on right side of Figure 4.

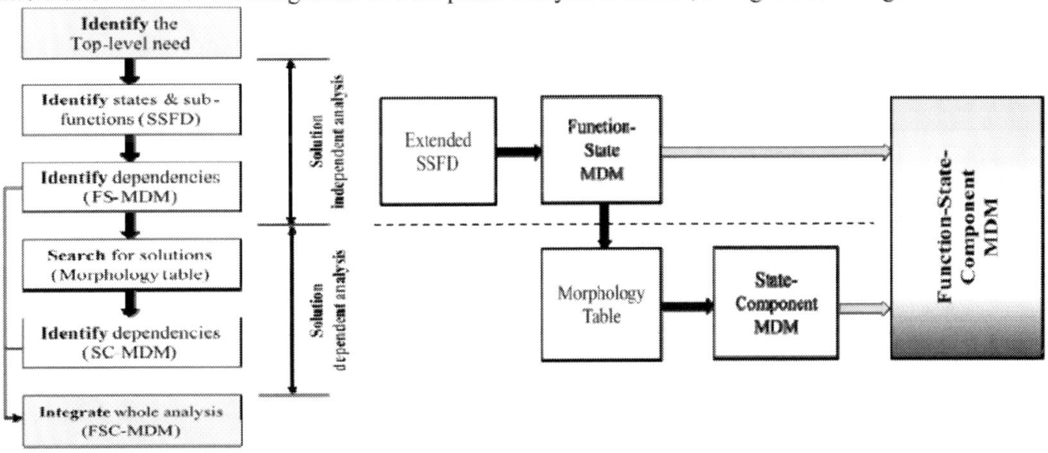

Figure 4. Proposed integrated approach process structure and information flow

The working of the proposed approach is discussed in this section, followed by its illustration with a case study in the next section.

4.1. Solution independent analysis phase

The solution independent phase is based on two key tools: System State Flow Diagram (SSFD) and Function-State (FS) MDM. The product definition and architecture analysis starts with solution independent stage underpinning a tool known as SSFD, introduced for the first time in [12] and discussed in detail in [13]. It is then followed by FS-MDM tool comprising of function DSM, state DSM and two FS-DMMs, driven by SSFD. Here, an SSFD is adapted to some extent by involving *constraint* elements on states' attributes. A distinct feature of SSFD is its ability to describe a main flowing object's states transition based on graphical representation. The solution independent MDM analysis is conducted on the modelling concepts and framework given by [23], however, the elements of our functional model are slightly different, particularly in the context of their *operation* and our *function* definition. Moreover, the interfaces in their [23] MDM between the domains are represented via qualitative and quantitative schemes such as mark 'x' and values ranging from 1-3. Our approach involves input-output flows specifications in the state-DSM interfaces and by keeping the same qualitative scheme in function-DSM in the solution independent MDM.

4.2. Solution dependent analysis phase
Once the functionalists complete the SSFD analysis and transform the knowledge into FS-MDM, it is then transferred to architect's team that explore the solutions and perform solution dependent analysis. In this phase, there are again two key tools available to engineers: the morphology table and the State-Component (SC) MDM. The state DSM within SC-MDM remains same as in the stage of solution independent phase whereas the interfaces between component DSM are represented via four types of interactions i.e. P/M/E/I with their flows exchanges and specifications.

In the end, the knowledge of both phases is integrated into a single framework named as Function-State-Component (FSC) MDM, shown in Figure 4's right side, which represents the whole product architecture in a structured manner. State domain is the linkage which is common between solution independent and dependent analysis. In the final FSC-MDM, this domain helps the architects to know components' capabilities in terms of delivering the desired input-output flows and in what capacity they interact with each other in terms of flows and spatial specifications. For the functionalists, the FSC-MDM manages to identify the required functions for states transition and the dependencies between functions. It can be observed that equal number of tools are available for both phases: the SSFD and FS-MDM tools in solution independent phase and the morphology table and SC-MDM tools in solution dependent phase.

5. Case Study
A domestic scale solar robot toy (SRT) multi-disciplinary case study is selected for illustrating the proposed framework. To keep the study simple, a specific unit (such as wave arm or leg) in SRT has been considered for analysis. Solar energy powers the SRT solar board and is absorbed via thermal collectors, and this energy is then used to drive mechanical systems. The SRT receives solar energy as an input and generates rotational energy to rotate its arm/leg as an output. In order to illustrate the process and to keep the case study 'simple', only the *energy flow* mechanism is considered here.

5.1. Identifying the top level need
At desired abstract level, the top function of SRT is 'convert solar energy into rotational energy' in order to rotate the robot arm full 360^0. It is represented by an extended-SSFD format comprising of constraint conditions in Figure 5.

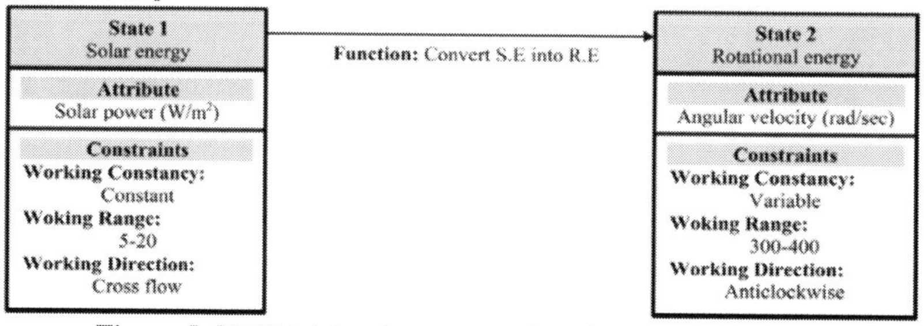

Figure 5. SRT high level representation via extended SSFD

5.2. Solution independent analysis

5.2.1. SSFD tool: high level function decomposition
The design team can decompose the high level function into sub-functions with the identification of intermediate states as shown in Figure 6. The functionalists at this stage perform many iterations for function structure development, thereby identifying intermediate states and corresponding functional requirements. Finally, the functionalists agree to the iterated function structure shown in Figure 6.

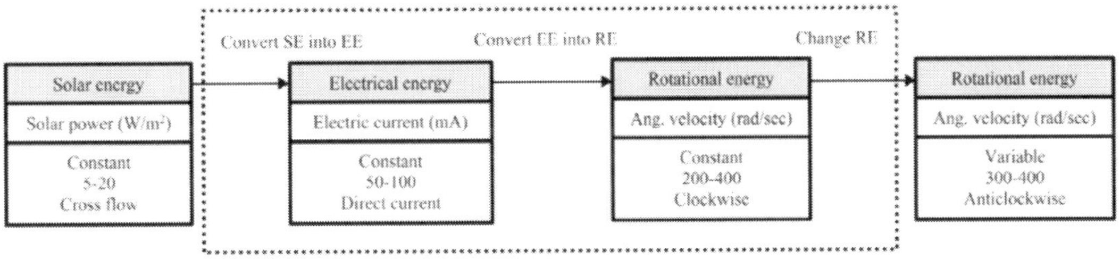

Figure 6. The main flow representation with sub-functions via SSFD

5.2.2. Knowledge transformation into FS-MDM

The knowledge from SSFD is then transformed into FS-MDM. In the FS-MDM, functional interfaces become visible while divorcing the states from functions. Moreover, flowing objects in state domain are also divorced from functions and their interfaces are also identified and managed in terms of input-output flows specifications.

5.3. Solution dependent analysis

After the solution independent analysis, the designers search for solutions that could satisfy the identified functions and relevant input-output states.

5.3.1. Morphology table

A morphology table, built-on function-state and constraint knowledge, is used to provide the feasible solutions, shown in Figure 7. The relevant solutions are adopted and taken forward for detail analysis.

Functions	States		Attributes		Constraints						Solutions
					Constancy		Direction		Range		
	I	O	Input	Output	I	O	I	O	I	O	
Change R.E	R.E	R.E	Ang. vel (rad/s)	Ang. vel (rad/s)	Constt	Constt	CW	ACW	[R]	[R]	Gear Pair Box
Change R.E	R.E	R.E	Ang. vel (rad/s)	Ang. vel (rad/s)	Constt	Var.	CW	To and fro	[R]	[R]	Crank Rocker
Convert E.E to R.E	E.E	R.E	Current (mA)	Ang. vel (rad/s)	Constt	Constt	DC	CW	[R]	[R]	DC-Motor
Convert E.E to R.E	E.E	R.E	Current (mA)	Ang. vel (rad/s)	Constt	Var	AC	CW	[R]	[R]	AC-Motor
Convert S.E to E.E	S.P	E.C	Solar p. (W/m^2)	Current (mA)	Constt	Constt	Cross flow	DC	[R]	[R]	Solar Board

Figure 7. Morphology table

5.3.2. Knowledge transformation into SC-MDM

The identified solutions are then placed in SC-MDM according to states input-output flows. This MDM helps the designers to capture the interface nature between the two components rather than brainstorming. The designers have to search for spatial or physical specifications between the interacting components.

5.4. Analysis in single framework

The FS-MDM and SC-MDM analysis are then combined in the FSC-MDM to provide a structured and coherent information flow from function to structure domains for product architecture. The whole architecture of SRT at system level, based on SSFD and morphology table, is transformed in FSC-MDM framework shown in Figure 8.

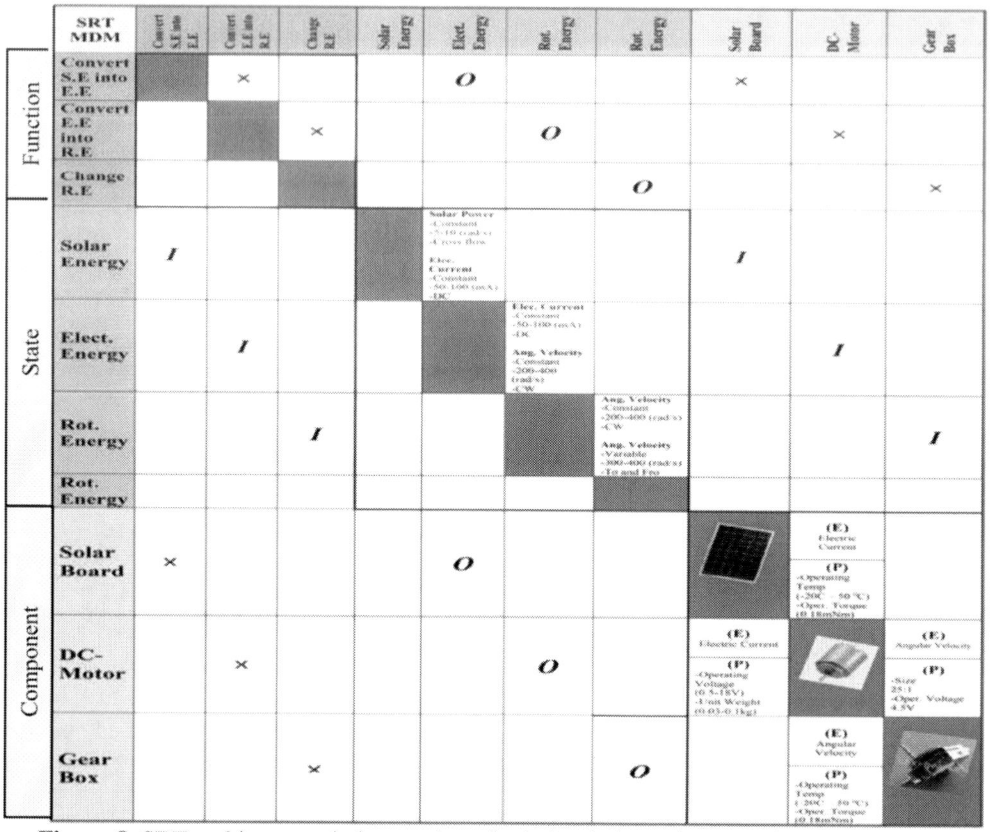

Figure 8. SRT architecture via integration of solution independent and dependent analysis

6. Approach Benefits and Limitations

As shown, the proposed approach meets its objectives highlighted in Section 3. The final MDM shows consistent information in a way that functional and structural elements share same form of input-output flows specifications. The state domain is the key linkage, binding the function and component domains in a coherent and structured manner. An application of the approach is also discussed from engineering change management and solution generation perspectives in the following paragraphs.

The functionalists and architects can both trace the changes from top-down and bottom-up in FSC-MDM. Firstly, from bottom-up, the exchanges between components are based on functions and states information rather than brainstorming. DC-Motor and Solar Board, both coupled, share electric current (energy flow) in which the current as an input to DC-Motor and output from Solar Board. Secondly, the same DSM shows that DC Motor *provides* specifications information to solar board and Gear Box or, inversely speaking, Solar Board and Gear Box *require* the specific information from DC-Motor such as operating voltage, motor weight, and motor size. This managed information helps the designers to visualise if they change or replace DC-motor then which specifications need to be carefully handled for other components. Any change in DC-motor will cause the solar board to change to meet the DC-motor specifications and similarly it will also impact on the corresponding functions and states all the way up. This distinct feature is observable in Figure 8. Hence these two sorts of interfacing management features (i.e. exchanges and specifications) in components DSM also help not only for final concept design but also for assembly design or change management.

Secondly, from top-down, if any functional requirement is changed at the top level then it will also impact all the way down to components and the interfaces between them. For example, the product architecture based on following components i.e. *Solar Board+DC-Motor+Gear Box* was selected as a final concept design. Now assume if the top level functional requirement associated with arm rotation is changed from 360^0 rotation to $0\text{-}180^0$. Moreover, it should return back from $180\text{-}0^0$, then the design team will look for another solution, e.g. 'Crank Rocker' from morphology table (see Figure 7) that meets the same sub-functional and states requirement but with different-constraint conditions, as it provides 'to and fro' motion. In that case, the product architecture design concept would be *Solar Board+DC-Motor+Crank Rocker* and it will require the designers to manage the specifications and interfaces between components accordingly. Thus, the proposed approach integrates solution independent and dependent analysis in a structured way.

Though the approach serves its primary objectives, however it has some key limitations which are discussed and considered for future work. The proposed approach has a systematic structure, however, it results in establishing and completion of large matrices. The input-output is described with only one attribute and its constraints. According to SSFD [13], a state is a generic object which can be described by a set of measurable attributes. In the proposed approach, a state domain is used as an intermediary and mapping element from functions to structures. The current approaches such as FBS, C&CM, and OPM offer many structured building blocks for function to structure mapping and recently, MDM-based approach also include behaviour domain within the function and structure domains [31]. The customer/market requirements are also not captured in the proposed approach. These limitations would be considered for future work.

7. Discussion and Conclusions

The main aim of this paper was to develop an integrated and structured design approach for complex product architecture in the context of solution independent and dependent analysis based on tools that are currently practice. The review of current methods and tools as well as the importance of product functional and structural modelling in the academic and industrial practice pointed out the need for the proposed approach. The proposed framework is built-upon the adaptations of principles of SSFD, morphological table and MDM tools. The SSFD tool helps for functional decomposition, morphological table for searching solutions and MDM for design architecture synthesis. The framework provides a systematic and balance procedure to perform both solution independent and dependent analysis. A core binding linkage between the function and structure domains is the state domain that keeps the information in a way that both functional and structural elements share same sort of input-output flows specifications.

The Solar Robot Toy case study illustrates the working of the proposed integrated approach. The approach strengths are also highlighted with the SRT case study in the contexts of identifying and comparing various solutions at same abstraction level and from engineering change management. The framework can be applied in various decomposition levels. The proposed approach is easy to understand and it employs the minimum graphical tools that help in improving the communication between functionalists and architects.

References

[1] Uddin A et al 2013 Function-based conceptual design expert (CDE) systems *Int. Conf. on Adv. in Mech and Manu. Engg.* Kuala Lumpur Malaysia
[2] Chmarra M K et al 2008 Revisiting the divide and conquer Strategy to deal with complexity in product design *IEEE/ASME Int. Conf. on Mechatronic and Embedded Systems and Application*
[3] Pimmler T U and Eppinger S D 1994 Integration analysis of product decompositions *ASME Design Theory and Methodology Conference*, September Minneapolis
[4] Jarratt T A W 2004 A model-based approach to support the management of engineering change PhD Thesis University of Cambridge, UK
[5] Pahl, P et al 2007 *Engineering design a systematic spproach* (Springer).
[6] Browning T R 2001 Applying the design structure matrix to system decomposition and integration problems a review and new directions *IEEE Transactions on Engg. Management*, vol 48 no 3 pp 292-306

[7] Danilovic M and Browning T R 2007 Managing complex product development projects with design structure matrices and domain mapping matrices *Int. J. of Project Management* vol 25 no. 3 pp3 00-314
[8] Campean et al 2013 Systems engineering excellence through design an integrated approach based on failure mode avoidance *SAE Int. J. Mater. Manf.*, vol 6 no 3 pp 389-401
[9] Tomiyama et al 2009 Design methodologies industrial and educational applications *CIRP Annals - Manufacturing Technology* vol 58 pp 534-565
[10] Lopez-Mesa B and Nicklas Bylund 2011 A study of the use of concept selection methods from inside a company vol 22 no 1 pp 7-27
[11] Ulrich K T and Eppinger S D 1994 *Product design and development* (McGraW Hill)
[12] Campean I F and Henshall E 2008 A function failure approach to fault tree analysis for automotive systems *SAE International Technical Paper* no 2008–01-0846.
[13] Campean I F et al 2013 A structured approach for function based decomposition of complex multi-discipinary systems *Proceedings of the 23th CIRP design conference* pp 113-123
[14] Albert A and Christian Z 2011 Interdisciplinary systems modelling using the contact and channel model for SysML *International Conference on Engineering Design*
[15] Albers A et al 2003 Engineering design in a different way : cognitive perspective on the contact and channel model approach *Visual and Spatial Reasoning in Design* pp 1-21
[16] Goel A et al 2009 structure behavior and function of complex systems The structure behavior and function of complex systems
[17] Suh N P 1998 Axiomatic Design Theory for Systems *Research in Engineering Design* pp 189-209
[18] Komoto H and Tomiyama T 2011 A theory of decomposition in system architecting *International Conference on Engineering Design*
[19] Umeda Y et al 1996 Supporting conceptual design based on the function-behaviour-state modeller *Artificial Intelligence for Engg Design, Analysis, and Manuf.* vol 10 no 4 pp 275-288
[20] Dori D 2002 *Object-process methodology a holistic systems paradigm* (Springer Berlin Heidelberg)
[21] Gero J 1990 Design prototypes a knowledge representation scheme for design *AI Magazine*, vol 11 no 4 pp 26-36
[22] Christophe et al 2010 RFBS A model for knowledge representation of conceptual design *CIRP Annals - Manufacturing Technology* vol 59 no 1 pp 155-158
[23] Deubzer F and Lindemann U 2009 Product architecture definition and analysis using matrix-based multiple-domain approach *Proceedings of ASME IDETC/CIE* pp 1197-1205
[24] Maurer M and Lindemann U 2008 The application of the multiple-domain matrix pp 2487-2493
[25] Lindemann et al 2009 *Structural Complexity Management* (Springer Berlin Heidelberg)
[26] Jarratt T A W et al 2004 Development of product model to support engineering change management *Proceedings of TMCE* April Switzerland
[27] Martin M V and Ishii K 2002 Design for variety developing standardized and modularized product platform architectures *Research in Enggg Design* vol 13 pp 213-235
[28] Campean et al 2011 A structured approach for function analysis of complex automotive systems *SAE Technical Paper* no 2011–01-1268 pp 1-14
[29] Konig et al 2008 Multiple-domain matrix as a framework for systematic process analysis 10th *Int. Design Structure Matrix Conf.* November Stockholm Sweden
[30] K. Oizumi et al 2011 Management of prod. development proj. through integrated modeling of product and process information *International Conference on Engg Design*, Denmark
[31] Hamraz B et al 2013 Requirements-based development of an improved engineering change management method *Journal of Engineering Design*

IOP Conf. Series: Materials Science and Engineering **65** (2014) 012006 doi:10.1088/1757-899X/65/1/012006

Simulated Energy Usage for a Novel 6 DOF Articulated Robot

A.A. Shaik[1], **N. Tlale**[2] **and G Bright**[3]
[1] Council for Scientific and Industrial Research, Pretoria, RSA
[2] Anglo American, Pretoria, RSA
[3] University of Kwa-Zulu Natal, Durban, RSA

E-mail: ashaik@csir.co.za

Abstract. The serial robot architecture is widespread in modern day manufacturing, and over the last few decades the technology has matured and settled to its current state. One drawback from the architecture however is the location of motors and gearboxes which are either at the joint it controls or close by. A novel hybrid 6 DOF robot was designed to move all the actuators to the robot base, and to control the desired axis through a set of connected links and gears, while maintaining the same workspace and dexterity. This would reduce the inertia of the movable part of the robot and some of the moment arms on the 3 axes required for translation of the 3 DOF spherical wrist. Doing so would decrease the energy requirements when compared to a 6 DOF serial robot. This paper focuses on the mathematical modelling and simulation of the novel hybrid machine design and compares it to an equivalent serial robot.

1. Introduction

Parallel robots save energy and increase speed when compared to serial robots due to the fact that they move less mass. A striking feature of parallel robots is that the motor/gearbox pairs remain stationary while motion of an end effector is controlled by connected links, in closed loop architectures. The main idea in this design utilizes that concept from parallel robots and fixes the position of all 6 motor/gearbox pairs to its stationary base. The inertia could be further reduced with the use of high strength, light weight linkages using composite materials which should maintain the arm's desired stiffness/rigidity.

Unlike parallel robots however a large workspace, which is equivalent to that of a serial robot is maintained. This goal is realized by using certain building block i.e. a concentric geared mechanism, torque transfer linkages and a 3 DOF (degree of freedom) spherical wrist, all of which are illustrated in figure 1. The concentric geared mechanism was a multi-geared component in which each geared section rotates independently of the others while sharing a common rotation axis. This component makes it possible to control 6 axes by 6 motors which remain stationary at the base of the machine. A practical model of the design is shown in figure 2. For additional information on the design please refer to [1].

A simplified dynamic model was created to model the dynamics of the robot, which would be used for energy usage calculations. These numbers were then compared to those of a typical serial robot to indicate if this idea was in fact useful and beneficial.

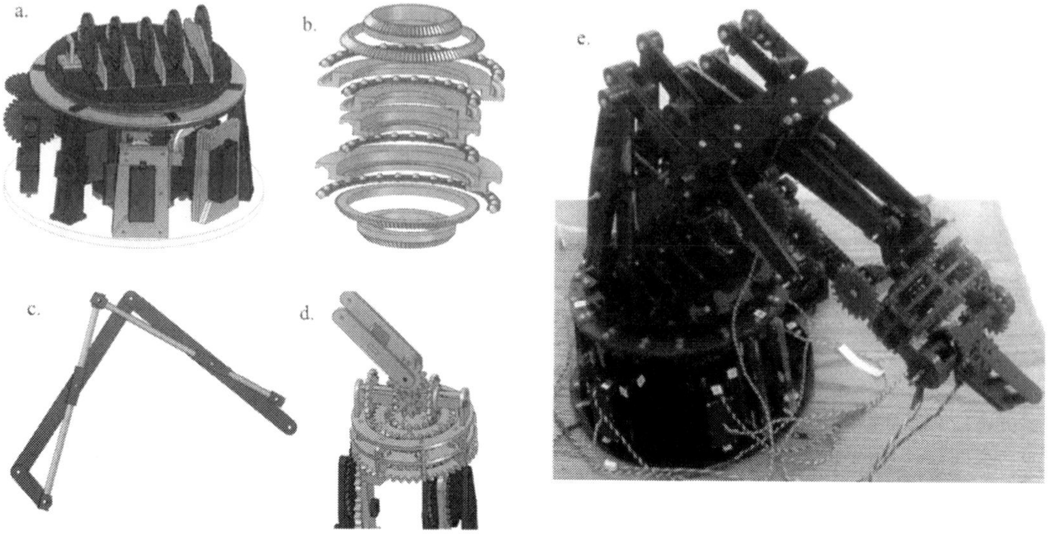

Figure 1. Model of a novel 6 DOF articulated robot. [1]

2. Experimental method

Joint torques for robot motion having a desired velocity profile while following a specific path are calculated in the machine level dynamics simulation. A standard recursive dynamics algorithm can be found in [2]. The algorithm requires the following data for implementation: joint angles for each pose along the path indicated, joint angular velocities and accelerations, robot model parameters which include inertia tensors, masses, centre of mass (COM) locations, velocities and accelerations. The work done by the robot is used as a measure of energy consumption which is given by the cumulative sum of joint work, for all joints at each pose along the path. Joint work is the multiplication of torque applied at the joint and the angular displacement it induces.

For the main purpose of the objective set out in this research, only the work done in translation of the end effector was of interest. The wrist was not controlled in any specific way and work done in orientation of the wrist was thus not considered for any of the models. For the robot design a large number of parameters could vary, taking into account all would not be practical, and for this reason one parametric model was used for the new design.

The serial robot model however had mass from its wrist, elbow and shoulder motors which it had to move, and for that 2 parametric models were used to highlight the effect of motor mass. The models used low mass (LM) and high mass (HM) motors respectively, and it was assumed that the motors would have sufficient ability to move as required in the path specification. Motor mass was taken as a percentage of all the mass it moved, and for high mass motors this was set at 35%.

The motor geometry was modelled as a block with varying edge length, whose density was estimated at 15% steel, 40% ferrite, 40% copper and 5% empty space, i.e. 6709.5 $kg.m^{-3}$. The density was essential to derive motor volume and the side lengths or edges were calculated from parametric relations between the edges. In total the motor mass for the light version of the serial robot was an additional 8.2% more than that for the new design. For the heavy version of the serial robot it was 71% heavier.

The simulations were carried at 2 speeds, the first being very slow so that the dynamic response of machine was close to its static state. This meant that the torques required for motion were slightly larger than that required to maintain a static pose. The second speed was excessive and may not be physically achievable but it was necessary to illustrate how the robot models compared with each other when the dynamic response was dominant. For each path the end-effector followed a simple velocity profile. It started at the beginning of the path with $0 m.s^{-1}$ velocity, and accelerated with constant positive acceleration to a maximum, after which it ramped down to $0 m.s^{-1}$ with constant negative acceleration. There must be a finite duration between changes in acceleration in real world situations, and so the acceleration graph should be an S-curve instead of a straight edged step curve. Since the derivative of acceleration was never used, that bit of real world detail was ignored.

A robot's work path could have a large number of segments either linear or curved which are connected in an infinite number of ways. It was decided to compare the dynamics over the simplest comprising elements of a complex path. These were straight line segments and circular curves. The straight line paths were vertical, horizontal and radial. Vertical lines are perpendicular to the xy plane, radial lines are perpendicular to the z axis and are parallel to the xy plane, and horizontal lines are perpendicular to both vertical and radial lines. Any other line path could be considered as a linear combination of the above paths, i.e. a diagonal line would be composed of a horizontal line with a vertical or a radial line with a vertical line, etc. For each straight line path type the dynamics were tested over a range of lines. This is illustrated in figure 2b for vertical line paths, and the discussion is the same for horizontal and radial paths. The vertical lines have the same length given by Δz, and are specified by the Start, Centre and End (SCE) coordinates. Motion for the vertical paths would start at S and end at E, with a path increment of ∂z. Once a path completes the y coordinate is incremented by ∂y. The robot then traces the next path from S to E incrementing the z coordinate only. This continues until the last vertical path is reached where the total change in y coordinate from the first to the last vertical path is Δy. This is done at 2 speeds first very slow, with a total path time of 5s and then very fast with a total path time of 0.25s. The straight line path profiles are captured in table 1.

Only the total energy consumed at the end of each path was taken, which was plotted against the coordinate that was advanced (the y value for the vertical and horizontal paths; and the z value for the radial paths).

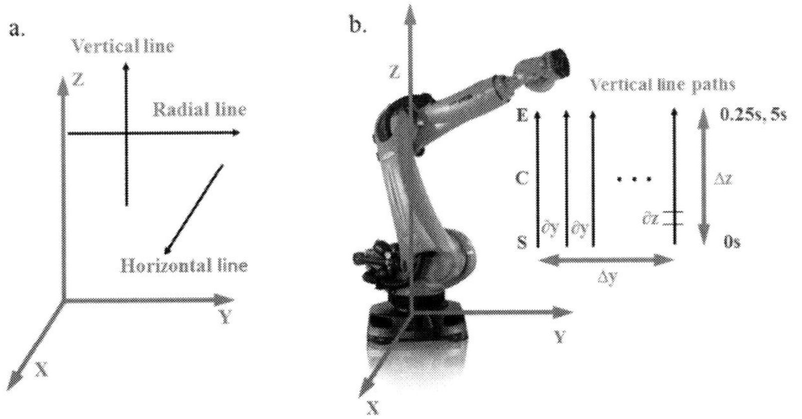

Figure 2. Illustration of straight line paths.

Table 1. Straight line path specification.

Path	SCE (m)	Variable coordinate (m)	Path length (m)	Time (s)
Vertical lines	$\begin{bmatrix} 0 & 0 & 0 \\ y_i & y_i & y_i \\ 0.213 & 0.746 & 1.280 \end{bmatrix}$	$0.32 \leq y_i \leq 1.706$ $\Delta y = 1.386$ $\partial y = 0.035$	$\Delta z = 1.066$	5 0.25
Horizontal lines	$\begin{bmatrix} -0.533 & 0 & 0.533 \\ y_i & y_i & y_i \\ 0.746 & 0.746 & 0.746 \end{bmatrix}$	$0.32 \leq y_i \leq 1.706$ $\Delta y = 1.386$ $\partial y = 0.035$	$\Delta x = 1.066$	5 0.25
Radial lines	$\begin{bmatrix} 0 & 0 & 0 \\ 0.213 & 0.746 & 1.280 \\ z_i & z_i & z_i \end{bmatrix}$	$0.533 \leq z_i \leq 1.599$ $\Delta z = 1.066$ $\partial z = 0.027$	$\Delta y = 1.066$	5 0.25

For the circular paths it was decided to isolate the first 3 axes responsible for wrist centre translation, i.e. only 1 of those 3 angles would change while the remaining 5 were kept constant. This allowed the wrist centre to move on a circular arc about the actuated axis.
Additionally this resulted in a large change in joint angle (difference between the start and end angular positions), unlike in the straight line paths. Each circular path, for each axis had a common start angle, which was given relative to that joint's coordinate frame axis. The end angle was varied between selected minimum and maximum values, with the increment given by $\Delta\theta$. The increment in each path was given by $\partial\theta$. The circular path specifications for angle 1 are illustrated in figure 3 and are given in table 2. In the table variable m sets the number of paths in the angular range for a joint, and variable n sets the angular resolution for a particular path. Again the dynamics for each model was tested at 2 different speeds (slow then fast).

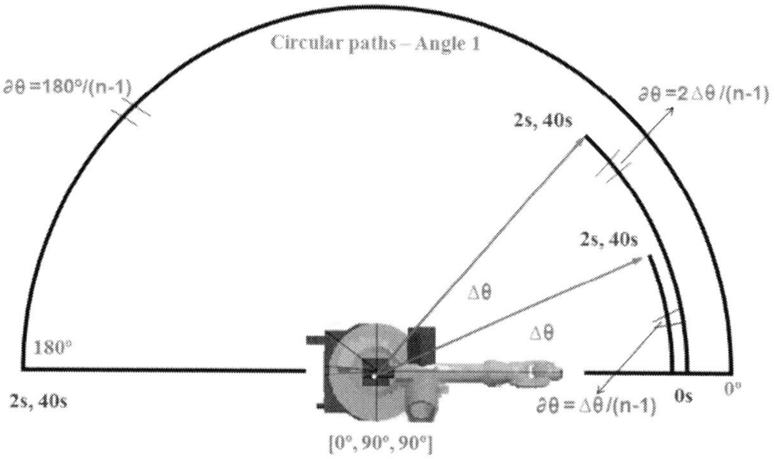

Figure 3. Illustration of circular path specification for angle 1.

For each of these 3 axes the arm was tested at the best and worst case mass moments about the individual axes, which would give the lowest and highest joint torque respectively. The worst moments are achieved when the arm mass is at its furthest location from the axis being tested, and the best case was the opposite. For rotation about the vertical z axis the worst case is when the arm points out and is perfectly horizontal, where the distal and proximal links are parallel. The best case is achieved when the arm points up vertically, and again both links are parallel. Those were singularity conditions but the aim was to move the wrist centre on the workspace boundary. There are similar cases for the remaining shoulder and elbow joint (axes 2 and 3).

Table 2. Circular path specification.

Angle varied	Arm configuration $[\theta_1\ \ \theta_2\ \ \theta_3]$; $[\theta_4\ \ \theta_5\ \ \theta_6] = [0°\ \ 0°\ \ 0°]$	Range $2 \le k \le m;\ m \ge 2$ $1 \le i \le n;\ n \ge 2$	Time (s)
Angle 1	$[\theta_{1(i)}\ \ 0°\ \ 0°]$	$\Delta\theta = \frac{180°}{m-1}$ $\Delta\theta \le \theta_{1(k)} \le 180°$ $\theta_{1(k+1)} = \theta_{1(k)} + \Delta\theta$	40 2
	$[\theta_{1(i)}\ \ 90°\ \ 0°]$	$\partial\theta = \frac{\theta_{1(k)}}{n-1}$ $\theta_{1(n)} = \theta_{1(k)};\ \theta_{1(1)} = 0°$ $0° \le \theta_{1(i)} \le \theta_{1(n)}$ $\theta_{1(i+1)} = \theta_{1(i)} + \partial\theta$	40 2
Angle 2	$[0°\ \ \theta_{2(i)}\ \ 0°]$	$\Delta\theta = \frac{90°}{m-1}$ $\Delta\theta \le \theta_{2(k)} \le 90°$ $\theta_{2(k+1)} = \theta_{2(k)} + \Delta\theta$	40 2
	$[0°\ \ \theta_{2(i)}\ \ 170°]$	$\partial\theta = \frac{\theta_{2(k)}}{n-1}$ $\theta_{2(n)} = \theta_{2(k)};\ \theta_{2(1)} = 0°$ $0° \le \theta_{2(i)} \le \theta_{2(n)}$ $\theta_{2(i+1)} = \theta_{2(i)} + \partial\theta$	40 2
Angle 3	$[0°\ \ 0°\ \ \theta_{3(i)}]$	$\Delta\theta = \frac{180°}{m-1}$ $-90° + \Delta\theta \le \theta_{3(k)} \le 90°$ $\theta_{3(k+1)} = \theta_{3(k)} + \Delta\theta$ $\partial\theta = \frac{\theta_{3(k)}}{n-1}$ $\theta_{3(n)} = \theta_{3(k)};\ \theta_{3(1)} = -90°$ $-90° \le \theta_{3(i)} \le \theta_{3(n)}$ $\theta_{3(i+1)} = \theta_{3(i)} + \partial\theta$	40 2

3. Multi-straight line path simulation results

The graphs for the new design are relative to the serial robot, and are displayed as a percentage thereof for both cases. The dash double dot and solid lines represent the low and high dynamics cases respectively for comparison against the low mass (LM) and high mass (HM) serial robot models.

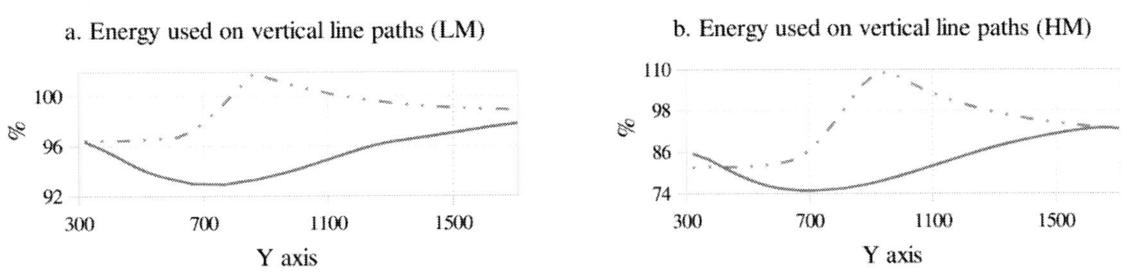

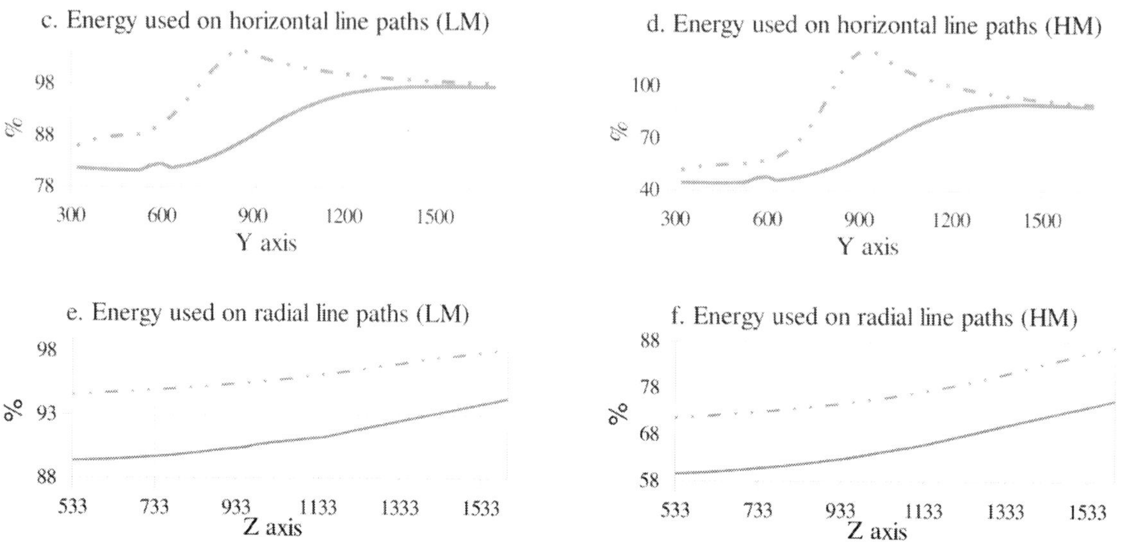

Figure 4. Total energy consumption data for multiple vertical, horizontal and radial line paths.

4. Circular path simulation results

The simulation results presented here were a comparison of total energy usage only (total energy consumed per model at the end of each circular path), and is illustrated in table 3. The tabular form is due to the fact that the percentage energy comparison against the serial robot for these paths and joint constraints were either constant throughout the range of motion or varied very slightly.

Table 3. Circular path energy usage results.

Varying angle	Configuration	Serial robot model	**% Energy Usage comparison** Low speed (40s)	High speed (2s)
Angle 1	θ_2=0°, θ_3=0°	LM	94.49	94.49
		HM	80.71	80.71
	θ_2=90°, θ_3=0°	LM	97.02	96.86
		HM	43.52	43.52
Angle 2	θ_1=0°, θ_3=0°	LM	94.23	94.23
		HM	76.04	76.06
	θ_1=0°, θ_3=170°	LM	90.86	90.86
		HM	56.55	56.6
Angle 3	θ_1=0°, θ_2=0°	LM	98.41	98.41
		HM	101.26	101.28

5. Conclusion

The data indicates that in most cases the new design is more energy efficient than a serial robot even when the serial robot's motor mass is low. During some parts in the vertical and horizontal line path motion of the low dynamics case the serial robot was more energy efficient, which included the high motor mass case. When the robot moves fast however the new design uses less energy in every case.

The new design performed better when moving axes 1 and 2 only in the circular path motion. At higher speeds there was a larger saving when compared to the higher motor mass serial robot. The situation was different for the elbow joint also known as axis 3. Here even though the serial robot was moving more mass there was a more significant balancing effect at that joint. This effectively brought the mass moment of that link closer to the axis and lowered its overall energy usage.

In conclusion a design that forces a static location of motor/gearbox pairs in a large workspace, articulated arm type structure can save energy, and in the long term it will be significant. The static balancing effect of a typical serial robot however is important and it should be applied to the new design to bring about an additional drop in energy usage. This will be investigated further.

6. References

[1] Shaik A A, Tlale N, and Bright G 2011 CARs&FOF proceedings, Novel 6 DOF hybrid machine design.
[2] Craig J J, 2005 Introduction to robotics, mechanics and control, Pearson Education.

IOP Conf. Series: Materials Science and Engineering 65 (2014) 012007 doi:10.1088/1757-899X/65/1/012007

Genetic Algorithm to minimize flowtime in a no-wait flowshop scheduling problem

Imran A. Chaudhry[1], **Riaz Ahmed**[2], **Abdul Munem Khan**[3]

[1, 2, 3]National Univ of Sciences & Technology, H-12, Islamabad, Pakistan

E-mail: [1]imranchaudhry-cae@nust.edu.pk, [2]dresearch@nust.edu.pk, [3]munem-cae@nust.edu.pk

Abstract. No-wait flowshop is an important scheduling environment having application in many industries. This paper addresses a no-wait flowshop scheduling problem, where the objective function is to minimise total flowtime. A Genetic Algorithm (GA) optimization approach implemented in a spreadsheet environment is suggested to solve this important class of problem. The proposed algorithm employs a general purpose genetic algorithm which can be customised with ease to address any objective function without modifying the optimization routine. Performance of the proposed approach is compared with eight previously reported algorithms for two sets of benchmark problems. Experimental analysis shows that the performance of the suggested approach is comparable with earlier approaches in terms of quality of solution.

1. Introduction

In a manufacturing environment, scheduling is the efficient use of resources over time. Scheduling is defined as the process of deciding what happens where and when. Any process that defines a subset of *what* x *when* x *where* can be said to "do scheduling" [1]. The objective in any scheduling problem is to determine a method to allocate and sequence the use of the shared resources such that production costs are minimized while satisfying the production constraints. In today's competitive world efficient scheduling schemes / algorithms are a key to achieve high production efficiency.

Flowshop is an important shop model that has been studied widely in the manufacturing community. Since its introduction by Johnson [2] in 1954, flowshop scheduling problem has received considerable attention. A recent detailed survey of flowshop scheduling has been given by Gupta and Stafford Jr [3]. No-wait flowshop is an extension of standard flowshop scheduling problem. No-wait flowshop does not have any intermediate storage between the machines and all jobs are to be processed from first to last machine without any interruptions between the operations. Different industries have no-wait flowshop applications that include concrete ware production, food processing, chemical processing, and pharmaceutical processing. No-wait situation also exist in production lines where the flow of the jobs is continuous and there is no in-process inventory e.g. JIT production lines. Hall and Sriskandarajah [4] gives a thorough analysis of applications and research of no-wait flowshop scheduling problem. No-wait flowshop scheduling problem is considered to be NP-hard even for a two machine case [5].

Minimization of flowtime or makespan are the two most commonly studied objective functions for this class of problem. In this paper minimization of flowtime for no-wait flowshop scheduling problem is considered using genetic algorithm to determine an optimal sequence of jobs.

2. Recent Literature Review

In this section a review of some of the recent studies related to no-wait flowshop is presented. The first instance of flowtime minimization in no-wait flowtime scheduling problem has been reported by Rajendran and Chaudhuri [6]. Since then numerous researchers have attempted to solve the problem with various heuristics. Rajendran and Chaudhuri propose two heuristic algorithms that produce near-optimal solutions for large sized problems.

Li *et al.* [7] present three composite heuristic by integrating forward pair-wise exchange-restart (FPE-R) and FPE with an effective iterative method. Framinan *et al.* [8] present a constructive heuristic for minimizing flowtime in a no-wait flowshop. The heuristic outperforms Pilot-1-Chins heuristic proposed by Fink and Voß [9]. Jarboui *et al.* [10] present a hybrid GA to minimize both flowtime and makespan. The authors use a variable neighborhood search in the last step of the GA to improve upon quality of the solution. Wang *et al.* [11] present three hybrid harmony search (HS) algorithms to minimize flowtime. Well-known benchmark problems are used to carry out computational experiments for the proposed algorithms.

Yagmahan and Yenisey [12] present a multi-objective ant colony system algorithm to minimize both total flowtime and makespan. Gao *et al.* [13] present a discrete harmony search algorithm based on well-known algorithm proposed by Nawaz *et al.* [14] to minimize flowtime. Laha and Sapkal [15] propose an efficient constructive heuristic for solving the problem. Sum of the processing times of a job on the bottleneck machine(s) is used to determine the priority of the job in the initial sequence.

Shafaei *et al.* [16] consider a two stage no-wait flowshop problem with multiple machines at each stage and propose six meta-heuristic algorithms to minimize flowtime. Simulation studies are conducted to evaluate the performance of the proposed approach. Gao *et al.* [17] propose four composite heuristics, improved standard deviation heuristic (ISDH) and improved Bertolissi heuristic (IBH), by combining the standard deviation heuristic [13] and Bertolissi heuristic [18] with the procedure of the constructive heuristic [19]. The authors then propose four composite heuristics, i.e., ISDH and IBH with local search and ISDH and IBH with iteration operator to improve the solutions of the ISDH and IBH. Guang and Junqing [20] propose an evolved discrete harmony search (EDHS) to minimize maximum completion time, total flowtime and total tardiness.

Gupta [21] present an algorithm to minimize weighted flowtime when processing times are uncertain. Gupta [22] also present an algorithm to minimize flowtime in two machine no-wait problem under the constraint of machine availability. Nagano *et al.* [23] present a GA and cluster search based evolutionary clustering algorithm for no-wait flowshop where setup times of jobs are separate from the processing time. Sapkal and Laha [24] also present an efficient heuristic method to minimize total flowtime. Laha and Sapkal [15] method is used to determine the job priority in the initial sequence.

3. Problems & Assumptions

A flowshop is a multi-stage scheduling where jobs are required to be processed through different stages. The no-wait flowshop problem is defined as: Each of the n jobs from a set $J = \{1, 2,....., n\}$ are required to be processed on m $\{M = 1, 2,....., n\}$ machines. Each job has m operations (o_{j1}, o_{j2},...., o_{jm}). All operations of job j are to be processed on machines without interruptions i.e., the earliest start time operation o_{jM+1} must be equal to completion time of o_{jM}. In other words there must not be any waiting between successive operations of job n. All jobs must follow a same sequence of operations on all the machines. The problem then is to find job sequence (permutation schedules) that minimizes total flowtime of n jobs.

Additionally following assumptions are considered in this research: (a) All jobs are available at the start of the planning horizon; (b) An operation once started cannot be interrupted; (c) All processing times are known and constant; (d) Setup time for setting up of machine to process a particular operation is included in the operation time; (e) Transportation time of jobs between machines is negligible; (f) At any given time, each machine can process only one job; (g) None of the machine can process more than one operation at a time; and (h) The machines does not have any intermediate storage for the unprocessed jobs.

4. Genetic Algorithm Details

Genetic algorithms are stochastic search technique inspired by natural evolution. The search process GAs are guided by Darwin's survival of the fittest principles. GAs are therefore able to traverse a large search space. GAs were first invented by Holland [25] and his colleagues in 1975 at the University of Michigan.

First application of GAs in job scheduling domain has been given by Davis [26]. In this research, GA optimization routine is implemented in a spreadsheet environment. We employ a commercial general purpose GA by Palisade [27], that works as an add-in to Microsoft Excel® spreadsheet. The schematic diagram for the integration of Microsoft Excel and GA are given in Figure 1.

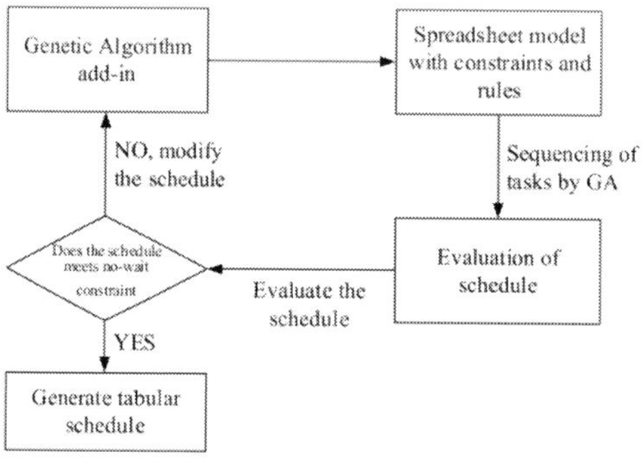

Figure 1. Spreadsheet - Genetic Algorithm Integration

5. Implementation Details

'No-wait' model proposed in this paper works in two phases. First phase calculates the delay in the starting of job p after q. Second stage delays job q after p by the delay factor that is calculated in first phase. This delay thus yields a no-wait schedule. The minimum delay in the start of job q after p i.e., $F(p, q)$ is calculated by the approach proposed by Reddi and Ramamoorthy [28]. Thus the minimum delay time between the completion of job J_p and the initiation of J_q, then $F(p, q)$ is given by:

$$F(p, q) = \max\,(t_{2,p} - t_{1,q},\ t_{2,p} + t_{3,p} - (t_{1,q} + t_{2,q}), \dots,\ t_{2,p} + t_{3,p} + t_{4,p} + \dots + t_{m,p} - (t_{1,q} + t_{2,q} + \dots + t_{m-1,q}), 0)$$

$$= \max_k\left(\sum_{i=2}^{k} t_{i,p} - \sum_{i=1}^{k-1} t_{i,q}, 0\right), \qquad 2 \le k \le m$$

The use of above equation is illustrated with the help of a sample problem. Consider 5-job 3-machine problem given in Table 1.

Table 1. Numerical Example Data

	Processing Times				
Machine	**Job 1**	**Job 2**	**Job 3**	**Job 4**	**Job 5**
Mch1	3	4	1	1	4
Mch2	2	5	4	3	3
Mch3	4	3	5	2	7

The Gantt chart for the data given in Table 1 for a job sequence of 5-3-2-1-4 is as show in Figure 2.

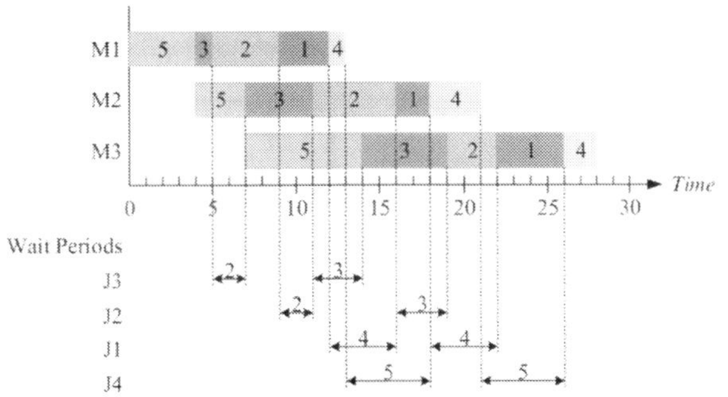

Figure 2. Gantt chart for sequence of job 5-3-2-1-4 and corresponding wait periods

The start delay for each of the job is calculated as describe above. By delaying the start of job q after job p for the schedule in Figure 2 by duration $F(p, q)$ would result in a no-wait schedule as given in Figure 3.

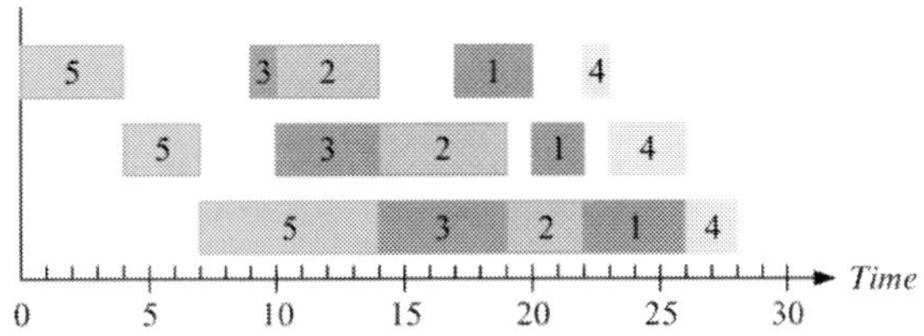

Figure 3. No-wait schedule for job sequence 5-3-2-1-4

6. Experimental Results & Comparisons

To test the performance of the proposed GA experimental analysis was carried out. The results of the proposed approach are compared with other heuristic solutions taken from the literature for well-known benchmark problems provided by Taillard [29]. The benchmarks consists of two set of problems with ten problems each for $n=20$ and $m=5$ and 10. Following heuristics are used to compare the performance of proposed approach:

ISDH : Improved Standard Deviation heuristic based on Gao *et al.* [13]
IBH : Improved Bertolissi heuristic based on Bertolissi [18]
Dipak : Constructive heuristic by Laha and Chakraborty [19]
AA_Omega : Heuristic algorithm by Aldowaisan and Allahverdi [30]
ISDHLS : Improved Standard Deviation with local search heuristic by Gao *et al.* [17]
IBHLS : Improved Bertolissi heuristic by Gao *et al.* [17]
ISDHIter : Improved Standard Deviation with with iteration operator Gao *et al.* [17]
IBHIter : Improved Bertolissi heuristic with iteration operator Gao *et al.* [17]

The best values for all above heuristics for $n=20$, $m=5$ and $n=20$, $m=10$ are given in Table 2 and Table 3 respectively. The results are based on forty runs for each instance. For $n=20$, $m=5$, the proposed the performance of GA was superior for 7 out of 10 problems while for $n=20$, $m=10$ the results were better for all instances as compared to eight heuristics mentioned above.

7. Conclusions

In this paper we presented a genetic algorithm approach for no-wait flowshop scheduling problem where the objective was to minimize total flowtime. The problem is categorized as NP-hard even for a two machine case. The GA routine is implemented in a spreadsheet environment. The proposed approach implements a general purpose GA for the optimization routine and as is easily customizable to address any objective function without modifying the basic optimization routine. The experimental results show that the proposed approach produced superior results for 17 out of 20 instances.

Table 2. Best flowtime values for *n=20, m=5*

Instance	ISDH	IBH	Dipak	AA_ Omega	ISDHLS	IBHLS	ISDHIter	IBHIter	GA	%age Improve
1	16553	16562	16421	16357	16414	16230	16381	16302	15674	3.4258
2	16749	16435	16551	16268	16164	16172	16220	16230	17250	-6.7186
3	15160	15197	14959	15258	14943	15024	15051	15018	15855	-6.1032
4	18989	18864	19048	18644	18732	18679	18788	18782	17970	3.6151
5	17293	16587	16570	16353	16684	16475	16385	16467	15317	6.3352
6	16268	15841	15974	15669	16109	15832	15620	15841	15501	0.7618
7	16302	16533	16538	16116	15990	15898	16117	16312	15693	1.2895
8	17836	17509	17277	17528	17403	17499	17340	17421	15955	7.6518
9	16802	17096	17186	16760	16551	16736	16802	16588	16394	0.9486
10	15693	15897	15776	15688	15785	15051	15208	15373	15329	-1.8471

Table 3. Best flowtime values for *n=20, m=10*

Instance	ISDH	IBH	Dipak	AA_ Omega	ISDHLS	IBHLS	ISDHIter	IBHIter	GA	%age Improve
1	27043	25664	26431	25410	26582	25657	25410	25664	25319	0.3581
2	26976	27037	26794	26847	26748	26774	26773	26586	26363	0.8388
3	25033	24509	24856	24377	24230	24509	24260	24277	22910	5.4478
4	23323	23353	23284	22905	22976	23120	22905	23138	22243	2.8902
5	24056	24185	23824	23779	23611	23838	24056	23998	23191	1.7788
6	23503	23416	23319	23743	23187	23016	23503	23380	22011	4.3665
7	24371	24236	24574	24344	24264	23967	24372	24500	21939	8.4616
8	24614	24416	24878	24294	24294	24315	24294	24294	24265	0.1194
9	24947	25128	25535	25799	25040	24663	24771	25107	23522	4.6264
10	27043	25664	26431	25410	26582	25657	25410	25664	25319	0.3581

References

[1] Van Dyke Parunak H 1991 Characterizing the manufacturing scheduling problem *J Manuf Syst* **10(3)** pp 241-259

[2] Johnson S M 1954 Optimal two- and three-stage production schedules with setup times included *Nav Res Logist Q* **1(1)** pp 61-68

[3] Gupta J N D and Stafford Jr E F 2006 Flowshop scheduling research after five decades *Eur J Oper Res* **169(3)** pp 699-711

[4] Hall N and Sriskandarajah C 1996 A Survey of Machine Scheduling Problems with Blocking and No-Wait in Process *Oper Res* **44(3)** pp 510-525

[5] Garey M and Johnson D 1979 *Computers and Intractability: A Guide to the Theory of NP-Completeness (Series of Books in the Mathematical Sciences)*

[6] Rajendran C and Chaudhuri D 1990 Heuristic algorithms for continuous flow-shop problem *Nav Res Log* **37(5)** pp 695-705

[7] Li X, Wang Q, and Wu C 2009 Efficient composite heuristics for total flowtime minimization in permutation flow shops *Omega-Int J Manage S* **37(1)** pp 155-164

[8] Framinan J M, Nagano M S, and Moccellin J V 2010 An efficient heuristic for total flowtime minimisation in no-wait flowshops *Int. J. Adv. Manuf. Technol.* **46(9-12)** pp 1049-1057

[9] Fink A and Voß S 2003 Solving the continuous flow-shop scheduling problem by metaheuristics *Eur J Oper Res* **151(2)** pp 400-414

[10] Jarboui B, Eddaly M, and Siarry P 2010 A hybrid genetic algorithm for solving no-wait flowshop scheduling problems *Int. J. Adv. Manuf. Technol.* **54(9-12)** pp 1129-1143
[11] Wang L, Pan Q-K, and Fatih Tasgetiren M 2010 Minimizing the total flow time in a flow shop with blocking by using hybrid harmony search algorithms *Expert Syst Appl* **37(12)** pp 7929-7936
[12] Yagmahan B and Yenisey M M 2010 A multi-objective ant colony system algorithm for flow shop scheduling problem *Expert Syst Appl* **37(2)** pp 1361-1368
[13] Gao K-Z, Pan Q-K, and Li J-Q 2011 Discrete harmony search algorithm for the no-wait flow shop scheduling problem with total flow time criterion *Int. J. Adv. Manuf. Technol.* **56(5-8)** pp 683-692
[14] Nawaz M, Enscore Jr E E, and Ham I 1983 A heuristic algorithm for the m-machine, n-job flow-shop sequencing problem *Omega-Int J Manage S* **11(1)** pp 91-95
[15] Laha D and Sapkal S U 2011 An Efficient Heuristic Algorithm for m-Machine No-wait flow shops in *International MultiConference of Engineers and Computer Scientists* Hong Kong pp 132-136
[16] Shafaei R, Moradinasab N, and Rabiee M 2011 Efficient meta heuristic algorithms to minimize mean flow time in no-wait two stage flow shops with parallel and identical machines *Int J Manag Sci Eng Manag* **6(6)** pp 421-430
[17] Gao K, Pan Q, Suganthan P N, and Li J 2012 Effective heuristics for the no-wait flow shop scheduling problem with total flow time minimization *Int. J. Adv. Manuf. Technol.* **66(9-12)** pp 1563-1572
[18] Bertolissi E 2000 Heuristic algorithm for scheduling in the no-wait flow-shop *J Mater Process Tech* **107(1–3)** pp 459-465
[19] Laha D and Chakraborty U K 2008 A constructive heuristic for minimizing makespan in no-wait flow shop scheduling *Int. J. Adv. Manuf. Technol.* **41(1-2)** pp 97-109
[20] Guang X and Junqing L 2012 Evolved Discrete Harmony Search Algorithm for Multi-objective No-wait Flow Shop Scheduling Problem in *The 2nd International Conference on Computer Application and System Modeling* Taiyuan Institute of Science and Technology, Taiyuan, Shanxi, China pp 791-794
[21] Gupta S 2012 Weighted mean flowtime in no-wait uncertain scheduling under various parameters *Int J of Comp Inte Tech* **3(2)** pp 99-101
[22] Gupta S 2012 TWO-MACHINE NO-WAIT FLOWSHOP SCHEDULING- A NEW APPROACH *Int. J. Manage.* **3(2)** pp 54-58
[23] Nagano M S, da Silva A A, and Lorena L A N 2012 A new evolutionary clustering search for a no-wait flow shop problem with set-up times *Eng. Appl. Artif. Intel.* **25(6)** pp 1114-1120
[24] Sapkal S U and Laha D 2013 A heuristic for no-wait flow shop scheduling *Int. J. Adv. Manuf. Technol.* **68(5-8)** pp 1327-1338
[25] Holland J H 1975 *Adaptation in natural and artificial systems* (Ann Arbor, MI)
[26] Davis L 1985 Job Shop Scheduling with Genetic Algorithms in *Proceedings of the 1st International Conference on Genetic Algorithms* L. Erlbaum Associates Inc. pp 136-140
[27] Palisade 1998 *Evolver: The genetic algorithm super solver Version 4* (New York, USA)
[28] Reddi S S and Ramamoorthy C V 1972 On the Flow-Shop Sequencing Problem with No Wait in Process *J Oper Res Soc* **23(3)** pp 323-331
[29] Taillard E 1993 Benchmarks for basic scheduling problems *Eur J Oper Res* **64(2)** pp 278-285
[30] Aldowaisan T and Allahverdi A 2004 New heuristics for m-machine no-wait flowshop to minimize total completion time *Omega-Int J Manage S* **32(5)** pp 345-352

IOP Conf. Series: Materials Science and Engineering **65** (2014) 012008 doi:10.1088/1757-899X/65/1/012008

Implementing gait pattern control and transition for legged locomotion

Zhijun Yang[1]*, Marlon V. Rocha[2], Priscila M.V. Lima[3], Mehmet Karamanoglu[1] and Felipe M.G. França[2]

[1]Department of Design Engineering and Mathematics, School of Science and Technology, Middlesex University, London NW4 4BT, UK

[2]COPPE, Federal University of Rio de Janeiro, Rio de Janeiro, Brazil

[3]Dept. of Math., Federal Rural University of Rio de Janeiro, Rio de Janeiro, Brazil

E-mail: z.yang@mdx.ac.uk

Abstract. In this work, a generalised central pattern generator (CPG) model is formulated to generate a full range of gait patterns for a hexapod insect. To this end, a recurrent neuronal network module, as the building block for rhythmic patterns, is proposed to extend the concept of oscillatory building blocks (OBB) for constructing a CPG model. The model is able to make transitions between different gait patterns by simply adjusting one model parameter. Simulation results are further presented to show the effectiveness and performance of the CPG network.

1. Introduction

The constituents of the locomotive motor system are traditionally modelled by nonlinear coupled oscillators, representing the activation of flexor and extensor muscles driven by, respectively, two neurophysiologically simplified motor neurons [1-4]. Different types of oscillators can be chosen and organised in a designed coupling mode, and usually with appropriate topological shape to allow simulating the locomotion of particular animals [5-9]. All internal parameters and weights of coupled synaptic connections of the oscillator network are controlled by the environmental stimulations, central nervous system instructions and the network itself. The nature of the parallel and distributed processing is a prominent characteristic of this oscillatory circuit that can be canonically described by a group of ordinary differential equations (ODE), which may also be an autonomous system. In other words, a complex biological pattern generator system such as the central pattern generators (CPG) can be simplified and implemented in a phenomenological model that uses the concrete artificial neural network dynamics.

Following our previous modelling [10-12] and implementation [13] works, a generalised locomotion CPG architecture is presented here not only to generate a range of legged gait patterns but also to make the transitions between different patterns. A mathematical formalism, extended from our previous works for gait pattern generation, is proposed to incorporate the gait pattern transitions. The CPG model uses an oscillatory building block (OBB) [12] as a pair of flexor and extensor motoneurons to drive individual joints. The interconnection of OBBs formulates a CPG model capable

* To whom any correspondence should be addressed.

of generating different gait patterns and their transitions. It is also shown that only one OBB parameter is used to control the creation of different gait patterns, and gait pattern transition is therefore implemented by changing this OBB parameter.

The proposed CPG model provides a reconfigurable architecture to integrate many observed gait patterns of any legged animals. The scalability and modularity features make the model particularly amenable to hardware or software implementation. A computer simulation shows that the model is able to run smoothly for both single pattern operation and pattern transitions provided that its initial state is properly configured.

The rest of the paper is organised as follows. Section 2 derives a mathematical framework for the OBB module and the CPG model, which is suitable for the neuronal network design. Some simulation results are presented in Section 3 to show its performance. Finally, Section 4 concludes the paper.

2. The model

In this section a graph dynamics is first introduced, which is followed by the dynamics of a generalised OBB module description.

2.1 Graph dynamics

Consider a neighbourhood-constrained system composed of a set of nodes and a set of shared resources represented by a connected graph G=(N,E) where N is the set of nodes, and E, the set of all resources between any pair of interconnected nodes. Between any two nodes i and j, $i, j \in$ N, there can exist e_{ij} resources, $e_{ij} \geq 0$. The reversibility of node i is r_i, i.e., the number of resources that shall be reversed by node i towards each of its coupled nodes, indiscriminately, at the end of its operation. A node will operate if and only if it possesses r_i resources from all of its coupled nodes.

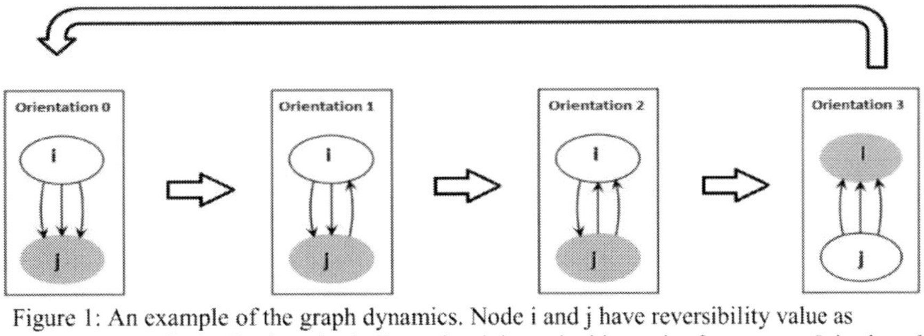

Figure 1: An example of the graph dynamics. Node i and j have reversibility value as 3 and 1, respectively. Dark nodes indicate the sinks and white nodes for sources. It is clear that a cycle of this graph dynamic system has 4 orientations. Node i becomes a sink exactly once, and node j becomes a sink 3 times in a cycle.

The reversibility value for each coupled node needs to be chosen together with a suitable number of resources belonging to each node. Two criteria exist for the arrangement of coupling parameters to avoid starvation or deadlock of the period operation: (1) $\max\{r_i, r_j\} \leq e_{ij} \leq r_i + r_j - 1$. (2) $f_{ij} = r_i + r_j - \gcd(r_i, r_j)$. where f_{ij} is the sum of the greatest multiple of $\gcd(r_i, r_j)$ that does not exceed the number of shared resources oriented from n_i to n_j, and from n_j to n_i, respectively in the initial orientation. The first rule stipulates a range of the number of the resources while the second decides the exact number of resources in the range and their directions. Based on the two rules a dynamic attractor can be made with flexible control of its active patterns, and be immune of the system halt due to deadlock or starvation [14][15]. Figure 1 illustrates the graph dynamics.

2.2 Dynamics of an OBB module

Inspired by the Hopfield Neural network model, the SMER graph dynamics can be described by a group of difference equations for computer simulation. Consider a pair of coupled neuron i and j with r_i and r_j as their reversibility, respectively. This coupled neuron pair is referred to as an OBB. The postsynaptic membrane potential of neuron i at t time instant, $M_i(t)$, depends on three factors, i.e., the potential at last instant $M_i(t-1)$, the impact of its coupled neuron output $v_j(t-1)$, and the negative feedback of neuron i itself $v_i(t-1)$, without considering the external impulses. The difference equation in the discrete time domain of this system can be formulated as follows: each neuron's self-feedback strength is $w_{ii} = -w_{ij}$, $w_{jj} = -w_{ji}$, respectively, and the activation function is a sigmoidal Heaviside type. Thus we have,

$$\begin{bmatrix} M_i(t) \\ M_j(t) \end{bmatrix} = \begin{bmatrix} M_i(t-1) \\ M_j(t-1) \end{bmatrix} + \underbrace{\frac{1}{r'} \times \begin{bmatrix} -r_i & r_j \\ r_i & -r_j \end{bmatrix}}_{W} \times \begin{bmatrix} v_i(t-1) \\ v_j(t-1) \end{bmatrix} \tag{1}$$

Where W is the weight matrix. We have the outputs of neurons as,

$$\begin{cases} v_i(t) = \max(0, \mathrm{sgn}(M_i(t) - \theta_i)) \\ v_j(t) = \max(0, \mathrm{sgn}(M_j(t) - \theta_j)) \end{cases} \tag{2}$$

The selection of system parameters, such as the neuron thresholds and synapse weights, are crucial for modelling the OBB module. In the model, let $r' = h(r)$, h is a function of getting highest integer level and multiplying by 10, e.g., if $r_i = 77$ and $r_j = 463$ then $h(r) = h(\max(77, 463)) = h(463) = 10^3$. The neuron i and j's thresholds θ_i and θ_j and their synaptic weights can be designed as $\theta_i = r_i / f_{ij}$, $\theta_j = 1 - \theta_i$, $W_{ij} = r_i / r'$, $W_{ji} = r_j / r'$. The model parameters can be arranged by comparing the two nodes' reversibility. If $r_i > r_j$, then $\theta_i > \theta_j$ and $w_{ij} > w_{ji}$ (i.e., asymmetric coupling), that means, a node with smaller reversibility, corresponding to a neuron with lower threshold in an OBB module, will oscillate at a higher frequency than its companion does.

The combination of the duty cycle (the ratio between the interval of the positive output and its associated oscillation period), the oscillation frequency and the phase latency of a coupled pair of neurons is the key set of joint parameters for modelling a one DOF joint. The oscillatory pattern transition, which is another important concept in addition to the pattern generation, can thus be understood as a transition from an old to a new set of the joint parameters. It is clear that the duty cycle of an extensor motor neuron plays an important role in deciding the locomotion speed of a legged animal [16-18]. In this model, the duty cycle of a neuron in a coupled two neuron system is dependent on the model parameters. The choice of reversibility of two coupled neurons thus dictates the transition between different patterns as it decides the model parameters, and hence the duty cycle. Therefore, the design of transition in patterns is simplified to the selection among different reversibility values.

Suppose both coupled neurons have their reversibility changed in the amount of r_i^d and r_j^d, respectively, the model formula (1) in a more general format involving pattern transition is as follows.

$$\begin{bmatrix} M_i(t) \\ M_j(t) \end{bmatrix} = \begin{bmatrix} M_i(t-1) \\ M_j(t-1) \end{bmatrix} + \left(W + \underbrace{\frac{s_i}{r'} \begin{bmatrix} -r_i^d & 0 \\ r_i^d & 0 \end{bmatrix} + \frac{s_j}{r'} \begin{bmatrix} 0 & r_j^d \\ 0 & -r_j^d \end{bmatrix}}_{\Delta W} \right) \times \begin{bmatrix} v_i(t-1) \\ v_j(t-1) \end{bmatrix} \tag{3}$$

where $s_k = \begin{cases} 1, & if\ r_k\ changed \\ 0, & if\ r_k\ not\ changed \end{cases}$, a transition control signal, $k \in (i, j)$ (4)

The model parameters are now changed to: $\theta_i^{new} = (r_i + r_i^d) / f_{ij}^{new}$, $\theta_j^{new} = (r_j + r_j^d) / f_{ij}^{new}$, $W_{ij}^{new} = (r_i + r_i^d) / r'$, $W_{ji}^{new} = (r_j + r_j^d) / r'$, $f_{ij}^{new} = r_i + r_i^d + r_j + r_j^d - \gcd(r_i + r_i^d, r_j + r_j^d)$. These equations indicate that, in theory, the pattern transition can be incurred by the reversibility change of any one of the two coupled neurons.

3. Simulation results

In this section, some case studies of the operations of the OBB modules, in the formats of a single OBB or a group of OBBs for the collective behaviours, are demonstrated in terms of the oscillatory patterns generation and transition.

3.1 Pattern generation

Let's suppose that in a simple experiment a hexapodal insect has only one joint in each leg. The coupling relation of 6 legs can be represented by an OBB architecture as shown in Figure 2.

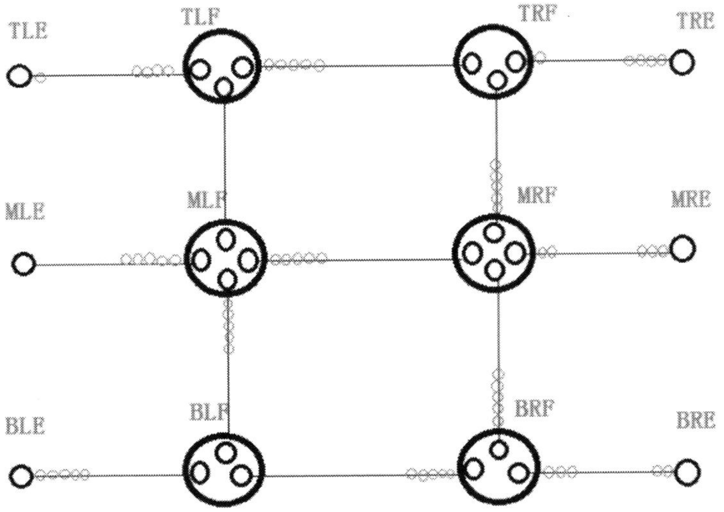

Figure 2: The OBB based architecture for the coupled legs of a hexapodal insect. Each leg has only one degree of freedom represented by a pair of flexor and extensor. The flexors are denoted by the big circles which contains 3 or 4 small nodes couping with neighbouring extensors or flexors. The smallest circles denote the resources shared by a pair of coupling nodes. Any two coupled nodes have an edge linking them. The significance of characters is: T: Top; M: Middle; B: Bottom; L: Left; R: Right; F: Flexor; E: Extensor, e.g., TLE represents for top left extensor.

Figure 2 configuration corresponds to a slow gait pattern of a hexapodal insect. Each flexor is regarded as a neuron population composed of several flexor neurons, whose reversibility is always one, coupling with the neighbouring flexor neurons or extensor neurons. The model parameters are shown in table 1.

Table 1. OBB module parameters

	Flexor neuron	Extensor neuron
Reversibility	1	5
Weight	0.1	0.5
Threshold	0	1
Initial value[a]	0.35	0.65

[a] Initial membrane potential values can be chosen randomly in the range of [0,1].

The oscillatory dynamics of the OBB module can be obtained by using Matlab Simulink, as shown in Figure 3. It is noticeable that the coupled neurons start with a self-organised period with the given initial membrane potentials. The system then undergoes a stable periodic oscillation. The duty cycle of

IOP Conf. Series: Materials Science and Engineering **65** (2014) 012008 doi:10.1088/1757-899X/65/1/012008

a neuron is decided by the model parameters, and thus indirectly related with the reversibility of two coupled neurons.

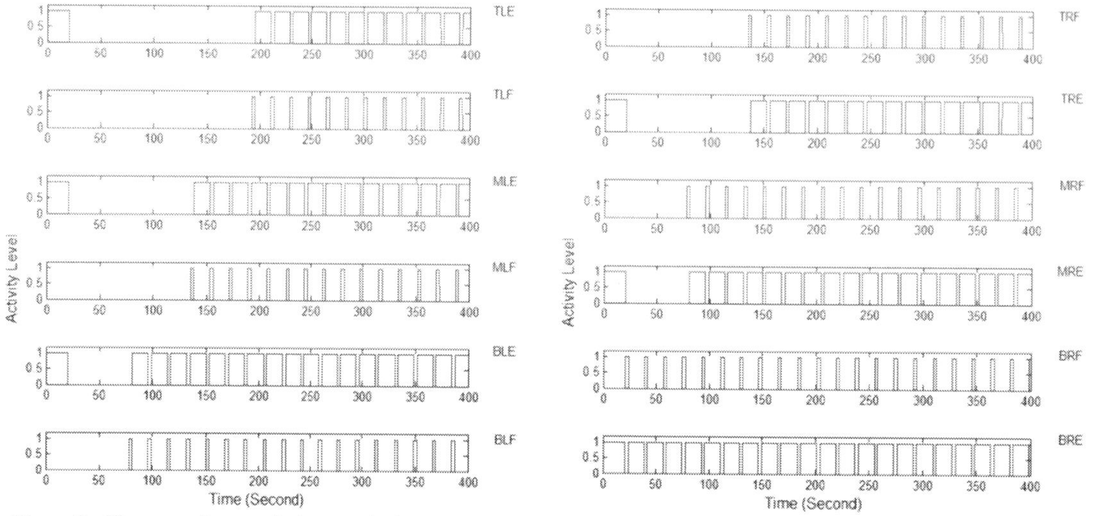

Figure 3 : The waveforms of a hexapodal insect OBB model in time domain. When the system becomes stable, the oscillatory period is 18 seconds and the duty cycle of extensor neurons is 15 seconds.

3.2 Pattern transition

As described in the coupled node equations in Section 2, a change of the reversibility of any one of two coupled neurons results in the change of model parameters, and hence the change of oscillatory patterns. Therefore, the pattern transition in the OBB model is straightforward. In Simulink simulation, a control signal, corresponding to the control signal in formula *3*, is used to switch between the previous and current model parameters derived from the previous and current reversibility of the coupled neurons. For instance, if we need to change the reversibility of all pairs of coupled flexor and extensor neurons from $\{r_{Flexor}=1, r_{Extensor}=5\}$ to $\{r_{Flexor}=1, r_{Extensor}=3\}$ and then to $\{r_{Flexor}=1, r_{Extensor}=1\}$, the dynamic model parameters are changed accordingly,

Table 2. OBB module parameters

	P1[a] Flexor	P1 Extensor	P2 Flexor	P2 Extensor	P3 Flexor	P3 Extensor
Reversibility	1	5	1	3	1	1
Weight	0.1	0.5	0.1	0.3	0.1	0.1
Threshold	0	1	0	1	0	1
Initial value	0.35	0.65	-[b]	-	-	-

[a] P1, P2 and P3 represent Phase 1, 2, 3, respectively. [b] This symbol means do not care.

Like a switch being used to control the pattern change, a transition between previous and current patterns can be achieved with some possible intermediate self-organisation period (see Figure 4).

It is clear that if no transition happens then neuron i will continue its first pattern, which becomes high at time instant 39 and lasts for 12 seconds till 51. The duty cycle for neuron i is 0.8 (and for j is 0.2 accordingly). As pattern transition occurs at 40, ideally the new pattern starts immediately after this time instant. Practically a self-organisation stage exists so the new pattern starts at the time instant of 51. This is because the membrane potentials of two coupled neurons are not ready (or, not as close as possible to their thresholds due to the operation of the old pattern) to make the transition to happen immediately. After a short period, though, the model will evolve into the desired new pattern with the duty cycle of neuron i as 0.5 (neuron j as 0.5). We argue that this phenomenon is biologically plausible as no real creatures will act immediately, i.e., zero delay, upon a command of action.

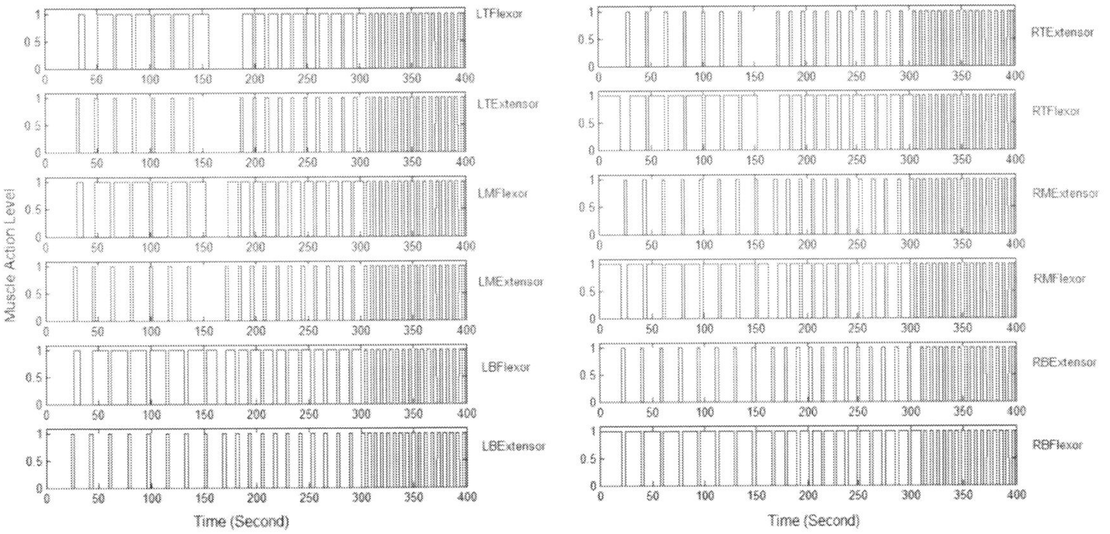

Figure 3: A pattern transition process. The transition occurs at the time instant of 160 and 300 seconds. It is obvious that there is a self-organised transition period between two patterns after the new pattern becomes stable.

4. Concluding remarks

An extended OBB model that is able to be configured to build up a tailor designed architecture for both model generation and transition has been proposed in this work. The simple OBB module constitutes a basis from which a complex, rhythm-producing model can be designed. Due to adoption of the OBB module, the whole model can be modular and scalable for design, prototype, manufacture and test. It is also an asynchronous and self-clocked system if the reversibility values and initial membrane potentials are chosen for individual OBB modules. Because of the simplicity of the system, the hardware version of a simple OBB module can be made such that a system with arbitrary complexity can be hopefully developed for real-time hardware implementation.

References

[1] Matsuoka K 1987 Mechanisms of frequency and pattern control in the neural rhythm generators *Biol. Cybern.* **56** 345-53

[2] Taga G 1995 A model of the neuro-musculo-skeletal system for human locomotion-I. Emergence of basic gait *Biol. Cybern.* **73** 97-111

[3] Ghigliazza RM and Holmes P 2004 A minimal model of a central pattern generator and motoneurons for insect locomotion *SIAM J. App. Dynamical Sys.* **3** 671-700

[4] Ijspeert AJ, Nakanishi J, Hoffmann H, Pastor P and Schaal S 2013 Dynamical movement primitives: Learning attractor models for motor behaviors *Neural Comp.* **25** 328-73

[5] Linkens DA, Taylor Y and Duthie HL 1976 Mathematical modeling of the colorectal myoelectrical activity in humans *IEEE Trans. Biomed. Engr.* **23** 101-10

[6] Tsutsumi K and Matsumoto H 1984 A synaptic modification algorithm in consideration of the generation of rhythmic oscillation in a ring neural network *Biol. Cybern.* **50** 419-30

[7] Bay JS and Hemami H 1987 Modeling of a neural pattern generator with coupled nonlinear oscillators *IEEE Trans. Biomed. Engr.* **34** 297-306

[8] Lewis MA, Tenore F and Etienne-Cummings R CPG design using inhibitory networks *Proc. IEEE Int. Conf. Robotics and Automation* 3682-7, Barcelona, Spain

[9] Still S, Hepp K and Douglas RJ 2006 Neuromorphic walking gait control *IEEE Trans. Neural Netw.* **17** 496-508

[10] Yang Z and França FMG 1998 Generating arbitrary rhythmic patterns with purely inhibitory neural networks *Proc. 5th European Symp. Art. Neural Netw.* 53-8, Brugges, Belgium
[11] França FMG and Yang Z 2000 Building artificial CPGs with asymmetric Hopfield networks *Proc. Int. Joint Conf. Neural Netw.* **4** 290-8, Como, Italy
[12] Yang Z, and França FMG 2003 A generalized locomotion CPG architecture based on oscillatory building block *Biol. Cybern.* **89** 34-42
[13] Yang Z, Cameron K, Lewinger W, Webb B and Murray AF 2012 Neuromorphic control of stepping pattern generation: A dynamic model with analog circuit implementation *IEEE Trans. Neural Netw. Learning Sys.* **23** 373-84
[14] Barbosa VC 1996 *An introduction to distributed algorithms.* Cambridge MA The MIT Press
[15] Barbosa VC, Benevides MRF and França FMG 2001 Sharing resources at nonuniform access rates *Theory of Computing Sys.* **34** 13-26
[16] Pearson KG 1976 The control of walking *Scientific American* **235** 72-86
[17] Patrick SK, Noah JA and Yang JF 2009 Interlimb Coordination in Human Crawling Reveals Similarities in Development and Neural Control With Quadrupeds *Journal of Neurophysiology* **101** 603-13
[18] Ivanenko YP, Labini FS, Cappellini G, Macellari V, McIntyre J and Lacquaniti F 2011 Gait transition in simulated reduced gravity *Journal of Applied Physiology* **110** 781-8.

IOP Conf. Series: Materials Science and Engineering **65** (2014) 012009 doi:10.1088/1757-899X/65/1/012009

Synthesis of Evolving Cells for Reconfigurable Manufacturing Systems

[1]J. Padayachee, [2]G. Bright

[1,2]Discipline of Mechanical Engineering, University of KwaZulu-Natal
Durban, South Africa, 4001

[1]padyacheej@ukzn.ac.za (corresponding author), [2]brightg@ukzn.ac.za

Abstract. The concept of Reconfigurable Manufacturing Systems (RMSs) was formulated due to the global necessity for production systems that are able to economically evolve according to changes in markets and products. Technologies and design methods are under development to enable RMSs to exhibit transformable system layouts, reconfigurable processes, cells and machines. Existing factory design methods and software have not yet advanced to include reconfigurable manufacturing concepts. This paper presents the underlying group technology framework for the design of manufacturing cells that are able to evolve according to a changing product mix by mechanisms of reconfiguration. The framework is based on a Norton-Bass forecast and time variant BOM models. An adaptation of legacy group technology methods is presented for the synthesis of evolving cells and two optimization problems are presented within this context.

Keywords: Reconfigurable Manufacturing Systems, Manufacturing System Design, Group Technology, Cellular Manufacturing Systems.

1. Introduction

The concept of Reconfigurable Manufacturing was first proposed by researchers at the University of Michigan in the late 1990s. A definition of Reconfigurable Manufacturing Systems was presented by Koren et al [1]: " A Reconfigurable Manufacturing System (RMS) is designed at the outset for rapid change in structure, as well as in hardware and software components, in order to quickly adjust production capacity and functionality within a part family in response to sudden changes in market or regulatory requirements."

The RMS concept has evolved out of the global manufacturing challenges of: economically managing the rapid rate of product development, frequent changes in parts and products, fluctuations in product demand and mix and changes in product and process technology [2]. In order to address these challenges RMSs are envisioned to exhibit transformable system layouts, reconfigurable processes, cells and machines. International research is currently being conducted in order to develop a new set of operating principles and methods for RMSs to be able to exhibit these characteristics; this research included the development of new group technology and factory floor design methods.

This paper presents a group technology framework for the development of evolving manufacturing cells that will enable RMSs to effectively deal with changes in products and the rapid development of new products. The paper proceeds as follows: section two presents relevant literature on recent developments in RMS design; section three develops the RMS cell design problem using a Norton-Bass forecast and BOM Models; section four presents the concept of evolving manufacturing cells and an adaptation of existing group technology methods for their design; section five presents two optimization problems within the scope of evolving cells and section six concludes with a discussion of future work.

2 Relevant Literature

A method for the design of a RMS was presented by Koren and Shpitalni [3]. The method extends to the design of a RMS line for a single part family. The method focused on generating feasible line configurations for the manufacturing of the part family and selection of the best configuration by means of throughput, investment cost, reconfigurability and scalability analysis. The use of the Analytic Hierarchy Process, (AHP), was proposed by Abidi [4] as an advanced decision making tool for the selection of the best configuration (machine placement) in a RMS line. The AHP required a set of RMS design alternatives and a set of design criteria as inputs to the algorithm. Abdi listed reconfigurability, cost, quality and reliability as the design objectives. The use of a Genetic Algorithm to design a Multiple Part Line (MPL) for RMSs was proposed by Tang et al [5]. Tang et al defined a MPL as a line that consists of several serial stages with a finite size buffer between every two stages. The objective of the optimisation was to allocate machines to the various stages of the MPL from a given library of available machines thereby optimizing the line configuration.

Galan et al [6] proposed a group technology method for part family formation for a RMS. The methodology starts by similarity analysis between pairs of products. The method then makes use of the AHP methodology and the Average Linkage Clustering algorithm in order to develop a dendogram that shows the diverse sets of product families that may be formed. The final selection of part families is then left to the discretion of the designer.

The literature has revealed that much of the RMS design methodologies being developed focused on the physical placement (configuration) of machines in a manufacturing line or cell. Little development was found on new group technology methods for RMS. Existing literature focused on the development of static machine-part groups for RMS. This paper proposes the concept of evolving machine-part groups for manufacturing cells that are able to respond more efficiently to changes in the part mix as dictated by the master production schedule.

3. Formulation of the Evolving Cell Problem

3.1 Long Term Forecasting and the Norton-Bass Model

A reconfigurable manufacturing system by definition is a system that is designed for a rapid change in structure as well as in hardware and software components. A RMS cannot change its structure rapidly and cost effectively unless the system design facilitates a transition from an existing configuration to new configuration. Pre-emptive knowledge is necessary such that the current manufacturing system configuration (physical design) can easily transition to a new configuration. A RMS can therefore be defined, from a design perspective, as a system whose current structural state supports a quick and seamless transition to a new structural state.

The Norton-Bass model is one such model that can provide pre-emptive information on how a RMS should evolve over time in order to meet market demand. The Bass Model was originally introduced by Frank Bass [7] to forecast the performance of a new product in the market place. The Bass model is noted as being one of the most widely used, frequently referenced and thoroughly researched marketing models in the world. The model was extended by Norton and Bass [8] to include the modelling of market performance for multiple generations of products. Significant information that may be derived from the Norton-Bass model includes timing for market launch of new products, rates of sales increase/decline, time of peak sales and the market growth and interplay of successive product generations. The Norton-Bass model for four successive generations of a single product is given by:

$$S_1(t) = F(t)m_1[1 - F(t - \tau_2)], \quad (1)$$
$$S_2(t) = F(t - \tau_2)[m_2 + F(t)m_1][1 - F(t - \tau_3)], \quad (2)$$
$$S_3(t) = F(t - \tau_3)\big[m_3 + F(t - \tau_2)[m_2 + F(t)m_1]\big][1 - F(t - \tau_4)], \quad (3)$$
$$S_4(t) = F(t - \tau_4)[m_4 + F(t - \tau_3)\big[m_3 + F(t - \tau_2)[m_2 + F(t)m_1]\big]] \quad (4)$$

Where $S_i(t)$ is the total sales at time t, m_i is the incremental market potential and τ_i is the time to market of the ith generation. The cumulative probability density of product adoption is given by:

$$F(\cdot) = \frac{1-e^{-(p+q)t}}{1+\frac{q}{p}e^{-(p+q)t}} \quad (5)$$

Where p is the innovation coefficient and q is the imitation coefficient. A full discussion of the Norton-Bass model and the estimation of its parameters are beyond the scope of this paper, however equations (1)-(4) provide the information required to determine the product mix and volume a given time. This will ultimately determine how the manufacturing cells should evolve over a period of time. Figure 1 shows a typical Norton-Bass model forecast for two products with two generations. The dashed lines indicate the times in the lifecycle of the manufacturing system where the part mix can be expected to change.

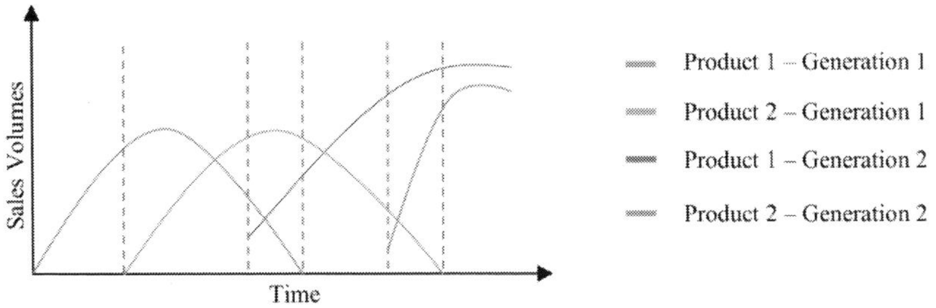

Figure 1: Norton-Bass Forcast for two Products in Two Generations

3.2 BOM Models and Property Maps

The design of a manufacturing system and investment in capital equipment is an investment that will influence its profitability and productivity over a long period of time. In order to pre-empt the changing part mix in a RMS, over a long period of time, the results of the Norton-Bass model must be coupled with a time varying Bill of Materials (BOM) trees as shown in Figure 2. A BOM Tree must be available for each curve that is shown on the Norton-Bass Model. Each product to be manufactured in the RMS over the design life cycle must modelled by a BOM Tree as shown. Each BOM Tree consists of entities E_i (vertices), where an entity represents either a part or a subassembly that is required for complete manufacturing of the product. With respect to product structure, each entity E_i with an in-edge is a parent entity to the entity E_j with the corresponding out-edge. This use of vertices and directed edges models a part to assembly/subassembly relationship. Property maps are the main link between the abstract nature of the BOM Tree and additional characteristics of product variety and manufacturing requirements.

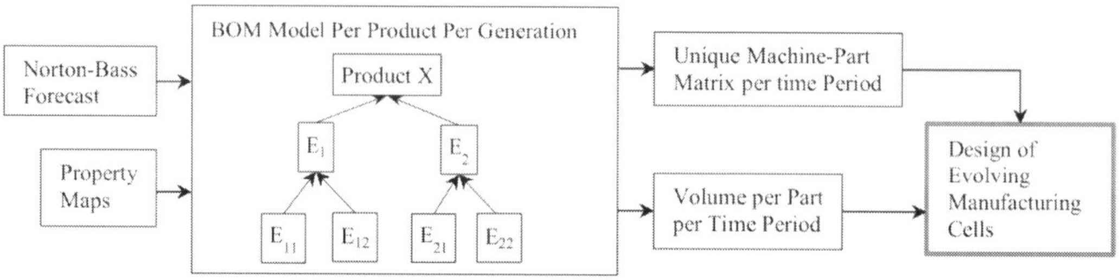

Figure 2: Conceptualization of the Evolving RMS Cell Design Problem

Variety Property Map
Considering that variety will exist within a product, an entity E_j, may be binary selective for integration into parent E_i, i.e $N_{ij} \in \{0,1\}$. The entity maybe also be selected for integration into the parent in different quantities according to the variety that is required, i.e. $n_{min} \leq N_{ij} \leq n_{max}, N_{ij} \in \mathbb{Z}^+$. In order to forecast the demand of an entity E_j over the design lifecycle of the manufacturing system, a probability distribution function $P_{i,j}$ is introduced over the range $[n_{min}, n_{max}]$. Every parent entity E_i will therefore contain a Variety Property (V_i) of constituent entities E_j coupled with a probability distribution that describes the likelihood and quantities of E_j being instantiated in E_i, i.e. $V_i = \{\langle E_j, P_{ij}\rangle\}$

Make or Buy decisions and Manufacturing Resource Map
Not every entity in the BOM tree will be manufactured in a RMS; some entities may be bought or the manufacturing may be outsourced. It is therefore essential to attach a Make or Buy property (MB_i) to each entity of the BOM tree in order to determine manufacturing system design requirements from this tree, $MB_i \in \{Make, Buy\}$. In addition to the make or buy property, a manufacturing resource property is required for each entity E_i. For those entities that have property $MB_i = Make$, an ordered list of machines $M_i = \{m_1, m_2, ..., m_n\}$ is attached to the node E_i, this is the manufacturing resource property. The list is ordered to represent the sequence of machine visitation such that E_i can be manufactured with the correct operation precedence. For those machines that have property $B_i = Buy$, $M_i = \emptyset$.

Time Varying Machine-Part Matrix and Product Demand
The output of propagating the results of the Norton-Bass model through each BOM Tree associated with a forecast curve will yield a unique machine-part matrix for various time periods in the lifecycle of a RMS. It will also yield a time varying demand characteristic for each entity E_i. It is essential that this information be available at the outset of the RMS cell design such the the evolution of the machine-part group associated with the reconfigurable cell can be optimized over a long term horizon.

4. Development of Evolving Manufacturing Cells

An evolving manufacturing cell is one in which the capacity of the cell changes, as well as the machine-part grouping. This section addresses the issue of formulating evolving machine-part groups. The primary objective of the cellular manufacturing paradigm is to reduce material flow complexity in the manufacturing system. The objective of system design for reconfigurable manufacturing is to provide the exact manufacturing functionality and capacity where it is needed, when it is needed. In order to achieve this, a slight variation applied to legacy cell formation methods. These are the steps:

1. a unique machine-part matrix is formulated for various time periods,

2. legacy algorithms are applied to block diagonalize the matrix. Methods used may include PFA [9], Rank Order Cluster Algorithm[10], etc.
3. the block diagonals in each matrix are paired using a similarity coefficient to propose evolving groups;
4. machine migrations from one group to another are identified and are submitted for a two stage filtration process (Discussed in Section 5)

Proposal of Evolving Machine-Part Groups- Illustration by Example
Consider the two machine-part matrices shown in Figure 3. These matrices correspond to the parts that are to be manufactured by a RMS in two different time periods or reconfiguration cycles:

Time Period One

	A	C	D	E
M_1	X		X	
M_2	X		X	
M_3	X		X	
M_4	X			
M_5		X		X
M_6				
M_7		X		
M_8		X		X

Time Period Two

	C	D	F	G	H	I
M_1		X		X		X
M_2		X				X
M_3		X			X	X
M_4				X		
M_5	X		X		X	
M_6			X		X	
M_7	X			X	X	
M_8	X		X		X	

Figure 3: Machine – Part Mix in RMS for Two Different Time Periods

If the parts and machines from these two time periods are combined into a single matrix as in legacy methods and groups are formed, the resulting groups are as shown in Figure 4. The resulting machine-part groups may be considered static as they prevail over the lifecycle of the associated manufacturing cells. The final solution contains three intra-cell movements as shown by the off-diagonal elements.

	A	B	C	D	E	F	G	H	I
M_1	X	X		X			X		X
M_2	X			X					X
M_3	X			X				X	X
M_4	X						X		
M_5		X	X		X	X		X	
M_6						X		X	
M_7		X	X				X	X	
M_8			X		X	X		X	

	A	D	I	G	B	H	C	F	E
M_1	X	X	X	X	X				
M_2	X	X	X						
M_3	X	X	X			X			
M_4	X			X					
M_7				X	X	X	X		
M_6						X		X	
M_5					X	X	X	X	X
M_8						X	X	X	X

Figure 4: Group Formation by Merging Part Mix

Time Period One

	A	D	C	E
M_1	X	X		
M_2	X	X		
M_3	X	X		
M_4	X			
M_5			X	X
M_7			X	
M_8			X	X

$G_{1.1}$ → M_3; $G_{1.2}$ → M_7

Time Period Two

	D	I	C	F	G	H
M_1	X	X			X	
M_2	X	X				
M_3	X	X				X
M_4					X	
M_5			X	X		X
M_8			X	X		X
M_7			X		X	X
M_6				X		X

$G_{2.1}$ → M_2; $G_{2.2}$ → M_8

Figure 5: Group Formation by Independently Assessing Part Mix

When the two smaller matrices are diagonalized independently per time period the results are as shown in Figure 5. Evolving cellular groupings are identified by pairing the block diagonals in the two different matrices using the Jaccard similarity index:

IOP Conf. Series: Materials Science and Engineering **65** (2014) 012009 doi:10.1088/1757-899X/65/1/012009

$$J(G_x, G_y) = \frac{|G_x \cap G_y|}{|G_x \cup G_y|} \quad (6)$$

The index is calculated on the commonality of the machines in each block divided by the union of the set of machines in the blocks being compared. Blocks $G_{1.1}$ and $G_{2.1}$ have a similarity index J = 0.75 and they propose an evolving machine group of {M_1, M_2, M_3, M_4} to { M_1, M_2, M_3}. Blocks $G_{1.2}$ and $G_{2.2}$ have a similarity index of J = 0.6 and they propose an evolving machine group of {M_5, M_7 M_8} to {M_4, M_5, M_6, M_7, M_8}. There are three distinct characteristics of this new solution versus the legacy solution:

1) there are now two intra-cell movements instead of three;
2) it is apparent that machine M_6 is only required in the second time period;
3) there is a migration of machine M_4 from one evolving cell to another.

This variation on legacy cell formation methods therefore has the potential to provide better work flow in a time period and promotes delayed capital investment in machinery by allowing cells to reconfigure or evolve their machine-part grouping.

5. Optimization of Initial Solutions

The concept of allowing cells to evolve their machine-part group over a period of time creates new optimization problems that must be solved for the concept to become economically feasible. Consider that over a period of 5 years and a quarterly reconfiguration cycle, up to 20 unique machine-part group solutions may exist. The challenge is to merge these solutions into viable evolving cells as all proposed evolutions may not make practical or economic sense.

Optimization Problem One – The immovable machine constraint
After evolving cells have been proposed in the immovable machine constraint must be applied. In order to determine the host cell for an immovable machine one must consider all machine-part groups visited by it in the preliminary solutions. The host cell can then determine by the cell that would otherwise incur the highest intra-cell material handing cost. The material handing cost of not having the immovable machine in a cell may be determined as follows:

$$CM_x = \sum_{T=1}^{n} k_T Q_T \quad (7)$$

Were CM_x is the intra-cell material handling cost of not locating the immovable machine in cell x over the lifetime of the system; the cost in each time period T is estimated by a material handling cost per part, k_τ , in that period and the volume of parts Q_τ making the related intra-cell. The volume V_τ may be forecast by relating the relevant curve on the Norton-Bass model to appropriate entities in the BOM. The material handling cost per part has also been described by time dependant variable k_τ, allowing the cost to be related to inflation over a number of years.

Optimization Problem Two – Minimization of machine migration
After the immovable machine constraint has been applied to the preliminary evolving machine-part groups, all machine migrations must be examined as some of these migrations may be eliminated for economic reasons. The optimization problem is to determine which migrations to eliminate such that the cost of machine migration (reconfiguration by machine movement) is at a best compromise with the cost of moving parts between cells. The solution method is illustrated with an example:

Example:
If the number of migrations for a machine M, is n in the proposed solution; the solution can be encoded in a binary string of length n. If the migration is approved the bit value corresponding to the

migration is 1, if it is denied the bit is 0. The number of possible solutions will therefore be 2^n. The best solution for small problems can be found by exhaustive search and by a genetic algorithm for larger problems. Genetic operators for binary strings are well documented and will not be discussed here. Consider the example shown in Table 1 below, which has seven migrations across eight time periods, therefore the number of possible solutions is 128. C_T is the cost of moving a machine in time period T, k_T is the material handling cost per part in time period T and Q_{*i} is the quantity of parts from each of three cells X, Y and Z that require access to the machine in question. The explicit quantities Q can be determined from the Norton-Bass forecast in each time period.

Table 1: Symbolic Example of Machine Migration Problem

The Machine Migration Problem – Migration of M							
T	**Proposed Resident Cell**	**Migration i**	C_T	k_T	Q_{xi}	Q_{yi}	Q_{zi}
1	X	0	C_1	k_1	Q_1		
2	Y	1	C_2	k_2		Q_2	Q_3
3	X	2	C_3	k_3	Q_4		
4	Z	3	C_4	k_4			Q_5
5	Y	4	C_5	k_5		Q_6	
6	X	5	C_6	k_6	Q_7		
7	3	6	C_7	k_7	Q_8	Q_9	
8	1	7	C_8	k_8	Q_{10}		

The solution is encoded in a binary string and the quality of the string is determined by decoding and evaluating it using equation (8). The solution is decoded using binary activity coefficients X_i, Y_i, Z_i and x_i, y_i, z_i for proposed and final solutions respectively. If the coefficient $X_i = 1$, this means that machine M is present in that cell (group) in the ith period, in the preliminary solution. If $x_i = 1$, this means that the machine is also present in that cell (group) in the final solution. Consider the decoding shown in Table 2, where areas shaded in grey are populated from Table 1. B_i is the bit value corresponding to the approval/rejection of the ith migration.

Table 2: Decoding Solution to Obtain Final Machine Migration

		Proposed Intra-cell Migrations of M			**Approved/ Rejected**	**Final Intra-Cell Migrations of M_x** *If $B_i = 1$, $<x_i, y_i, z_i> = <X_i, Y_i, Z_i>$* *If $B_i = 0$ $<x_i, y_i, z_i> = <x_{i-1}, y_{i-1}, z_{i-1}>$*		
T	i	X_i	Y_i	Z_i	B_i	x_i	y_i	z_i
1	0	1	0	0	1	1	0	0
2	1	0	1	0	0	1	0	0
3 - 4	2	1	0	0	1	1	0	0
5	3	0	0	1	1	0	0	1
6-7	4	0	1	0	0	0	0	1
8	5	1	0	0	0	0	0	1
9	6	0	0	1	1	0	0	1
10	7	1	0	0	0	0	0	1

From the example above, the binary string 0110010 ultimately yields two migrations for machine M. The machine remains in group (cell) X for time periods 1-4 and then migrates to group (cell) Z for time periods 5-10. This reduction in machine migration will be at the expense of increased intra-cell movement of parts. The quality for this solution is evaluated against the objective function below, which aims to minimize the total cost considering machine movement and intra-cell movement of parts:

$$min \sum_{i=1}^{n} B_i C_i + k_i(x_i Q_{Xi} + y_i Q_{Yi} + z_i Q_{Zi}) \quad (8)$$

IOP Conf. Series: Materials Science and Engineering **65** (2014) 012009 doi:10.1088/1757-899X/65/1/012009

Once the immoveable machine constraint and the optimization of machine migrations has been performed on preliminary solutions, the final result is a set of evolving machine part groups that may be used to design reconfigurable manufacturing cells. The concept of an evolving manufacturing cell has shown the potential to reduce intra-cell movements and possibly eliminate the duplication of capital equipment (machines and material handling equipment) in a manufacturing system.

6. Conclusion

This paper has presented the concept of evolving machine-part groups as a type of group technology for cell design within the scope of Reconfigurable Manufacturing Systems. A RMS is a system that changes its structure in order to meet changes in production requirements. The concept of an evolving machine-part group provides the underlying framework that guides how the structure should be reconfigured, taking into account changes in part mix, while optimising the system for better part flow and utilization of available capital equipment. The framework utilizes a variation on existing group technology methods to propose the migration of machines from one manufacturing cell to another to provide the reconfiguration. Two methods were presented on how to filter proposed machine migrations for practicality and economic viability. Future work includes the development of line analysis tools for the physical layout design of manufacturing cells taking into consideration the foreknowledge on how the machine and part groups will evolve in that cell over time.

References

[1] Y. Koren, *et al.*, "Reconfigurable Manufacturing Systems," *Annals of the CIRP,* vol. 48, pp. 527-540, 1999.

[2] V. Malhotra, *et al.*, "Reconfigurable manufacturing system: an overview," *International Journal of Machine Intelligence,* vol. 1, pp. 38-46, 2009.

[3] Y. Koren and M. Shpitalni, "Design of Reconfigurable Manufacturing Systems," *Journal of Manufacturing Systems,* vol. 29, pp. 130-141, 2010.

[4] M. R. Abdi, "Selection of a Layout Configuration for Reconfigurable Manufacturing Systems using the AHP," *presented at the 8th International Symposium on the Analytic Hierarchy Process Multi-criteria Decision Making*, Honolulu, Hawaii, USA, 2005.

[5] L. Tang, *et al.*, "Concurrent Line-Balancing, Equipment Selection and Throughput Analysis for Multi-Part Optimal Line Design," *presented at the CIRP 2nd International Conference on Reconfigurable Manufacturing*, Ann Arbor, MI, USA, 2003.

[6] R. Galan, *et al.*, "A Systematic Approach for Product Families Formation in Reconfigurable Manufacturing Systems," *Robotics and Computer-Integrated Manufacturing,* vol. 23, pp. 489-502, October 2007 2007.

[7] F. Bass, "A New Product Growth for Model Consumer Durables," *Management Science,* vol. 15, pp. 215-227, 1969.

[8] J. A. Norton and F. M. Bass, "A Diffusion Theory Model of Adoption and Substitution for Successive Generations of High-Technology Products," *Management Science,,* vol. 33, pp. 1069-1086, 1987.

[9] J. L. Burbidge, "A manual of Production Flow Analysis," *Production Engineer* vol. 56, pp. 34-38, 1977.

[10] R. J. King, "Machine-component grouping in production flow analysis: an approach using rank order clustering algorithm," *International Journal of Production Research* vol. 18, pp. 213-232, 1980.

IOP Conf. Series: Materials Science and Engineering **65** (2014) 012010 doi:10.1088/1757-899X/65/1/012010

The use of Fuzzy expert system in robots decision-making

Mehdi Jamaseb[1]
Shahram Jafari
Farshid Montaseri
Masoud Dadgar
Robotic Lab, Computer Department, Scholl of Computer and Electronic, Shiraz University, Mollasadra street, Shiraz, Iran

E-mail: jamasb@cse.shirazu.ac.ir

Abstract. The main issue that is investigated in this paper, is a method for decision making of mobile robots in different conditions for this purpose, we have used expert system. In this way, that the conditions of the robot are analyzed by on expert person a special issue (like following a ball) using knowledge base and suitable decisions will be mode. Then, using this information fuzzy rules well be built, and using its rules, robots decisions can be implemented like an expert person. In this study, we have used delta3d base for implementing expert systems and CLIPS and also we have used NAO for simulation rcssserver3d robot and 3d football simulation have been used for implementing operation program.

1. Introduction

Humanoid soccer robot is widely used to extend the robotic science. Because of football's being dynamic, the main attention of the robot could be on football game [1, 2].Because of football's being dynamic the main focus of the robot would be on this game. Humanoid soccer robots have been entered to the robot research to extend the robotic and artificial intelligence research.

3d soccer simulation includes one simulator server , that simulates environmental and physical realities, and two teams that every team represent one executive program that is executed according to the number of the players(robots).this executive program should analyze the received message from the server including environmental information and the information of the robot itself, and regarding these data , it should make the necessary decisions and send them to the server in the command format [3].in this plan ,receiving sending , analyzing , updating the information , mathematic calculation and implementing robot's capabilities are the responsibilities of the program written in C++ language.

CLIPS tool is a tool of expert system that works using inference engine of leading chain. CLIPS is designed for communicating with other languages like C++ and java [4, 5].

In this paper, we want to implement decision-making of robot in special cases using expert systems. A state that is considered for decision making is the case of tracking the ball (intercept).

The robot should decide what to do in every two hundredths of the second.in order to reach the ball in the best condition and the highest speed.

A good state in this case is that the robot locate in the direction of the ball and the goal of opponent.

[1] Phone No. +989373563060

2. Procedure

For this purpose, we need fuzzy rules, because we should make the conditions like human football players system, because we need to support inaccurate data [7].

Characteristics considered in this work include four criteria of distance and angle of agent with the ball and also angle and distance of the agent with the rival gate. But we can consider other properties like distance and angle of agent with other agents, that because of high complexity, it hasn't been regarded in this paper.

2.1. priorities of system rules

For priorities of angle kind, nine fuzzy sets and for priorities of distance kind, we have defined four fuzzy sets.

Our priorities include the following:
X1= Distance to ball
X2= Distance to opponent goal
X3= Angle From ball
X4 = Angle from opponent goal

$$gaussian(x;c,\sigma) = \exp(\frac{-1}{2}(\frac{x-c}{\sigma})) \quad (1)$$

Where c is the center of membership function and σ is the width of membership function.

$$bell(x;a,b,c) = \frac{1}{1+\left|\frac{x-c}{a}\right|^{2b}} \quad (2)$$

In the second formula, b is considered a positive number. In variables of distance kind, we have considered the following fuzzy sets regarding (1) and (2) formula [6].
1-Very near distance (V Near): V Near = Gaussian (0, 0.4247)
2-Near distance (Near): Near = Gaussian (1, 0.4247)
3-Medium distance (M): M= Gaussian (2, 0.4247)
4-Far distance (Far): Far= Gaussian bell (53.88, 107.8, 57.13)

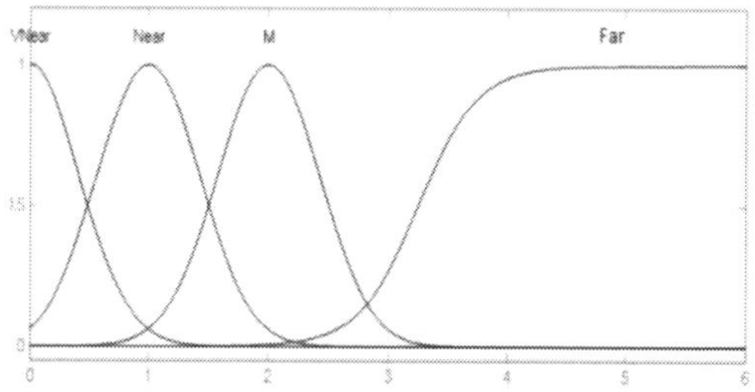

Figure 1. Membership function of define fuzzy sets on the inputs of distance kind

And in variables that are of angle kind (angel can be between -180 to +180) we have considered the following fuzzy sets:
1-Back right direct direction (BR): BR = Gaussian (-180, 9.555)
2-A little leaning to back and right (LBR): LBR = Gaussian (-157.5, 9.555)
3-Right direction (HR): HR = Gaussian (-90, 28.66)
4-Slightly leaning to right and forward direction (LFR): LFR = Gaussian (-22.5, 9.555)
5-Forward direction (DF): DF = Gaussian (0, 9.555)
6-Slightly leaning to left and forward (LFL): LFL = Gaussian (22.5, 9.555)
7-Left direction (HL): HL = Gaussian (90, 28.66)
8-Slightly leaning to the left and back (LBL): LBL = Gaussian (157.5, 9.555)
9- Left back direction (BL): BL = Gaussian (180, 9.555)

IOP Conf. Series: Materials Science and Engineering **65** (2014) 012010 doi:10.1088/1757-899X/65/1/012010

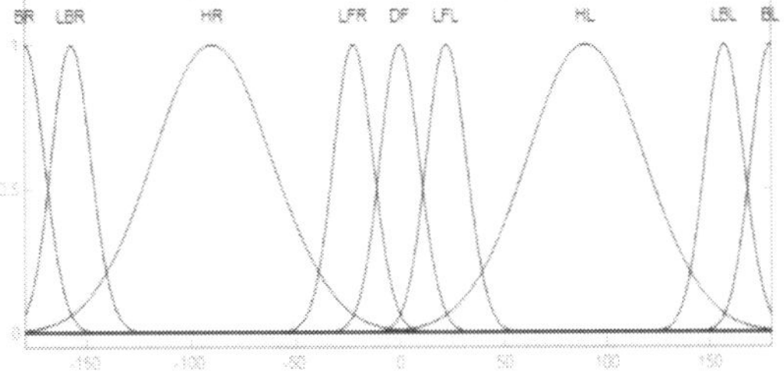

Figure 1. Membership function of defined fuzzy sets on inputs of angle kind.

2.2. Consequents of system rules

In Delta3D base Code, some of the basic operation of robot has been implemented. In this article, we have consequents in rules used for robots' decision-making.

Walking straight forward
Walking straight backward
Rotating to the right
Rotating to the Left
Walking in the direction of the ball
Step to the right
Step to the left

2.3. Creating system rules

Regarding that we can merge Br Fuzzy sets that are used in X3 and X4, finally we can implement the system by maximum 1024 rules. On the one hand, all the inputs are not required in many of the rules, We have rules with length of 2, 3 and 4, and by this assumption , the number of our rules are very low[8].for the example when the angle of ball and opponent goal with agent is very low, it's not needed to consider distances. In many cases ,rules can be combined, for example, when the distance with the ball opponent goal is low for small angle and very small angle of the agent with the ball or opponent goal, the same orations a similar be done.

Finally, we have covered all the cases with 58 rules. We can reduce the complexity of rules and system design, using the mentioned mechanisms, so the rules well be more clear and transparent. Finally using such an approach will result in easier development of the system. For establishing the rules, we should also pay attention to the conditions of the agent in the fields. Agent conditions are divided into two parts, general conditions and special conditions. General conditions refer to the conditions that in the most pant of the play happen for one football player agent, like when the ball is in a small angle from the agent and the opponent goal is also in front of the agent. These conditions usually happen in a far distance from the ball.

But special conditions are conditions in which the football player agent less possibly face them during one play, for example conditions in which the ball is beside the agent and the agent is close to the opponent goal.

General conditions constitute a small number of our rules and we have many rules for special conditions.

After building a knowledge base of the rules in WXCLIPS and regarding that WXCLIPS is fully compatible with C++ language, Expert systems can be implemented in operating program, and expert system can be used in robots decision-making

Figure 3. A general Condition

Figure 4. a special Condition

2.4. System evaluation

In order to evaluate the presented method, we will compare the recommendations of the expert system to the robot with the recommendations of expert persons.

Expert persons include two persons that are in expert simulator competition (members of Delta3d team), one person from the base football players of the country and two person from semiprofessional players and a normal person.

For this purpose, we place the football player agent and the ball separately and randomly in a spot of the simulated field, then expert system and the mentioned persons should suggest a combination of suitable actions to reach to the ball by considering the angle of agent with the opponent goal. After suggestion every action, the agent do the defined action, till the agent reach the ball and actions' continuation end.

The mentioned expert persons send the continuation of the action using a controller that is designed to control the robot, that by using it, they guide the football player agent to the desired point. In fact the Expert person can announce every one of the 7 possible actions to the robot. Every repetition is in the way that we announce random coordinate of the agent and the ball to the expert persons in order to give the value of primary coordinate of the agent and the ball and the expert and whenever regarding a good angle with the opponent goal. The ball is in the good situation, the expert person end the controller. It's note Worthy that the interval of the agent arriving to the ball by pressing stop button is very short and doesn't affect the evaluation of suggested continuation by the expert person.

For example you can see the way in which the expert person guide the robot in figures 5, 6 and 7.

Figure 5. Detection of the backward movement by expert person

Figure 6. Detection of right rotation by expert person

Figure 7. Detection of forward movement by expert person

The evaluation criteria that we considered are the average of the time of reaching the ball and the average of final direction of the robot in 100 times. Direction is defined as follows: ((difference of the agent angle with the agent to the opponent goal)). The final evaluation criteria include the time to reach the ball and the required time for reaching to zero direction that is calculated from formula (3):

$$\text{Time}_{\text{total}} = \text{time}_{\text{dis}} + \frac{\text{angle}}{20} \quad (3)$$

In formula 3, <u>time dis</u> is the average of the time of reaching to the ball, angle is the value of direction and <u>time total</u> is the final evaluation value, that lower the final evaluation value is, the robot is quicker in this issue. The results are shown in table 1.

Table 1. The results of comparison between expert persons, previous system and expert system.

Evaluator	Time dis (s)	Angle	Time total
Expert system	85.5 (±19.8)	23(±12)	48.95
Expert person 1	42.1(±24.3)	34(±13)	44.80
Expert person 2	44.0(±25.2)	38(±10)	46.90
Football player 1	51.3(±18.7)	36(±14)	54.10

Football player 2	52.2(±22.5)	39(±13)	55.15
Football player 3	54.5(±23.0)	40(±12)	57.50
Normal person	59.4(±19.8)	31(±18)	61.95
Delta3d 2010	91.4(±31.8)	9(±3)	91.85

Regarding table 1, we conclude that although a small number of expert system in comparison whit the previous system ha was implemented in delta3d 2010.

3. Conclusion

In this article, a method was presented for decision making of the football player robots in a simulated environment.

The proposed system used Fuzzy expert systems for deciding the best action for following the ball and to a large extent act like an expert person. This method has improved the speed of the robot about 87.64% in the case following the ball in comparison whit the previous model that was presented in delta 3d 2010.

4. References

[1] C. Zhou, "Rules for the RoboCup humanoid league", Advanced Robotics, vol. 18(7), pp. 721-724, 2004.
[2] H. Kitano, and M. Asada, "The RoboCup humanoid challenge as the millennium challenge for advanced robotics" , Advanced Robotics, vol. 13(8), pp. 723-736, 2000.
[3] Wiki Page. (2012)The SimSpark Website. [Online]. Available: http://simspark.sourceforge.net/wiki.
[4] G. Riley. (2012, May 10). Clips a Tool for Building Expert Systems [Online]. Available: http://clipsrules.sourceforge.net/
[5] J. Giarratano and G. Riley, Expert System Principles and Programming, Brooks/Cole, pp. 328-329, 1998.
[6] J.S. Roger Jang, C.T. Sun and E. Mizutani, Neuro-Fuzzy and Soft Computing: A Computational Approach to Learning and Machine Intelligence, Prentice Hall, pp. 24-27, 1997.
[7] L.X.Wang, A Course In Fuzzy Systems and Control.Prentice Hall, pp. 48-72, 1996.
[8] W.Siler and J.J.Buckley, FUZZY EXPERT SYSTEMS AND FUZZY REASONING.John Wiley & Sons, pp. 15-51, 2005.

IOP Conf. Series: Materials Science and Engineering **65** (2014) 012011 doi:10.1088/1757-899X/65/1/012011

EXPLORING THE POTENTIAL HIGH ENERGY LOCATIONS AND INTENSITIES IN CONFINED WORK SPACES OF WAVEGUIDE DIMENSIONS

Ricardo Rodriguez and Winston G. Lewis

Department of Mechanical and Manufacturing Engineering, The University of the West Indies, St. Augustine Campus, Trinidad and Tobago, West Indies. Email: ricardoj.rodriguez@yahoo.com

Department of Mechanical and Manufacturing Engineering, The University of the West Indies, St. Augustine Campus, Trinidad and Tobago, West Indies. Email: wlewis@uwi.edu.tt

Abstract: Adequately identifying and managing hazards at the workplace can be a tedious task which extends into the realm of uncertainty, probability and prediction models in order to fully comprehend the nature of the hazard. As such, organizations cannot be blamed for knowledge gaps in the training of personnel they contract to ensure a safe and healthy work environment, especially where there are latent hazards. Electromagnetic wave propagation at frequencies in the SAR (specific absorption rate) region is a special concern to authorities involved in setting RF (radiofrequency) and microwave exposure guidelines. Despite that there is no conclusive evidence to suggest that non-ionizing electromagnetic radiation causes adverse health effects other than thermal, no effort should be lost to ensure that workers and the public at large are adequately protected from unnecessary exposure to radiation. Standards however set exposure limits for free space, plane wave propagation but fall short in compiling information on intensities of these waves after they undergo reflection and diffraction from wall surfaces. Waveguide technology has managed to constrain microwaves to remain within set boundaries, with fixed frequencies that force the waves to behave differently to if they were moving in free space. This technology has offered the ability to transport more energy for communication purposes other than transmission lines. The size of a waveguide may be to the order of a few centimetres and can guide RF of wavelengths of the order of centimetres also but what if spaces of larger dimensions are capable of being waveguides and can guide waves of larger wavelengths such as those that correspond to frequencies between 30MHz to 300MHz? Such RF waves belong to the SAR region of the spectrum where strict exposure limits are set for health and safety protection since a standing man acts as a dipole antenna for this radiation and can absorb maximum energy from propagating RF waves. This review visits the likelihood for potential energy build-up due to RF propagation in confined spaces that are of waveguide design but with larger dimensions. Such confined spaces include silos, tanks, pipes, man-holes, air-condition ducts, tunnels, wells, engine rooms and operator rooms on board vessels. In these confined spaces waves reflect off of the walls and combine constructively or destructively with incident waves producing reinforcement or cancellation respectively. Where there is reinforcement, the intensity of the wave for a particular distance in accordance with the standard, may exceed the exposure limit for this distance from the source thereby exposing the worker to larger intensities than the accepted limit and presenting a potential health and safety threat.

Keywords: RF hazard, SAR, confined space, reflection, waveguide

1. Introduction

In the past, workers in organizations were expected to complete their tasks in sub-standard conditions within a given time frame which contributed to worker health and safety issues. Of special interest to worker safety is work performed in confined spaces. A confined space is one which has limited access and egress and accommodates one employee or two to perform some work function that may take a period of time unsuitable to the health and safety of the employees exposed. Such spaces include: tanks, a/c ducts, storage bins, pits, vessels, silos, and man-holes. (NCDOL, 2012).

Confined spaces pose potential threats that are not readily visible, (latent hazards), to the health and safety of workers and may lead to fatal injuries because there is failure to recognise and treat the hazard and the suitability of the emergency response (NCDOL, 2012). These spaces may accumulate gases and toxic fumes, poor lighting, and may be a hub for harnessing electromagnetic fields. RF fields can be generated by propagating electromagnetic waves in the SAR region of the spectrum (30MHz to 1000 MHz) and can be a potential hazard to a standing man. International standards set stringent exposure limits for RF in this frequency range, siting restrictive measures of monitoring, maintenance and reporting for sites using equipment within this range (IEEE C 95.1 1999). The OSHA of the United States has set a maximum exposure level of 10 mW/cm^2 for frequencies between 10 MHz and 100 GHz averaged over a 6 minute period, while IEEE sets at 1mW/cm^2 and E (electric) field at 61.4 V/m over the same period.

2. Exposure Standards

International exposure limits are set for thermal effects and possible electro-stimulation, shocks and burns for controlled and uncontrolled environments. Standards are however not fault proof and may from time to time not fully address all the risks to the hazard present since hazards may change depending on their location, presence with other hazards or other perilous conditions and interaction with artefacts. The standards therefore give a probability of safety but do not guarantee it, which should prompt safety practitioners to engage in methods of determination and analysis of the risks posed by hazards in a more inclusive and holistic manner. (IEEE C95.1 1999).

There has been public concern that electromagnetic waves coming from the various sources cause persons to have headaches, develop rashes, cancer, nervous disorders, irritability, moodiness, breathing difficulties, muscle pain, mental depression, among other claims (Cember, 1996). While the thermal effects have been substantiated there is no conclusive evidence of non-thermal effects despite the findings and reviews from many scientists and researchers who claim that there are adverse health effects. To balance the claims, a neutral position is taken in the IEEE C 95.1 standard with strict limits set at 30 MHz to 300 MHz where the SAR (specific absorption rate) region is most dangerous to a standing man. (Rodriguez 2008).

The standards set guidelines generally for RF propagation in free space for plane waves. It is difficult to accept that RF exposure in free space is the same as that within a confined space where the energy is confined within boundaries. The standards have not strictly focussed on RF propagation in the 30 – 1000 MHz region or SAR region, inside of rectangular and cylindrical life-size structures likened to communication waveguides for microwave propagation. According to IEEE, a RF hot spot is a highly localized area of relatively more intense radio-frequency radiation that manifests itself in localized areas in which there exists a concentration of radio-frequency fields caused by reflections and/ or narrow beams from antennae.

There are three major concerns then, up for consideration at this point if we are to address this RF propagation in confined spaces:

i. whether there is any difference between RF propagation in free space (unbounded) as opposed to confined space (bounded) propagation;
ii. can we use waveguide theory to treat smaller frequencies (longer wavelengths) of RF propagation in large structures (confined spaces) in the same way we do with the larger frequencies (smaller wavelengths); and
iii. whether or not there are high energy points (RF hot spots) inside confined spaces where waves reflect and interfere constructively.

The modelling of wave propagation is essentially for the determination of the probability of accepted performance for communication systems and is a major factor when planning network systems. There are a variety of models to suit applications of which many do not suit the applications they were designed for and as such there is much room for RF propagation modelling. In this regard, RF modelling has its genesis in statistics as we try to determine the location and intensity in the environment in which it is contained. In the words of Seybold, '*the selection of a model to be used for a particular application often turns out to be as much art (or religion) as it is science*'. (Seybold, 2005). Between 30 MHz and 1GHz for RF waves, we can expect refraction and reflection as dominant factors but as the frequency increases, the skin effect produces more attenuation which is readily observed at microwave frequencies (Seybold, 2005). This is in support for the methodology proposed here in this article for RF hot spots, since reflections of RF in the frequency range 30 MHz to 1000 MHz (below microwave frequencies) off of the inner walls of rectangular confined spaces are under consideration.

When electromagnetic radiation strikes a surface some of the radiation reflects off the surface while some is transmitted through the medium it strikes. The amount of radiation reflected is dependent on the coefficient reflection of the material or reflectance, i.e. the ratio of the reflected to the incident flux. For high reflectance materials or specular reflection, we expect maximum reflection with small transmission and the angle of reflection to be equal to the angle of incidence. (Hecht 1975).

3. Waveguides

RF energy can be transmitted via transmission lines, antennae or waveguides as sources to receiver antennae. Waveguides are usually small with dimensions just larger than that of the wavelength of the wave travelling along it; these may be about 'a' = 6 cm x 'b' = 3 cm in rectangular cross sectional area, see Fig 1.

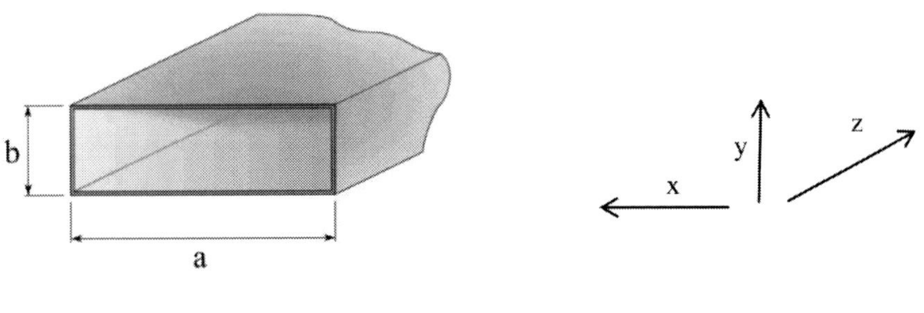

Fig 1. Picture of typical waveguide [<u>*Source:*</u> *http://en.wikipedia.org/wiki/File:Waveguide.svg*]

Table 1 shows typical waveguide dimensions for microwaves. The last column, column 6, shows the inner dimensions of width and height and it can be observed that the ratio is 2:1. For this condition, propagation of waves can exist without much loss in energy. For such dimensions there can be a number of possible modes by which energy can be transferred with each being characterized by a distinctive field configuration.

An electromagnetic field can propagate along a waveguide in various ways. Two common modes of propagation are known as transverse-magnetic (TM) and transverse-electric (TE). In TM mode, the magnetic lines of flux are perpendicular to the axis of the waveguide.

Table 1: Standard sizes of waveguides

Waveguide name			Frequency band name	Recommended frequency band of operation (GHz)	Cutoff frequency of lowest order mode (GHz)	Cutoff frequency of next mode (GHz)	Inner dimensions of waveguide opening (inch)
EIA	RCSC *	IEC					
WR2300	WG0.0	R3		0.32 — 0.45	0.257	0.513	23.000 × 11.500
WR2100	WG0	R4		0.35 — 0.50	0.281	0.562	21.000 × 10.500
WR1800	WG1	R5		0.45 — 0.63	0.328	0.656	18.000 × 9.000
WR1500	WG2	R6		0.50 — 0.75	0.393	0.787	15.000 × 7.500
WR1150	WG3	R8		0.63 — 0.97	0.513	1.026	11.500 × 5.750

[Source: http://www.bing.com/search?q=table+of+standard+waveguide+sizes&form=MSNH14&refig=99af3dee9e594fd3af12565b4683d5b2&pq=table+of+standard+waveguide+sizes&sc=0-3&sp=-1&qs=n&sk=]

In TE mode, the electric lines of flux are perpendicular to the axis of the waveguide. Either mode can provide low loss and high efficiency as long as the interior of the waveguide is kept clean and dry. (Bhattacharya et al, 1999).

To function properly, a waveguide must have a certain minimum diameter relative to the wavelength of the signal. If the waveguide is too narrow or the frequency is too low (the wavelength is too long), the electromagnetic fields cannot propagate. At any frequency above the cut-off (the lowest frequency at which the waveguide is large enough), the feed line will work well, although certain operating characteristics vary depending on the number of wavelengths in the cross section. Guided waves are used where high efficiency is required and despite losses that occur which are not normally found with free space waves, these are offset since the waves are no longer divergent as an expanding ball but confined to a rigid path of fixed cross-section (Mathews and Stephenson, 1968).

Waves of frequencies between 20 – 3000 MHz are used for line of sight communications in aircrafts, FM, TV and amateur radio broadcasting and contain wavelengths comparable to the dimensions of large trees and buildings which provide good reflecting surfaces. Communication can be cancelled or enhanced depending on whether there is destructive or constructive interference between direct and reflected radiation. (Whitaker 2002).

4. Confined space philosophy

Buildings, houses, vehicles, tanks, silos are just some of the everyday structures we design for ourselves that in fact may perform the same function as a waveguide used in communication for the efficient propagation of microwaves.

The formulae for wave propagation inside a rectangular wave guide are as shown:

1. $c = 3x10^8$ m/s (c= speed of light in vacuum)
2. $\lambda_{cmn} = 2/\sqrt{[(m/a)^2 + (n/b)^2]}$: (cut-off wavelength for m, n modes for m=1,2,3…, n = 0,1,2,3..
3. $\sin\theta = \lambda/\lambda_c = f_c/f$: Θ is angle of incidence at entrance of guide
4. $f_{cmn} = c/2\ \{\sqrt{[(m/a)^2 + (n/b)^2]}\}$: cut-off frequency for m, n modes

[Source: http://www.ece.msstate.edu/~donohoe/ece3323waveguides.pdf]

For any wave propagating down a wave guide, it does so in the Z direction with a group velocity given by U_g. In so doing, a stationary wave pattern is set up as the waves bounce off of the walls of the guide in a 'zig-zag' manner in the XZ plane with velocity in direction of wave travel, U_u.

There is therefore a component of wave motion in both Z and X directions. The Y direction addresses the oscillations of the electric field E only and there is no wave motion there just as there is no component of E in either Z or X directions. The E field spreads across the width of the guide and is half wavelength of the propagating RF for the TE10 mode with maximum field strength at the centre and falls off to zero at the walls. For the TE20 mode, the width is two ½ wavelengths and the E field distribution is E^+ and E^- across the width corresponding to maxima at top and bottom.
http://www.ece.msstate.edu/~donohoe/ece3323waveguides.pdf

5. Sample Scenario (Fig 2)

Consider that a rectangular high reflectance air-condition duct of width a = 1.75m and height b = 0.88m is in the line of sight of a RF broadcasting antenna. The incident plane waves emanating from the antenna strike the open rectangular duct (waveguide) at 60 degrees to the Z axis of the duct. Determine which frequencies can propagate through this duct if the space is to operate as a waveguide for the TE10 mode in single mode operation.

Single mode operation occurs when the guide allows only one frequency within the dominant mode to propagate, in this case it is 98.9 MHz. Recall the cut-off frequency $f_{cmn} = c/2\ \{\sqrt{[(m/a)^2 + (n/b)^2]}\}$; m=1 n=0 for TE10 mode which is 85.7 MHz for this guide dimension. For width 'a' = 1.75m and angle of incidence 60°, then $f = f_c / \sin 60$ which is equal to 98.9 MHz. If the antenna emits at 98.9 MHz, then since this frequency is higher than the cut-off, it should be able to propagate down the guide, air-condition duct in this case, once excited. For guide operating as a single mode operation, i.e., for the TE10 mode only, waves with frequency 98.9 MHz that strike the entrance at 60° will propagate and fill up between the points A to Z. Incident waves strike the wall of the duct at Z and G and at 30° to the normal (n) at the wall's surface and reflect at the same angle 30° and continue to E and F. These will then reflect from this surface to strike the first surface at H and L, see Fig.2.

Interference occurs within triangles: OZG, OEF and OHL where incident and reflected rays cross. These are considered potential RF hot spots and the triangles that house them are hot zones. The hot zones are all of same area and repeat periodically and give the probability of RF hot spots, where the waves interfere constructively. These hot spots will have higher field intensities than the waves that do not interfere constructively and are therefore potential spots of higher exposure.
The field intensity for an incoming wave is $E_i = E_0\, e^{j\omega t}.\, e^{-\gamma z}$, (unbounded) where $\gamma = \alpha + j\beta$ and γ is imaginary if (attenuation factor) $\alpha = 0$. i.e. there is no attenuation of the wave. Inside the confined space the wave equation becomes: $E_i = E_0\, e^{j(\omega t - \beta z)} \sin(k_x x)$ for the time varying wave propagation in the z direction and the stationary wave set up in the x direction, where ω is the frequency of the wave, ε is the phase change upon reflection, β is the phase constant of the wave, E_i, E_0, E_r are the incident, initial and reflected electric field intensities.
The reflected wave is $E_r = E_0\, e^{j([\omega t+\varepsilon] - \beta z)} \sin(k_x x)$. For constructive interference, $E = E_i + E_r$. The area of each triangle by geometry is the same. For triangle OZG then, the area is: 1/2 (ZG) (a/2), which is a/4 ZG.

IOP Conf. Series: Materials Science and Engineering **65** (2014) 012011 doi:10.1088/1757-899X/65/1/012011

ZG = AE = EF = HL, a/ ZG = tan Θ = tan 60. So ZG = a/ tan 60, = 1.01 units. Area of triangle is therefore: $(a^2/4)$ 1/tan 60, i.e., $(1.75^2/4)$ 1/tan 60 = 0.44 units^2. It is within this confined area we can expect RF hot spots to exist. http://www.ece.msstate.edu/~donohoe/ece3323waveguides.pdf

Table 2: Cut-off frequencies in TE10 mode for varying rectangular confined space widths 'a' in Single Mode Operation

Confined space name	a/ m	b/m	f_c/ MHz	Single mode freq. at 60^0 /MHz	SAR
CS200	2.00	1.10	75	86.6	Yes
CS175	1.75	1.40	85.7	98.9	Yes
CS150	1.50	0.75	100	115.5	Yes
CS125	1.25	0.65	120	138.5	Yes

If the angle of incidence of the waves at the guide entrance is smaller, then larger frequency waves will propagate and those below will not. If we are trying to stop the FM band between 86.1 and 107.1 MHz for example, then 'a' can be varied to achieve this. Table 2 gives some widths and corresponding cut-off frequencies for the TE10 mode. Table 2 unlike Table 1 addresses the larger guide dimensions and has been constructed in this presentation as an equivalent informative chart using the layout prescribed in Table 1. The Table will therefore assist in being a first observation point for knowing what frequencies will propagate in our design of a confined space and whether they are within the SAR region. These are the frequencies to be weary of since constructive interference within the confined space could lead to RF build-up and hence RF hot spots.

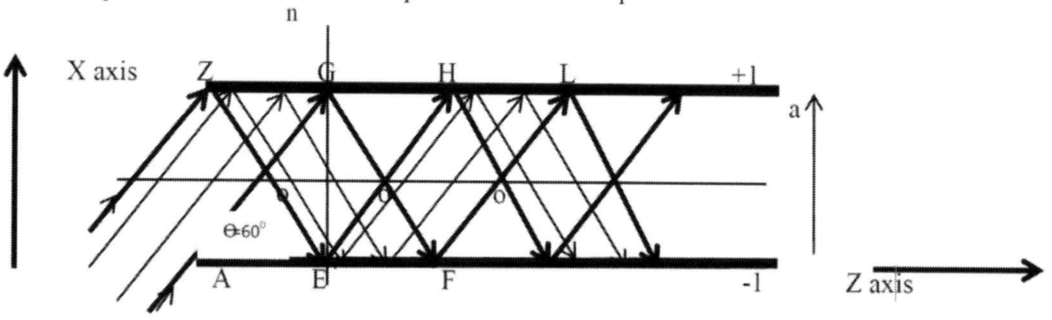

Fig 2: Confined space of width 'a', as a single mode operation waveguide for cut-off frequency in TE10

6. Conclusion

This article presents a thought – provoking notion of large waves in SAR region propagating down rectangular confined spaces just as microwaves would in waveguides setting up standing waves due to interference of incident and reflected waves within the space. These RF hot spots possess the potential to contain higher field intensities than if the waves were propagating in free space. When we therefore consider safety hazards due to noise, lighting, toxic and gas build-up within confined spaces, accumulation of RF along the confined space must also be given consideration.

References

[1] A Guide to Radio Frequency Hazards With Electric Detonators: N.C. Department of Labor Occupational Safety and Health Division,1101 Mail Service Center Raleigh, NC 27699-1101, Cherie Berry, Commissioner of Labor

[2] Bhattacharya. A., S.Gupta and A. Chakraborty. 1999. Analysis of rectangular waveguides and Thick windows as EMI sensors. Progress in Electronics Resears. PIER 22. 231-258

[3] Cember, H, 1996. Introduction to Health Physics. Mc Graw Hill, New York

[4] Hecht E, 1975. Optics. Mc Graw Hill, New York, USA

[5] http://www.ece.msstate.edu/~donohoe/ece3323waveguides.pdf

[6] Institute of Electrical and Electronics Engineers. 1992. IEEE Standard for Safety Levels with respect to Human Exposure to Radio Frequency Electromagnetic Fields, 3 KHz to 300 GHz (CP5.1-1995). New York: Institute of Electrical and Electronics Engineers.

[7] Mathews, P and I Stephenson, 1968. Microwave Components, Chapman and Hall Ltd, 11 New Fetter Lane, London EC. Ch 1, pg 3-27.

[8] Rodriguez, R. 2007. A HSE Radiation Management Model for the Workplace. Mechanical Engineering Department, University of the West Indies, Trinidad, West Indies.

[9] Seybold, John S. 2005. Introduction to RF Propagation.pg11. John Wiley and Sons, Inc., Hoboken, New Jersey,

[10] Whitaker, Jerry. 2002. The RF Transmission Systems Handbook. CRC Press LLC, 2000 N.W. Corporate Blvd., Boca Raton, Florida 33431. Pg 15-13.

Numerical Study of Conjugate Natural Convection Heat Transfer Using One Phase Liquid Cooling

F A Gdhaidh[1], K Hussain[2] and H S Qi[3]
[1] PhD student in Faculty of Engineering and Informatics at Bradford University, BD7 1DP, UK
[2] Senior Lecturer in Mechanical Engineering and the Head of the School of Engineering at Bradford University, BD7 1DP, UK
[3] Senior Lecturer in Mechanical Engineering at Bradford University, BD7 1DP, UK

E-mail:F.A.Gdhaidh@bradford.ac.uk

Abstract. A numerical study in 3-D is performed using water as a cooling fluid to investigate the one phase natural convection heat transfer within enclosure. A heat source representing a computer CPU mounted on one vertical wall of a rectangular enclosure is simulated while a heat sink is installed on the opposite vertical wall of the enclosure. The air flow inside the computer compartment is created by using an exhaust fan, and the flow is assumed to be turbulent. The applied power considered ranges from $15 - 40\ W$. In order to determine the thermal behaviour of the cooling system, the effect of the heat input and the dimension of the enclosure are investigated. The results illustrate that as the size of the enclosure increase the chip temperature declined. However the drop in the temperature is very small when the width increased more than $50\ mm$. When the enclosure was filled with water the temperature was reduced by 38%. Also the cooling system maintains the maximum chip temperature at 71.5°C when the heat input of 40 W was assumed and this is within the current recommended computer electronic chips temperature of no more than 85°C.

1. Introduction

The natural convection is considered as an effective and most economical cooling strategy for computer electronic devices [1]. However, the conventional method for free convection using air as the medium is not sufficient enough to remove the high heat flux from the modern computer electronic chips [2]. This is due to the low heat transfer rate of air when compared to other high rate fluids.

The results of natural convection inside rectangular enclosures filled with air to cool electronic chips show the limitation of air natural convection cooling [3]. There is a promising alternative way to keep the electronic devices at the safe operation temperature by using liquids. This is justified by the fact that the liquids have relatively high removal of heat when compared to air. In natural convection the fluid motion inside enclosures whether these enclosures were vertical or horizontal occurs due to the temperature gradient change, therefore the isothermal boundary condition is adopted [4-6]. Sathe et al. [7] studied numerically the effect of installed square enclosure filled with dielectric liquid on the chip temperature. They noticed that the highest chip temperature occurred when the upper wall of the enclosure was at fixed temperature and other walls adiabatic. Numerical and experimental studies were carried out by Heindel et al. [8] where two different coolant fluids (water and FC-77) were used within the enclosure to cool heaters where the cold wall is fixed at room temperature. They concluded that as the heat flux increases, the convection coefficients along the heater face and vertical velocity also increases. Further work by Heindel et al. [9] showed that using fins on the heaters enhanced the heat transfer as much as 24 and 15 times for vertical and horizontal cavity orientations respectively when the cold wall kept isothermally at 15°C. In [10, 11] an experimental studies were reported using water as the cooling fluid. Their results indicate that the Nusselt number based on heat source length increases as the heat flux increased. Other investigations on natural convection cooling, using ethylene glycol as fluid to cool heat sources which attached to rectangular enclosures were carried out by Keyhani et al. [12] and Ju and Chen [13]. In both of these studies the cold wall is maintained isothermal at a fixed temperature. They concluded that the enclosure width influences the heat transfer process but this influence is weak when the ratio of enclosure width to heater height was around 4.0.

Desktop computers are complex in terms of heat transfer control and it is impossible to analyse using exact solution methods. Therefore CFD prediction tools are adopted to explore various designs quickly with acceptable accuracy. In the thermal design there are some successes for analysing complex electronics system using CFD. The detailed temperature and flow fields have been simulated through computer chassis with two fans using Flotherm code by Lee and Mahalingam [14].

From the previous studies the far end of the enclosure has been assumed isothermal. However In this study, a desktop computer heat transfer characteristics is examined with the aim to remove excessive heat generated by the electronic chips using one phase water cooling system with adopting a realistic boundary conditions for the cold wall by assuming heat sink with air flow created by exhaust fan.

2. Problem description

A numerical study of coupling between natural convection heat transfer inside enclosure and forced convection within the computer covering case has been investigated. The heat source is mounted on a vertical substrate and the substrate is connected to the enclosure. A heat sink without fan is attached to the other end of the enclosure and the enclosure was filled with water. The air flow inside the computer covering case is created by a fan installed at the end of the computer case. Figure 1 shows a schematic diagram of the three-dimensional system used for the analysis. The heat input chosen was in the range of $15 - 40W$ with $5W$ increment and the value of power remains constant during the simulation. Table 1 lists all the required dimensions for the simulated model.

Table 1. Physical Model Dimension in mm.

Components	Height	Length	Width
Computer case	352	418	171
Hard drive	23	145	100
Heater	37.5	37.5	5
Enclosure	77	68	50
Substrate	77	68	3
Main board	245	200	1
Power supply	89	138	148

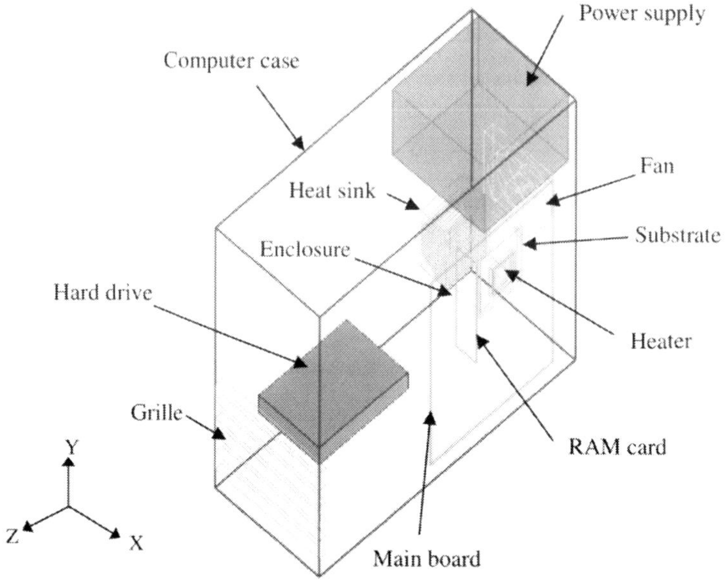

Figure 1. Schematic diagram of the physical model.

The fan is installed at the back of the computer case with 90 mm diameter and a constant flow rate of $0.026\,\mathrm{m^3/s}$ was assumed.

3. Mathematical model

The mathematical model was constrained by the following assumptions: 3-D steady state heat transfer, contact resistance between the heater/substrate, substrate/enclosure and enclosure/heat sink interfaces equal to $0.18°\mathrm{C}/W$. The radiation effects are neglected and an incompressible fluid flow was assumed. The flow through the computer chassis was assumed to be turbulent and hence the $k-\varepsilon$ type model was used.

Applying the above assumptions on the governing differential equations of continuity, momentum and energy, yield the following equations:

Continuity:

$$\nabla . V = 0 \quad (1)$$

Momentum equation:

$$\rho(V.\nabla V) = -\nabla P + \mu \nabla^2 V + F \quad (2)$$

Energy equation:

$$V.\nabla T = \alpha \nabla^2 T + S \quad (3)$$

where ρ is the density, V is the velocity components, μ is the viscosity, P is the pressure, α is the thermal diffusivity, T is the temperature and F and S are the body force and the source term respectively.

In order to complete the mathematical model, the following boundary and initial conditions were used: the flow rate inside the computer case which is created by the fan was assumed fixed. The air inlet temperature was set at 22°C, air physical properties are calculated at the same temperature and the isoflux condition was applied for every test.

The average Nusselt number for substrate/water interface is determined by integrating the local Nusselt number along the substrate surface. Therefore the average Nusselt number ($\overline{N_u}$) could be written as:

$$\overline{N_u} = \frac{\bar{h}l}{K_w} \tag{4}$$

where $\bar{h}$ is the average heat transfer coefficient, l is the substrate height and K_w is the thermal conductivity of water.

4. Solution procedure

In present study the commercial (ANSYS –CFD) software which has heat transfer capability was used to examine the system. This software is based on a Finite-Volume scheme to convert the governing equations to algebraic equations that can be solved numerically. The "SIMPLE" algorithm (semi-implicit method for pressure linked equations) is used to handle the coupling between pressure and velocity. To avoid the divergence in the iteration process, Under-relaxation techniques are used to slow down the changes between iterations.

Although there are different regions (solids and fluids), and these have different thermal properties and equations, however, the numerical solutions within the computational domain for continuity, momentum and energy equations are obtained simultaneously in all regions.

The effects of the elements number on the results have been examined. The final optimised mesh data points of 311,204 were used, and this gave stable solution.

5. Results and discussion

A numerical study was carried out to examine the thermal behaviour of the proposed system. Results and analysis of the flow and the thermal field are presented in this section. During the design of the cooling system, a number of limitations/constrains were considered, such as the available space and the maximum allowed operation temperature of the system. These constraints are applied to obtain the optimum width of the enclosure when the applied heat input was set at 40 W.

The second test was to examine the effect of using different heat input level (Q). Heat sources in the range of $15 - 40\ W$ with $5\ W$ increment were used. According to deSorgo [15] when two surfaces in contact, less than 1% of the surfaces is in touch because all the solid surface have certain roughness regardless of how smooth they are. Therefore, a thermal resistance value of $0.18\,°C/W$ between any two solid surfaces in contact was assumed.

5.1 Enclosure width effects

Figure 2 illustrates the maximum heater temperature for different enclosure widths. A range of $0 - 60\ mm$ with $10\ mm$ increments was used, and the heat input was fixed at $40\ W$. In the case when the enclosure width is $0\ mm$, the heater temperature surface was high and reaches a temperature of 115.4°C.

An enclosure of 20 mm wide reduced the temperature sharply to 76.4°C. A further reduction can be achieved by increasing the width of the enclosure. A 60 mm enclosure reduces the temperature to 70.6°C. . The change in the temperature is due to the increase in the heat transfer area of the enclosure wall surfaces and also the effect of the water thermal properties. Moreover the water flow inside the enclosure helps to cool regions closer to the hot wall.

The results show that only 1% change in temperature between the 60 and $50\ mm$ enclosure widths. Hence a $50\ mm$ enclosure was used as the optimum dimension for our enclosure to carry further analysis. Moreover a $50\ mm$ enclosure filled with water reduced the initial temperature without enclosure by 38%. Note that the volume of water inside the enclosure can be obtained using the dimension of enclosure and this result in a total volume of water of 0.207 litre.

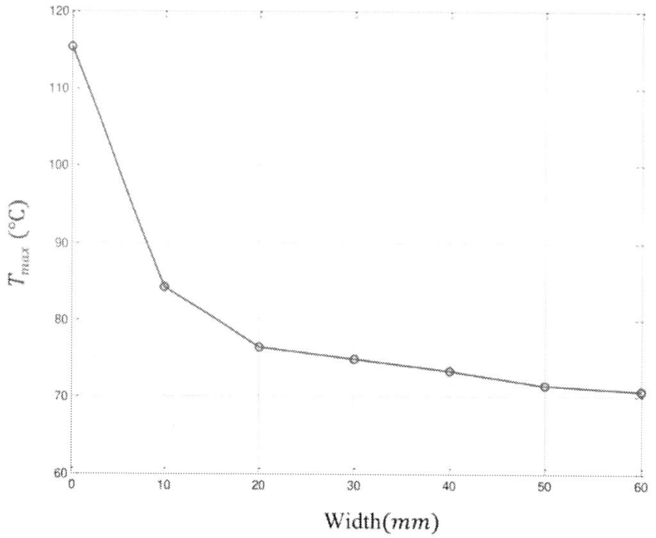

Figure 2. The enclosure width effects on the maximum heater temperature.

5.2 *Heat input effects*

In this investigation the heat input applied to the heater varies between $15 - 40W$. The heater material used corresponds to extruded aluminium with thermal conductivity of ($k_h = 205\ W/mK$), producing constant heat input. The substrate material chosen was copper with thermal conductivity ($k_s = 387\ W/mK$). Further, the fluid Prandtl number was assumed to be 5 corresponding to water. The fluids and the material properties are assumed constant. The effects of different values of heat energies on the maximum system temperature and average Nusselt number are presented.

The maximum system temperature. In this section the maximum temperature in the system which occurs in the heater is studied. As mentioned earlier, the recommended electronic chip temperature should not exceed 85°C in most electronic devices [16]. However, the ideal maximum working temperature is in range of 65 − 70°C. Figure 3 illustrates the maximum temperature in the system used in this study for different values of heat input.

Figure 3 illustrates that the maximum temperature of the cooling system increased as the heat input increased. The curve is almost linear and that is because the temperature is a function of heat input when the other conditions are fixed.

Figure 4 shows the temperature distribution for two values of heat input 15 and 40W. The maximum temperature increased from 40.02°C at 15W to reach 71.48°C at 40W. The contour lines of temperature are nearly vertical in the fluid region because the heat transfer in the enclosure is controlled by conduction due to the slow water motion within the enclosure. Further, the maximum temperature is located at the centre of the heater as shown. For clarity the temperature in the enclosure, substrate and heater are also illustrated in Figure 4.

Figure 5 shows the substrate/water interface temperature for the full range of the applied heat input. The heat input flows from the heater and the substrate towards the water. The flat part of the curve denotes zero temperature gradient where the temperature is maximum. This region corresponds to the size of the heater. The zero gradient region is due to the higher heater and substrate thermal conductivity. The positive gradient region of the curve represents the increase in the thermal boundary layer while the negative gradient represents the decrease in the thermal boundary layer.

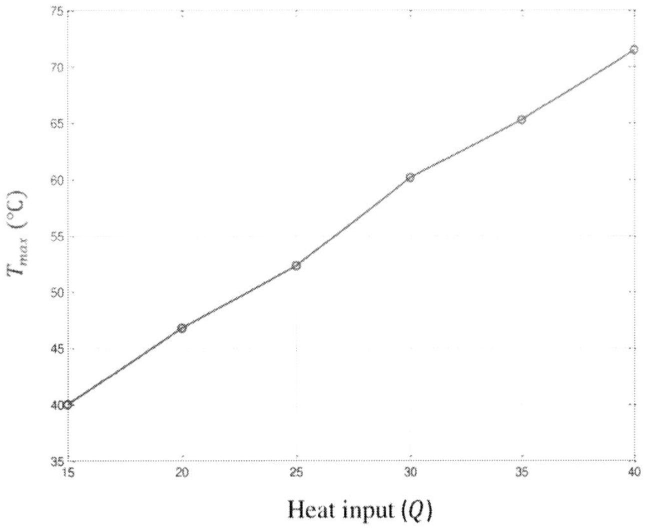

Figure 3. The variation of maximum temperature with different values of heat flux.

The increase in the thermal boundary layer is due to the reduction in the convection coefficient and the increase in the local bulk fluid temperature. While for the negative temperature gradient the thermal boundary layer start to decrease. The effects of the enclosure walls thickness are clearly shown at the substrate interior corners having steeper gradients. This is again due to the high thermal conductivity for both enclosure and substrate.

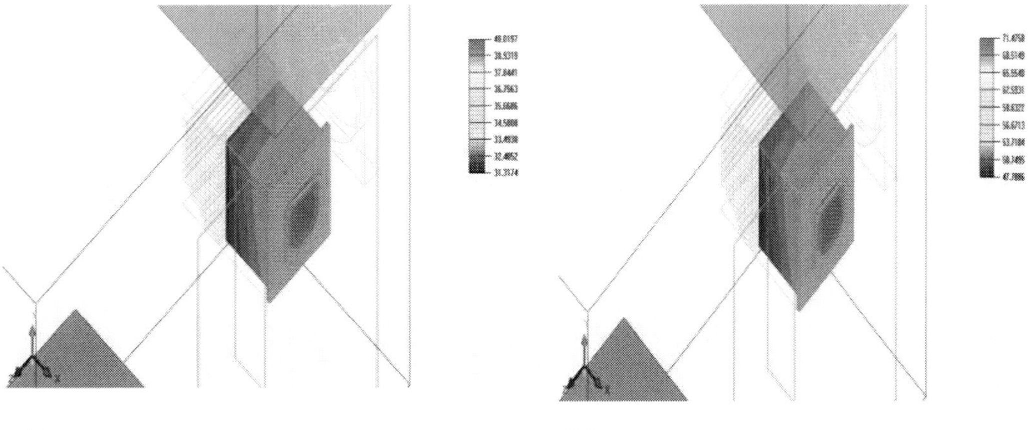

(a) (b)

Figure 4. Temperature distribution within enclosure, substrate and heater for (a) 15W and (b) 40W.

The different curves denote the change of temperature for various heat input levels. It is clearly shown that as the heat input increases the temperature level also increases.

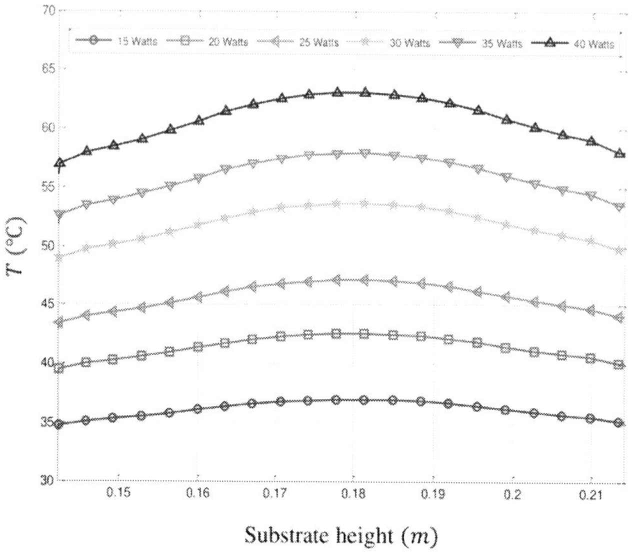

Figure 5. The local substrate/water interface temperature with different applied heat input.

5.2.1 Average Nusselt number. The Nusselt number illustrates the ratio between the convection and conduction heat transfer, and hence a larger Nusselt number corresponds to a more active convection. In this study the average Nusselt number is calculated along substrate/water interface.

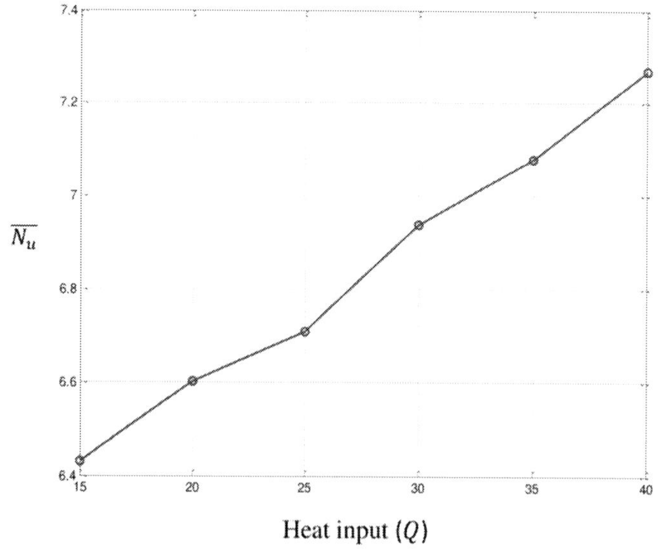

Figure 6. Average Nusselt number with different values of heat input

Figure 6 illustrates the variation of mean Nusselt number for different values of heat input. It is evident from the figure that, the average Nusselt number increases as the heat input increases due to the increase in buoyancy driven flow inside the enclosure. As a result the average heat transfer coefficient increases and the thermal boundary layer decreases.

6. Conclusion

Steady state 3-D natural conjugate convection analysis in a rectangular enclosure and forced convection within a computer covering case has been conducted numerically. One can conclude from the results of this research the following:

- The size of the enclosure has an effect on the value of the maximum temperature. However, it was also found that the temperature variation remains constant beyond certain enclosure size.
- The temperature at substrate/water interface in the place corresponding to the heater remains constant due to the high thermal conductivity of the heater and the substrate.
- The values of mean Nusselt number rises as the heat input increase from $15W$ to $40W$. That is due to the thermal boundary layer being thin and also the average convection coefficient increased.
- It was found that the addition of an enclosure with water had a greater effect on the value of the maximum temperature when compared to the case without enclosure. A reduction of 38% of maximum temperature was reached when the enclosure with water was added.
- The introduction of the water system reduced the temperature to 71.5°C which is much below the recommended manufacture temperature for the desktop computers.

7. References

[1] Ridouane EH, Campo A and Chang JY 2005 Natural convection patterns in right-angled triangular cavities with heated vertical sides and cooled hypotenuses *J. Heat Transfer*.**127**1181–1186.

[2] Bar-Cohen A1991Thermal management of electronic components with dielectric liquids*Proceedings of the ASME/JSME Thermal Engineering Joint conference*.**2**xv-xxxix.

[3]KelleherMD, Hnock R H and Yang K T1987 Laminar natural convection in a rectangular enclosure due to a heated protrusion on one vertical wall, Part 1: Experimental investigation *Proceedings of the Second ASME/JSME Thermal Engineering Joint Conference*. **2**169-177.

[4] Hasnaoui M, Bilgen E and Vasseur P1992 Natural convection heat transfer in rectangular cavities heated from below*J. Thermophysical Heat Transfer*.**6** 255-264.

[5] Valencia A and Frederick R1989 Heat Transfer in square cavities with partially active vertical walls*Int. J. Heat and Mass Transfer*.**32**1567-1574.

[6] Selamet E, Arpaci V S and Borgnakke C1992 Simulation of laminar buoyancy driven flows in an enclosure*Numerical Heat Transfer*.**22**. 401-420.

[7] Sathe S B and Joshi Y1991 Natural convection arising from a heat generating substrate-mounted protrusion in a liquid-filled two-dimensional enclosure*Int. J. Heat Mass Transfer*. 342149-2163.

[8] Heindel T. J, Ramadhyani S and Incropera F P 1995Conjugate natural convection from an array of discrete heat sources: part 1 two- and three-dimensional model validation*Int. J. Heat and Fluid Flow*.**16** 501-510,

[9] Heindel T J, Incropera F P and RamadhyaniS1996 Enhancement of natural convection heat transfer from an array of discrete heat sources*Int. J Heat and Mass Transfer*.**39**479- 490.

[10] Bhowmik H, Tso C P, Tou K W and Tan F L2005 Convection heat transfer from discrete heat sources in a liquid cooled rectangular channel*Applied thermal engineering*.**25** 2532-2542.

[11] Bhowmik H and Tou K W2005 Experimental study of transient natural convection heat transfer from simulated electronic chips*Experimental Thermal and Fluid Science*.**29** 485- 492.

[12] Keyhani M, Chen L and Pitts D R1991 The aspect ratio effect on natural convection in an enclosure with discrete heat sources*J. Heat Transfer*.**113**883– 891.

[13] Ju Y Land Chen Z Q1996 Numerical simulation of natural convection in an enclosure with

discrete protruding heaters*Numer. Heat Transfer.***30** 207–218.

[14] Linton R and Agonafer D1995 Coarse and Detailed CFD Modeling of a Finned Heat Sink*IEEE transactions on COMFQNENTS, packaging and manufacturing technology- part A.***18**517-520.

[15] Miksa deSorgo 1996 Thermal Interface Materials *Electronics Cooling Magazine* September, 1, No 3.

[16] RamaswamyC, Joshi Y, Nakayama W and Johnson W2002 High-speed visualization of boiling from an enhanced structure*Int. J. Heat and Mass Transfer.***45**4761- 4771.

Experimental and numerical evaluation of the fatigue behaviour in a welded joint

P. Almaguer[1] and R. Estrada2

Centro de Estudios CADCAM, Facultad de Ingeniería. Universidad de Holguín. Ave. XX Aniversario s/n. Piedra Blanca. Holguín. GP 57 .CP 80100. Cuba, Tel. (53)(24) 482678

Email: [1] pavel@facing.uho.edu.cu , [2] roberto@cadcam.uho.edu.cu

Abstract: Welded joints are an important part in structures. For this reason, it is always necessary to know the behaviour of them under cyclic loads. In this paper a $S-N$ curve of a butt welded joint of the AISI 1015 steel and Cuban manufacturing E6013 electrode is showed. Fatigue tests were made in an universal testing machine MTS810. The stress ratio used in the test was 0,1. Flaws in the fatigue specimens were characterized by means of optical and scanning electron microscopy. *SolidWorks* 2013 software was used to modeling the specimens geometry, while to simulate the fatigue behaviour *Simulation* was used. The joint fatigue limit is 178 MPa, and a cut point at 2 039 093 cycles. Some points of the simulations are inside of the 95% confidence band.

Keywords: Welded joint, fatigue, $S-N$ curve, AISI 1015, E6013, numerical simulation

1. INTRODUCTION

Welded joints have an important role at the present time. Practically in all the structures are these types of joints. Manual metal arc is a process commonly used to make the joint. This is characterized by the application of heat focused in the area to join. In the welding bead, the temperature reaches high values that are able to fuse the metal. However, the temperature decreases from the welding bead to the base metal. This causes a variable heat treatment in the welded joint that influences in the final metallographic structure of each area according to the maximum of temperature that has reached, and also in its mechanical properties. In this sense, it is possible to define three different zones in the welded joint: contribution material (CM), heat affected zone (HAZ) and base material (BM). HAZ has a special interest. In this, the BM doesn't melt, but that is due to the high temperatures reached during the welding, in it occurs important metallographic transformations and also in its mechanical properties. Some investigations [1] define the HAZ as metallurgical notch, due to the reduction of the tenacity that happens there.

Fatigue is one of the phenomena that influence the most in the useful life of the elements of machines and structures requested by cyclis loads. Fatigue cracks begin in a local flaw of the structure. In fact, the own geometry of the bead already constitutes a geometric discontinuity that turns out to be

IOP Conf. Series: Materials Science and Engineering **65** (2014) 012013 doi:10.1088/1757-899X/65/1/012013

a stresses concentrator. The "local approach", which is based on the location of the denominated hot spots, has great application in the study of the welded joints. According to this approach, if in any point of the joint the stress value are equal or higher that the yield limit, a fatigue crack begins.
The S-N curves is based on the medium fatigue life or in a given failure probability. The generation of the S-N curve of a material requires many tests. Due to the high statistical dispersion, the results are grouped in a confidence bands. A part of this dispersion can be attributed to test errors, but it is a property of the physical phenomenon that which provokes to carry out a great number of tests in order to determine the confidence band with enough precision. Therefore, to characterize a material to fatigue supposes a very high cost, due to the quantity of necessary test specimen, the complexity of the test machines and the necessary time to carry out each experiment.

Design modern techniques are supported in the CADCAM technologies. Thanks to the quick development of the computers, today is possible to carry out calculations and to simulate complex experiments that reproduce the real phenomenon with certain approach grade.
SolidWork software uses the theory of the accumulated damage (also called Miner´s lineal damage rule) for the studies to fatigue[2]. In this it is supposed that a load cycle causes a permanent damage that can be measured[2].

Lineal damage rule doesn't consider the effects of the load sequence. This way, it predicts that the damage caused by the stress cycle is independent from where it takes place in the load history. It also supposes that the rate of accumulation of damage is independent of the stress level. This can take to differences with the experimental values, because like it is known the cracks of fatigue begin and develop with different behavior at differents stress levels. Another difficulty that it can be observed to this rule it is that it doesn't consider the different stages of the development of the cracks of fatigue. In a general way, this is a factor that affects all the fatigue classic approach. In this sense, Fracture mechanics can offers better results.

In this study, it is pursued as objective to show the S-N for a butt welded joint in which is used AISI 1015 steel as BM and the CM is the E6013 electrode of Cuban manufacturing, besides a methodology to simulate by means of Finite Element Method (FEM) the answer of the joint under the action of the variable loads.

2. MATERIALS AND METHODS

To carry out this study AISI 1015 steel was used as BM. This is steel that has a good soldability.
The BM for the welding was in a plate of dimensions 500 mm x 600 mm x 4 mm. The joint was carried out without edges preparation, with beads to both sides assisting to the dimensions proposed in the Figure 1.

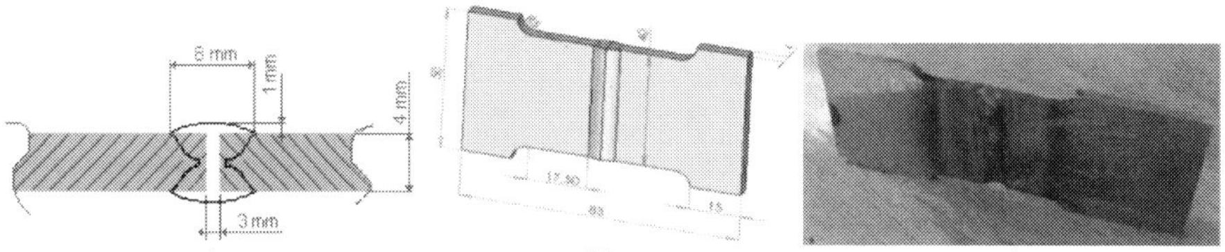

a) b) c)

Figure 1. Welded specimen geometry. a) The joint is carried out without edges preparation. b) Dimensions of the welded specimen. c) Specimen welded for fatigue test.

Chemical analysis of a square sample of side 5 cm and thickness 4 mm to check that the material used in the experiments was AISI 1015 steel was carried out in an ARL 3460 cuantometer. The results of this analysis is shown in the Table 1. On the other hand, the mechanical properties of the material according to the bibliography are in the Table 2.[3]

Table 1. Chemical composition of the AISI 1015 steel[3].

AISI 1015 steel	**Elements**				
Composition:	Carbon (%)	Manganes (%)	Silicon (%)	Sulphur (%)	**Phosphurus (%)**
Typical	0,13- 0,18	0,30 – 0,60	0,17 – 0,37	≤0,05	**≤0,04**
Measure	**0,166 090**	**0,504 210**	**0,027 460**	**0,005 780**	**0,013 850**

Table 2. Physical and mechanical properties of the AISI 1015 steel[3].

Properties							
σ_B **(MPa)**	σ_f (MPa)	σ_{-1} (MPa)	μ	ρ (kg/m^3)	α (1/°C)	δ (%)	ψ **(%)**
420	**315**	**230**	**0,29**	**7 870**	**119 x 10^{-7}**	**39**	**61**

Welded joint was carried out by means of Manual Metal Arc. E6013 Cuban manufacturing electrode was used. Table 3 shows the mechanical properties of the deposit.

Table 3. Mechanical properties of the CM[Manufacturer data].

Properties			
σ_B **(MPa)**	σ_f (MPa)	δ (%)	ψ **(%)**
431 – 510	**392**	**20 - 28**	**35 – 60**

The values of the welding process parameters to carry out the joint are in Table 4.

Table 4. Welding process parameters.

Current (A)	**Voltaje (V)**	**Number of beads**	**Welding velocity (mm/min)**	**Number of electrodes$^{(*)}$**	**Total time$^{(*)}$ (s)**
90 – 120	24 – 26	2	59,57 – 79,42	0,421	81,22 – 60,91
TRI-ARC 504. Rectificator with capacity between 70 A -500 A y 23 V – 40 V.					

(*)For one specimen

The welded joint studied was tested to fatigue. For it was necessary to determine the values of the parameters of the load cycle. The regime of load was asymmetric tensile.The first parameter was the maximum stress (1). The maximum stress of the cycle in a first load level it is considered a fraction 2/3 of the AISI 1015 steel yield limit.

$$S_{max} = \frac{2}{3} S_y \qquad (1)$$

The value of the maximum stress calculated by means of the previous expression is equal to S_{max} = 210 MPa to which corresponds a maximum load P_{max} = 33,60 kN.

Tests will be carried out at different load levels. The following load levels will be calculated according to the results of previous level. This is due to the great dispersion that exists in the fatigue test results. The logic that is used to determine the following load levels is the following one: the load value is diminished until the specimen tested reach equal or overcome the value of 1 500 000 cycles, for those correspond the fatigue limit in case of the joints according to ASTM E 739 - 91[4] standard. The asymmetry ratio of the cycle R = 0,1. The frequency of application of the load is 14 Hz. As failure criterion is used "total fracture". Test is interrupted when 1 500 000 load cycles are applied. The temperature of the laboratory was of 20 °C ± 2 °C. Fatigue test were carried out in a universal machine test MTS810.

IOP Conf. Series: Materials Science and Engineering **65** (2014) 012013 doi:10.1088/1757-899X/65/1/012013

The geometric performance was made in *SolidWorks* 2013 software and for the simulations the package *Simulation*. To make the simulations the methodology that appears in the Figure 2 is proposed.

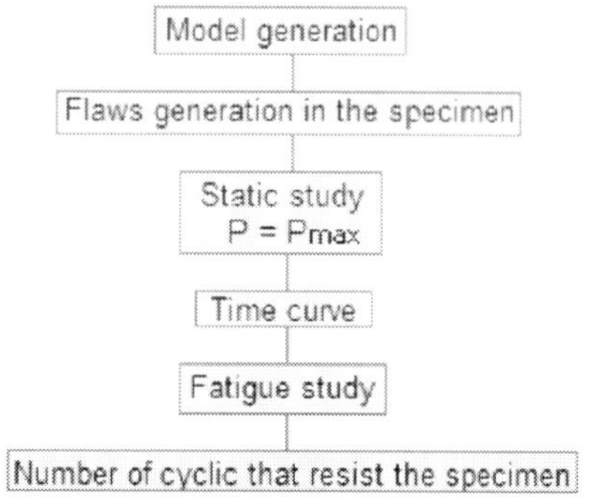

Figure 2. Proposed metodology to the joint fatigue simulation.

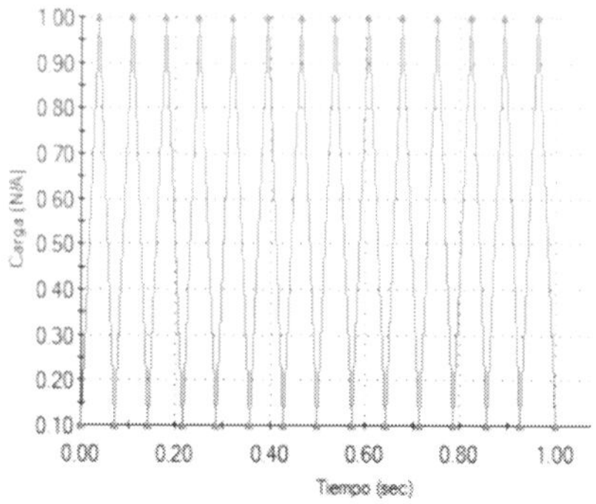

Figure 3. Load history curve.

The geometric model is the same that appears in the Figure 1. The second point of the methodology proposes the flaws creation. This is because simulations were carried out in the specimens without flaws, the software showed that the specimen failure occured in the section change, however, in the BM, what doesn't happen this way in the real welded joints. Next one runs a static study, considering that the applied load is the maximum of the load cycle. The time curve (Figure 3) models the variation of the load in the time. After the load curve a fatigue study is carried out which shows that it is possible to obtain the number of cycles that should fail the specimen.

3. RESULTS

According to what was expressed in the previous section experiments at 4 load levels were carried out. In the first load level a great dispersion of the number of cycles that the specimens failed was obtained. In the second level each specimen reach 1 500 000 cycles or higher that it is the value to stop the experiment. Then the third and fourth levels were used to adjust the intermediate values of the curve. Maximum stress for each load level are in Table 5.

Table 5. Maximum stress of each fatigue test load level.

Level	1	2	3	4
Maximum stress S_{max} (MPa)	210,00	178,1	196,9	190,6

4. DISCUSSION

4.1. Fatigue test statistical analysis

For the study of the results of the fatigue test the ASTM E 739 - 91 standard [4] is applied in this work. In a general way, it is possible to relate the fatigue life with the stress by means of the following equation:

$$\log N = A + B(S) \qquad (2)$$

Where A and B are the coefficients to fit the equation and are equal to $\hat{A}$ = 11,775 716 240 and $\hat{B}$ = -0,031 058 404 respectively. However, equation (2) now is:

$$\log N = 11{,}77571624 - 0{,}031058404 \times S_{max} \qquad (3)$$

To obtain equation (3) is necessary to asume a lineal model. If the lineal model is right or not is proved by means of the statistical test, based on the distribution F that appears in the Table 2, page 5 of [4]. For a significancy level of the 95% F_p=3,6823, the calculated parameter has a value of 3,4577 being smaller than F_p. Therefore the hypothesis of lineality of the model is accepted.

4.2. Fractographic analysis

In this work was carried out a fractographic analysis to all the specimens tested to fatigue, by means of optic microscopy (MO) and scanning electron microscopy (SEM) using an Estereomicroscopic Zeiss Stemi SV11 and an electronic microscope JEOL JSM-6460LV, respectively. In a general way the cracks always begin in the flaw and develope themselves with a semieliptic front. The area of fragile break that happens because of the static overload of the specimen shows an extriction, characteristic of the break by tensile loads.

The observations of the specimen 20 are shown. It fails in the HAZ. The crack originates in the detail pointed out in the Figure 4 a) that it was identified by means of the penetrant liquids. From this point it began to develop that was growing by means of the well-known semieliptic front until reaching the critical crack size. An observation by means of SEM of the area where is this flaw are showed in the Figure 4 c). In its, the flaws is appreciated that has an initial lengthened shape with a longitude of 0,6923 mm and an approximate width 0,0385 mm.

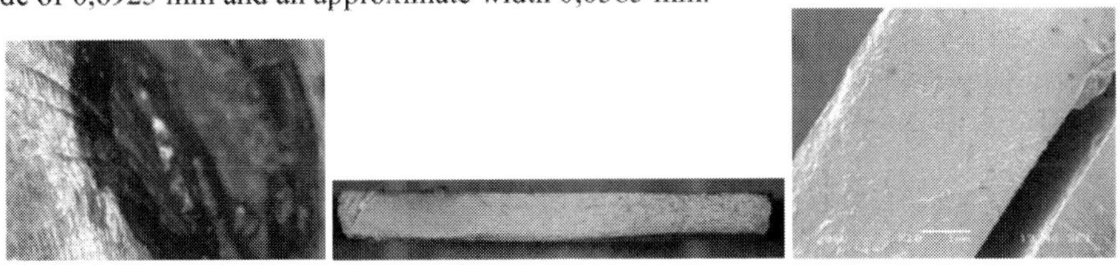

Figure 4. Specimen 20 fractographic analisys. a) Flaw identified by penetrant liquids. In tis point the crack begin. b) Observation by means of MO. c) Observation by means of SEM.

These defects and others that are not shown in this work are those that were modeled in the second step of the proposed methodology to carry out the welded joint fatigue behavior simulations.

4.3. Numeric simulation of the welded joint fatigue behavior

To carry out the study a solid mesh was used. Material model was lineal elastic and a type of finite elements was quadratic of high order. To check the model convergence several studies are carried out varying the size of the element until the results do not vary more than 5% of the previous analysis.
In the Figure 5, the specimen with static load, constraints and mesh is shown.

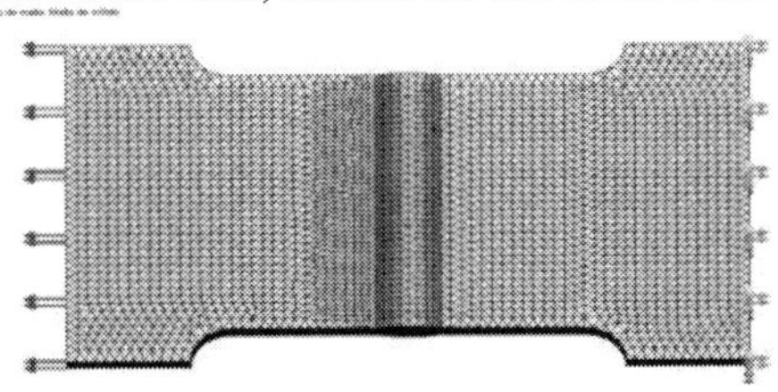

Figure 5. Specimen with load, constraints and mesh. Elements size: 1,233 mm.

As it was already said before, the different specimens were modeled with flaws that constituted the places where the fatigue crack began; and therefore, where it should the fatigue failure happen. Flaws size was taken from characterization by means of SEM.

Twenty four specimens were modeled. The fatigue study is based on running a static study to a certain value of load (in the case of this investigation is carried out to the value of maximum load) and later on a load curve is generated as that of the Figure 3 and execute a fatigue study. Results are the accumulated damage and total fatigue life in the specimen. The last one is expressed in load blocks. The load blocks means how many times it is necessary to repeat the applied load event to the fatigue failure occurs. As in the studied specimen a load curve was generated to one second, the number of load blocks for the times that one cycle are in each second, how many cycles resists the specimen until the fatigue failure is obtained. The frequency of application of the load is of 14 Hz, and then each cycle repeats 14 times in each second.

As example of the obtained results, the simulated specimen 9 is shown in Figure 6. In this specimen, a flaw it is like that is shown in the Figure 4.

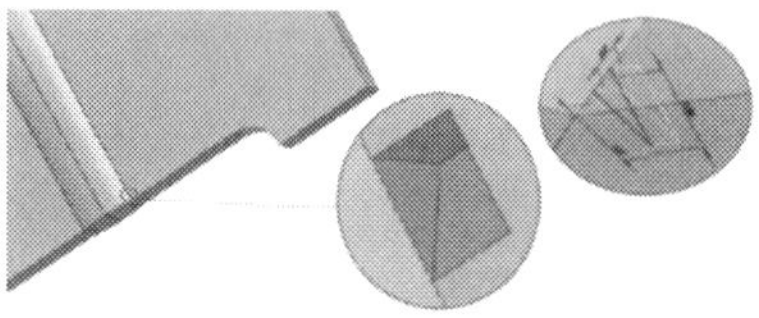

Figure 6. Flaw location and size in simulated specimen 9.

In the Figure 7 a) the results of the static study are shown. The biggest stress value is equal to 1 181 MPa. It is located in the stress concentrator, with a bigger value that the yield limit. This is in agreement with the local approach, in that point a fatigue crack begins.

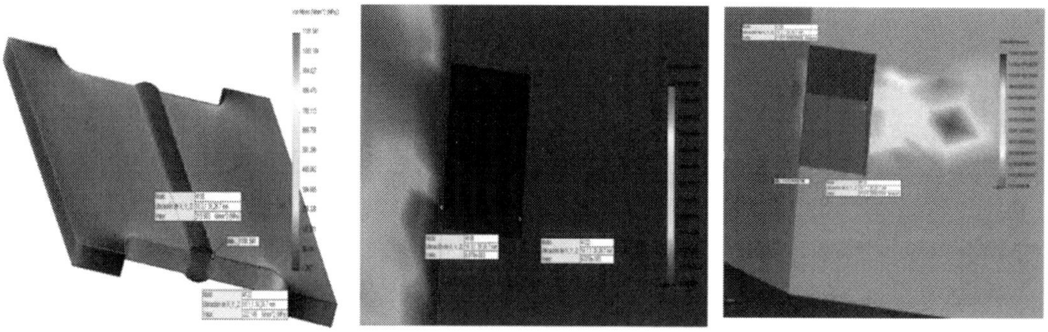

Figure 7. Specimen 9 simulation results. a) Maximum stress. b) Damage 9. c) Total life.

In the Figures 7 b) and c) are the values of damage and the total life for the simulated specimen 9. As it is appreciated, in the concentrator is where the fatigue failure occurs.

4.4. Welded joint S - N curve

The S - N curve of the welded joint is shown in the Figure 8. The experimental curve was obtained from the experimental data, processed according to [4] like it has been described in the epigraph 4.2. The cut point is for a stress value of 178 MPa and 2 039 093 cycles. The simulations results are in this figure too. As we can see, some of them are between the 95% confidence bands.

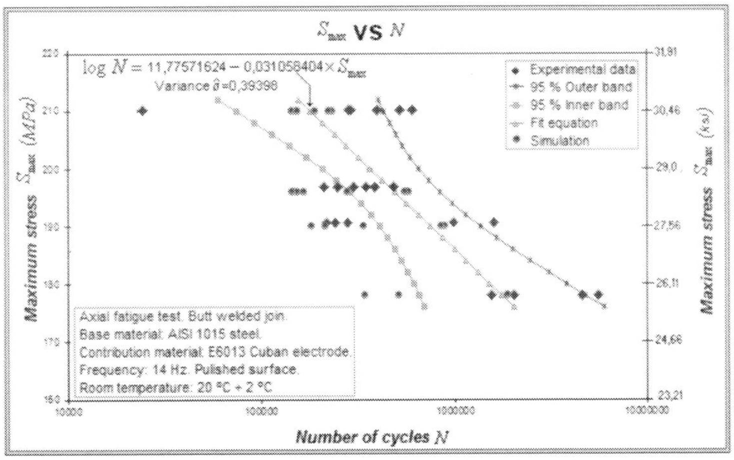

Figure. 8. S - N curve of the welded joint.

5. CONCLUSSIONS

1. A S - N curve was obtained. The equation that relates stress and number of cycles is $\log N = 11{,}775\,716\,240 - 0{,}031\,058\,404 \times S_{max}$. The fatigue cracks fatigue always began in a flaw, and then it grew following a semieliptic front crack until the break. Due to the great dispersion of the experimental results in each load level, is necessary to increase the quantity of tested specimen, with a better fit of the curve.
2. The proposed methodology can be used to simulate the welded joint fatigue behaviuor by means of FEM. As we can see in Figure 8 some simulated points are inside the 95 % confidence bands, even on the fit equation. With several it doesn't happen this. First, it is possible to attribute this to the characteristic of the fatigue, where great dispersion always exists in the experimental results and second, to the classic fatigue theory, that do not consider the different fatigue stages. It is considered that to obtain better results it is necessary use the Fracture Mechanics in next research.
3. With the use of the numeric simulation software it is achieved a saving of time and resources, diminishing the costs of the experimentation considerably.

6. REFERENCES

[1] Reina Gómez, Manuel. *Soldadura de los aceros. Aplicaciones.* 3ra edición. Gráficas Lormo. Madrid, España. 1994. p. 474. ISBN 84-605-14-75-7.
[2] Tutorial Software *SolidWorks* 2013.
[3] http://www.matweb.com. See on april, l 2011.
[4] ASTM E 739 – 91. Standard Practice for Statistical Analysis of Linear or Linearized Stress-Life (S-N) and Strain-Life (ε-N) Fatigue Data. 2004

27th International Conference on CADCAM, Robotics and Factories of the Future 2014 IOP Publishing
IOP Conf. Series: Materials Science and Engineering 65 (2014) 012014 doi:10.1088/1757-899X/65/1/012014

Analysis on the Load Carrying Mechanism Integrated as Heterogeneous Co-operative Manipulator in a Walking Wheelchair

I S Rajay Vedaraj[1],
Associate Professor, SMBS, VIT University
Ritika Jain
Final year B.Tech., CSE, VIT Universtiy
B V A.Rao
Senior Professor, Advisor to Chancellor(retd.), VIT University.

Abstract:

After industrial robots came into existence during 1960, the technology of robotics with the design and analysis of robots in various forms in industries as well as in domestic applications were developed. Nowadays, along with the automotive sector the robots are producing a great impact in the form of quality and production rate to register their existence reliable in various other sectors also. Robotic technology has undergone various phase translations from being tortured as humanoids to the present day manipulators. Depending upon the various forms of its existence, robot manipulators are designed as serial manipulators and parallel manipulators. Individually both types can be proved effective though both have various drawbacks in design and the kinematic analysis. The versatility of robots can be increased by making them work in an environment where the same work volume is shared by more than one manipulator. This work volume can be identified as co-operative work volume of those manipulators. Here the interference of manipulators in the work volume of other manipulators is possible and is made obstacle free. The main advantage of co-operative manipulators is that when a number of independent manipulators are put together in a cooperative work envelope the efficiency and ability to perform tasks is greatly enhanced. The main disadvantage of the co-operative manipulators lies in the complication of its design even for a simple application, in almost all fields.

In this paper, a cooperative design of robot manipulators to work in co-operative work environment is done and analysed for its efficacy. In the industrial applications when robotic manipulators are put together in more numbers, the trajectory planning becomes the tough task in the work cell. Proper design can remove the design defects of the cooperative manipulators and can be utilized in a more efficient way. In the proposed research paper an analysis is made on such a type of cooperative manipulator used for climbing stairs with three leg design and anlaysis were also done on the mechanism integrated to the system. Kinematics of the legs are analysed separately and the legs are designed to carry a maximum of 175kgs, which is sustained by the center leg and shared by the dual wing legs equally during the walking phase. In the proposed design, screwjack mechanism is used as the central leg to share the load and thus the analysis on the load sharing capability of the whole system is analysed and concluded in terms of failure modes.

1.Introduction:

Designing a wheelchair to get integrated with walking mechanism is possible only in the form of co-operative manipulators. In co-operative manipulators the load is shared uniformly to almost all the components involved in the motion while the other components will be sharing the function in a passive way, like a counter balance. Designing this needs a thorough knowledge of both the serial and parallel manipulators.

After the design of the wheel chair and the legs are done separately, they are integrated to work in the cooperative environment to follow the trajectory planned. The proposed wheelchair is made up of three slots provided with V-grooves. The middle slot is assigned for the telescopic screw jack mechanism and the two other slots are provided for the wing legs as described in Figure 1.

All the three legs are interfaced with the base of the wheelchair by hip joints. The wing legs are composed of three components, the hip joint and two serial links. And the screw jack leg is composed of the normal screw jack components joined to the base of the wheelchair by the hip joint.

The slots provided are of 230mm length and 45mm breadth with V-grooves. These details are provided in the wireframe and solid model specified below. Analysis on the wing legs are separately done applying the maximum load that is going to act on it during the walking and stair climbing phase. Also the screw jack leg is subjected to the same form of analysis allowing the load acting on it to be a uniformly distributed load. Special attention is made towards the screw jack leg to understand the angle it can make with the floor during walking and stair climbing. Since the angle of inclination of the screw jack leg in the walking phase is maximum, it is designed in such a way that the frictional forces at the foot of the screw jack leg should not allow it to slip. And the angle is restricted to 68° from various analysis made. The stress and strain created during the analysis beyond 68° show abnormal features which may support failure of the screw jack mechanism or any component of the screw jack leg. Slip is another factor when the angle is increased beyond 68°. A lead screw link is provided on the wing leg slots to drive the wing legs horizontally on the base of the wheelchair. During this motion only the weight of the legs are carried along the slots, which is easier than carrying the whole load and moving forward. The screw jack leg is fixed at the middle and the slot of the screw jack leg is only to reduce the weight of the base of the wheelchair. Thus pulling the screw jack leg forward will make the base of the wheelchair to move forward and this makes the whole mechanism move forward and the mass transfer form screw jack leg to the wing legs. This has been explained in the simulation of the mechanism and has been animated.

Two triangular components are provided on both sides of the base of the wheelchair facing upwards. They are the base level of the seating arrangement. On the top of these triangular components the seat is mounted. With all the components discussed here and mentioned above the exact model of the proposed walking wheelchair is shown in the figure 1.

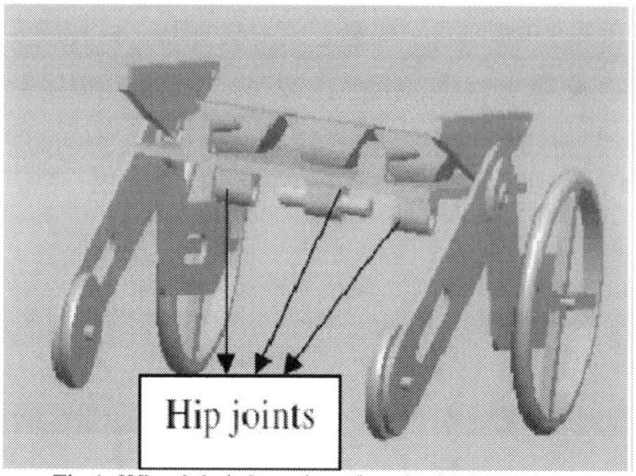

Fig.1. Wheelchair base interfaced with hip joints.

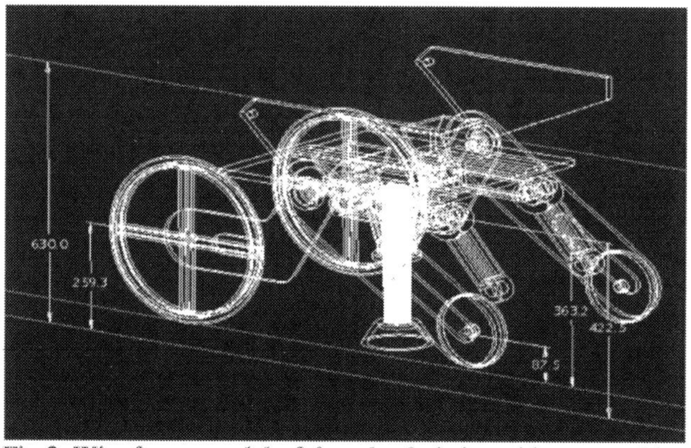
Fig.2. Wire frame model of the wheel chair and leg assembly.

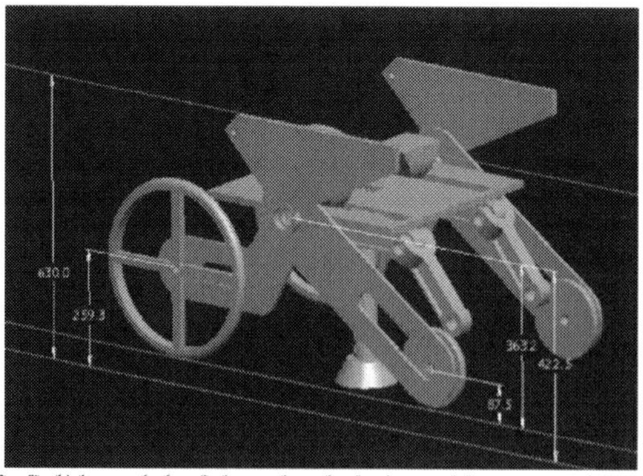
Fig.3. Solid model of the wheel chair and leg assembly with dimensions

In the co-operative environment, interference of manipulators are possible and are designed productive instead of destructive. This is done by the proper trajectory planning. In walking, the transfer of potential energy and kinetic energy is utilized in the walking wheelchair mechanism.

In the stair climbing process an ultrasonic sensor is used to analyse the dimensions of stairs in the form of raise and plinth. Then the controller used decides the position of wing legs when the whole mechanism is being lifted by the screw jack. During the first raise the stress and strain intensities are analysed to be high than the following steps. The length and size of the legs are designed in such a way that the next stair will not hinder the stair climbing process and similiarly the previous stair will not hinder the process of stair climbing down process. In all the cases the platform of the wheelchair is kept horizontal and the wheels are lifted above the platform of wheelchair level to act as a counterbalance to support the walking/stairclimbing process.

Figure explaining the wireframe model of the integrated mechanism identifies the intensity of co-operative manipulators to take and share the load among the mechanism. Any interference of the components into the work envelope of the other component can be easily identified by the wire frame model. Figure showing the solid model of the integrated mechanism and some isometric views of the walking wheelchair is shown in above figures. The triangular elements facing upwards fixed on the base of the wheelchair provides the base of the wheelchair seating arrangement. This is explained in the isometric views of the mechanism.

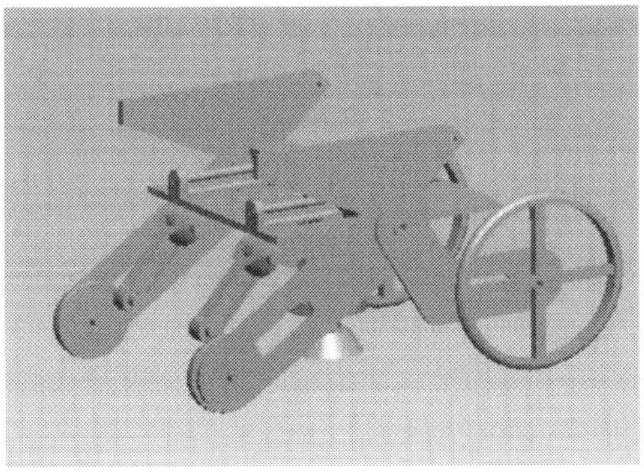

Fig.4.1 Isometric RHS view of the proposed mechanism

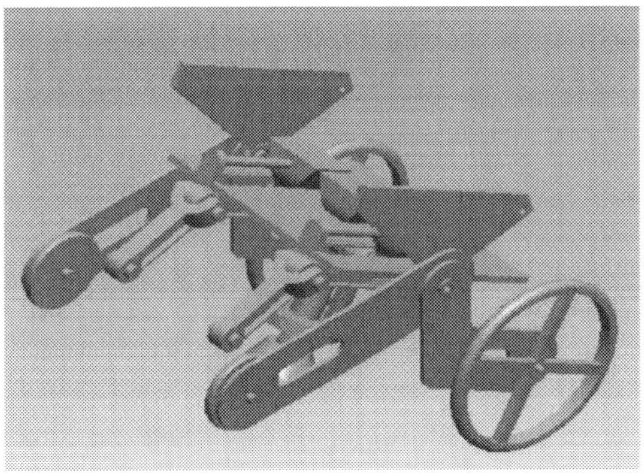

Fig.4.2 Isometric tilted top view of the proposed mechanism

Figures shown above infers the walking wheelchair designed with less weight and simple components to assemble the legs through the hip joints provided on the slots of the base of the wheelchair . This can be used as a normal wheel chair which can move with very less energy consumption. Either manually or propelled wheelchair, this can perform its task in a reliable manner. Figure salso infers the assembly of the legs. Wheels and the wheel connecting links are removed for clear visibility. The screw jack leg will be over hanging when the wheel are grounded. The height of the base of the wheelchair is 363.2mm and the height of the screw jack at fully retracted condition is 351.2mm. This provides a minute 12mm clearence of the screw jack leg touching the ground and the weight of the screw jack leg also will be carried by the wheelchair during wheeling process. The reason for the clearance given to the screw jack leg given is to reduce the jerkless operation of the lifting of the whole system. If the clearance in more, when the walking command is given the screw jack will extend to touch the ground. This may produce a impact load on the screw jack leg, if the clearnance is more and will be ultimately transferred to the whole mechanism. The wing legs will kept in the retracted condition.

From the base of the wheelchair to the seat arrangement base there will a space of 120mm. This space can be used to keep the electrical and electronics components like battries or motors as explained before. The whole system weighs roughly about 65kgs and can accomadate an average load of 80kgs and just above on it during both wheeling and walking mechanism. Since the results are oriented purely on the mechanism of the skeleton of mechanical components, any inclusion of components both electrical and electronics oreinted should be considered accordingly as per the individual properties of included components. And these components should be placed syemmtrically within the golden ratio of the wheelchair discussed before. Care should be taken in placing the components, since the major concept of this walking mechanism is based on center of gravity.

2.Analysis:

In the analysis phase the vital areas are identified on the basis of maximum load bearing position, maximum load bearing time and maximum failure zone. Depending upon the movement it differs. Like for stair climbing all the legs will be active parts where as wheels are passive parts. Wheels acts as counter balance to support walking.

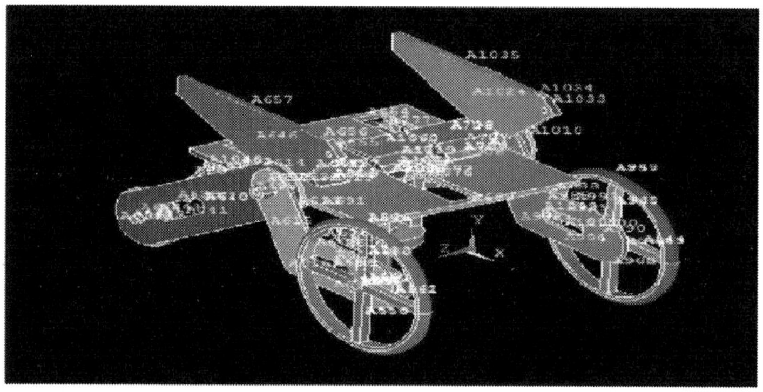

Fig.5. Identification of stress concentrating areas

Then the wheelchair is meshed with triangualar components for nodal analysis. In this analysis a thorough checkup is done on the elements of the wheelchair to avoid failure in any mode during the walking or stair climbing process. Since the wheels and most of the components connected with the platform is being lifted above the stress/strain intensity to cause failure is very minimal.

Fig.6. Meshed sequence of wheelchair

The mechanism integrated with the co-operative manipulators are then analysed separately and some of the results are tabulated below:

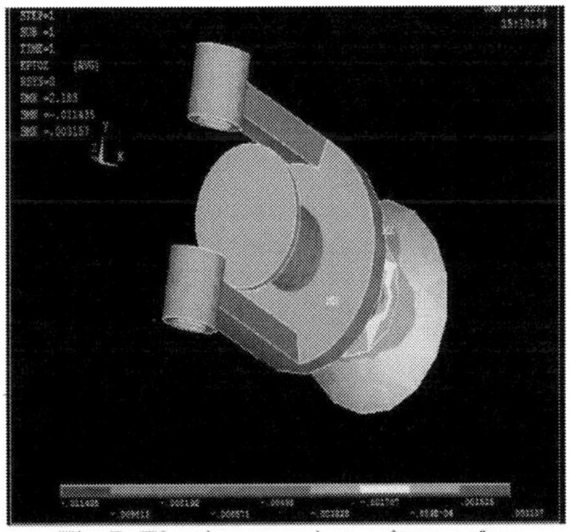

Fig.7. Elastic stress intensity on the screwjack mechanism

From figure.7, it can be inferred that the elastic stress intensity is concentrated much towards the stem region of the screwjack mechanism than the other areas. As the load carrying capacity of screwjack is well known, it can be justified that the whole system can be lifted and moved by the screwjack mechanism without failure. It is also analysed that the failure at joints are highly restricted during even eccentric loads, jerks and abnormal vibrations.

The hub of the screwjack mechanism is separately tested for the maximum load carrying capacity and its found to be more than the maximum load of the system with eccentricity caused by it. Diagram below shows the distribution of stress intensity during the working phaseof the whole system. It is understood that the connecting region pays a less failure intensity and thus provides the ability of the system in load carrying capability as co-operative manipulator.

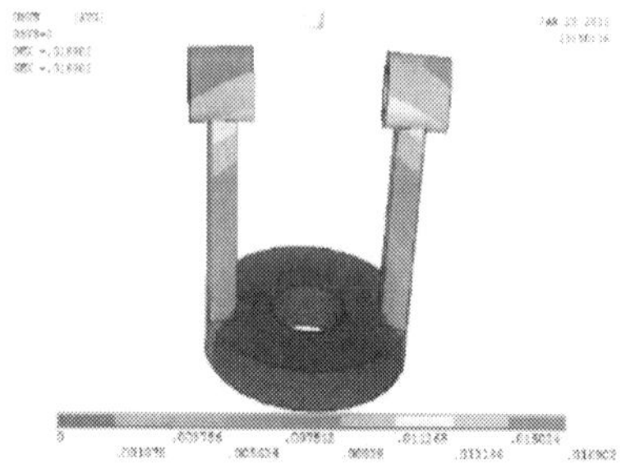

Fig.8. Mechanical stress intensity on the hub of screwjack mechanism

Finally the whole active co-operative manipulators components are analysed for the failure rate after elemental analysis the following figure shows the result:

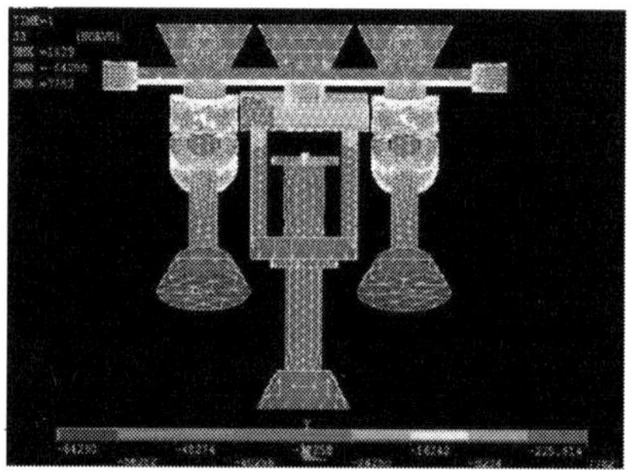

Fig.9. Nodal analysis of stress concentration on the screwjack mechanism during raise

It is analysed that the stem and joint region of the whole leg mechanism is subjected to the maximum deformation range. Though the deformation range is maximum it is much less than the failure range for the functional phase of the whole system.

3.Conclusions:

During the various functional phases the cooperative manipulators are subjected to follow various kinematics of serial and parallel manipulators. The results are obtained after subjecting each component of the cooperative manipulators and were tabulated. The efficiency of the proposed mechanism is thus proved by the results that have been produced. It is found in through the results that the components that work individually are subjected to more stress and deformations for the same load and constraints made than that of the components which are cooperatively performing the same task. Even the ability of the components that are jointly doing the task has been proven good. Thus the designed co-operative manipulators integrated wheelchair mechanism can be applied for walking and stair climbing process with a normal person of 80Kgs with out any jerk or vibrations. Further the whole system can be enhanced by proving more sensors to obtain feed back from the walking/stair climbing phase and can be real time based working co-operative environment. But more care should be taken in the tracjectory planning and other functions related to walking of the mechanism than the wheeling process of the mechanism. Study for providing the angles for various staircase heights and velocity needed for the foot to reach the target as inputs and applying artificial neural networks the stair climbing process can be made autonomous within the limits of the wheelchair leg dimensions.

References:

1. Hassan. S., Anwer. N., Khattak. Z. and Yoon. J. ,2010, 'Open architecture dynamic manipulator design philosophy', *Robotics and Computer-Integrated Manufacturing*, **Vol. 26**, pp.156–161.
2. Derek C. Schuurman and David W. Capson, Robust Direct Visual Servo Using Network-Synchronized Cameras, *IEEE Transaction on Robotics and Automation*, **Vol. 20**, NO. 2, pp.319-334.
3. Lawn M.J. and Ishimatsu T. 2003, 'Modeling of a stair-climbing wheelchair mechanism with high single-step capability', *IEEE Transactions on Neural Systems and Rehabilitation Engineering*, Vol. 11, No. 3, pp.323–332.
4. Gaetano Paviglianiti, Francesco Pierri, Fabrizio Caccavale and Massimiliano Mattei, 2010, Robust fault detection and isolation for proprioceptive sensors of robot manipulators, *Mechatronics* **Vol.20**, pp.162–170.
5. Jay Lee, 1987, Computer-Aided Robot Modelling, Programming and Simulation, *Elsevier Science Publishers, Robotics* **Vol.3**, pp.399-408.
6. Gao.Z., Zhang. D and Ge.Y, 2010, 'Design optimization of a spatial six degree-of-freedom parallel manipulator based on artificial intelligence approaches', *Robotics and Computer - Integrated Manufacturing*, **Vol. 26**, pp.180–189.
7. J. Angeles and O. Ma, "Performance evaluation of four-bar linkages for rigid-body guidance based on generalized coupler curves," *ASME J. Mech. Des.*, **vol. 113**, no. 1, pp. 17–24, 1991.
8. J. Angeles, "The design of isotropic manipulator architectures in the presence of redundancies," *Int. J. Robot. Res.*, **vol. 11**, no. 3, pp. 196–201, 1992.
9. M. Jamshidi, H. Seraji and Y.T. Kim, 1987, Decentralized Control of Nonlinear Robot Manipulators, *Elsevier Science Publishers, Robotics* **Vol.3**, pp. 361-370.
10. Renato Tinos and Marco Henrique Terra, 2002, Control of Cooperative Manipulators with Passive Joints, *Proceedings of the American Control Conference*, pp 1129-1134.
11. Nilanjan Chakraborty and Ashitava Ghosal, 2004, Kinematics of wheeled mobile robots on uneven terrain, *Mechanism and Machine Theory* **Vol.39**, pp.1273–1287.
12. Vitiello N, Lenzi T, De Rossi, S.M.M, Roccella S. and Carrozza M.C., 2010, 'A sensorless torque control for antagonistic driven compliant joints', *Mechatronics*, **Vol. 20**, No. 3, pp.355–367.
13. L. J. Stocco, S. E. Salcudean, and F. Sassani, "On the use of scaling matrices for task-specific robot design," *IEEE Transactions on Robotics and Automation*, **vol. 15**, no. 5, pp. 958–965, Oct. 1999.

Design of an Amphibian Exploring Robot

Atanu Maity[1] and Somajyoti Majumder

Central Mechanical Engineering Research Institute, M.G. Avenue, Durgapur-713209, West Bengal, India, www.cmeri.res.in

[1]E-mail: maity@cmeri.res.in

Abstract. To design and develop an amphibian exploring robot capable of operation in constrained mine environment puts a tremendous challenge to the system developers from both scientific and engineering perspective. Very few attempts have been made to fulfil these criteria of versatility in design, communication and control. The CSIR-CMERI developed amphibian subterranean robotic explorer (SR) is capable of moving over fairly rough terrain. It can swim as well as crawl over basin floor effortlessly. It is capable of operating at a maximum depth of 10m and can swim at 1 knot. A number of field trials have been carried out for performance testing of the system to ascertain its capability in underground flooded mine tunnels. This paper presents the insight on the design of an amphibian exploring robot for mine safety and disaster mitigation with special features of low power consumption vis-à-vis high mission time.

1. Introduction

A few mobile robots have been developed and applied to support the monitoring and mapping of active mine environments. Notable among these are 'Numbat' [1] - mine emergency recovery vehicle developed at Commonwealth Scientific and Industrial Research Organization (CSIRO) which is equipped with gas detection sensors, thermocouples etc. The mapping of 3-D mine tunnel was also investigated by Carnegie Mellon University [2]. Robotic mine mapping systems 'Groundhog' and 'Cave crawler' developed at CMU can recognize topological features of mine tunnels [3, 4, 5]. Besides, a few attempts have been made to introduce robotics for other mining related activities, e.g. a robotic system for underground coal mining has also been developed by CMU which can maneuver in highly constrained environments and cut coal without human operator onboard [6]. Researchers of Monash University - Gippsland Campus, are carrying out a project which aims to provide a computer assisted remote operation of mining machines and services [7]. This research also aims at developing an intelligent unmanned mining system using robotics technology for coal harvesting at a major Australian open cast mine. Forster & Schraft [8] have described a semi-automated coal extraction system with longwall drum ranging shearer. Another class of specially designed underwater robots developed and deployed in various tasks related to undersea applications. These robots are called Remotely Operated Vehicles (ROV). The un-tethered version of such robots with very limited payload, endurance and capability but self contained power and communication capability is called Autonomous Underwater Vehicle (AUV). Both ROVs and AUVs are used in deep sea activities rather than shallow turbid waters of underground coal mines.

This clearly shows that only a limited number of research works have been reported in literature for development of robotic system capable of exploring both dry and underwater environments. In general, chances of flooding is a serious threat to Indian coal mines because of higher groundwater table as well as the presence of flowing stream in the vicinity of coal-rich areas. Several accidents have been reported in the recent past [9]. Comprehensive accounts of mine inundation have been presented in [13] and [14] for Indian as well as international scenario. It is therefore essential for developing a robotic system which can be deployed within shortest possible notice to gather most recent information of the flooded mine galleries so as to plan and launch any emergency intervention in post flooding scenario. The SR, as depicted in figure 1 is an amphibian explorer which is capable of moving over both dry land and partially or fully submerged mine tunnels and simultaneously can collect data from the surroundings. While under water it is similar to an AUV and does not have any physical link with the control station. It is not designed to carry heavy payload and sole purpose of this system is to explore and gather data. The gathered data are stored in local storage for further analysis. A part of which with very low bandwidth-overhead is transmitted back online to assist navigation. Though most of the navigational parameters are locally handled at client-end, it can be manually intervened for failsafe operation.

Figure 1. The CSIR-CMERI Sub-terranean Robot (SR).

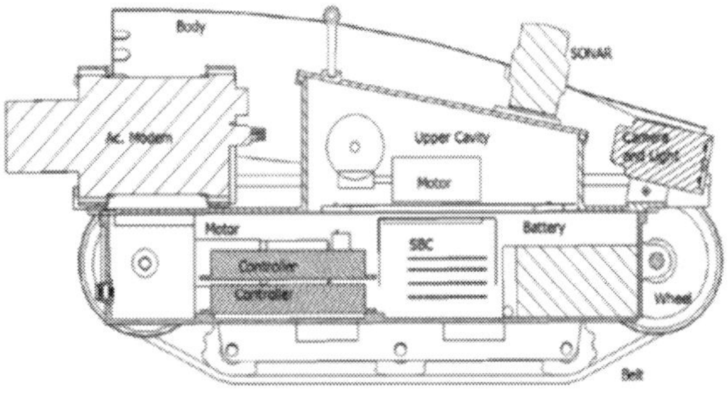

Figure 2. Layout showing relative positioning of major equipments.

2. System Design

The proposed system is aimed towards operation in partially or fully flooded mine tunnels and this calls for special considerations to be undertaken in its design so that the system can effortlessly move over a variety of terrain surfaces under varying conditions. Another interesting behavior is that the proposed system should exhibit swimming capability against moderate current. Foregoing discussion clearly shows that a variety of issues need to be considered critically to arrive at an acceptable system design. In view of above the following wish list has been created. It is desired that the system design must satisfy the above features and capabilities to a large extent.

2.1. Shape and Size

The key issues that influence the shape and size of the amphibian explorer is the equipments that need to be mounted on it, depth of operation, mission time and power requirement. A major share of the total power is consumed by the propulsion system, such as, motors and thrusters. Various equipments and sensors such as- acoustic modem, forward looking sonar, camera with light, drive system, radio modem etc. are packed within a very limited space. The hull made up of two watertight compartments enclosing various equipments like motors, Controllers, on-board battery banks, single board computer (SBC), necessary electronics and cabling. The system is provided with two modular track-wheel arrangements on both of its sides. The track modules are driven by two different motors from within the hull, the shafts of which are extended to two drive-wheels outside the hull. In addition to an underwater camera the SR is mounted with short range pencil beam sonar for obstacle avoidance and safe navigation through turbid water.

2.2. *Equipment Layout*

Arrangement of various equipments and placements of sensors were determined based on their *field of view* and *beam angle* and are shown in figure 2, 3 and 4. The underwater camera used has a built in light and having field of view 54°x30°. The sonar has a maximum scanning field of 360° and the acoustic modem with 17.8 kbps acoustic link has a range of 1.5km in narrow beam mode which is a 60° cone. Lower cavity of the hull is comparatively larger and houses most of the heavier equipments.

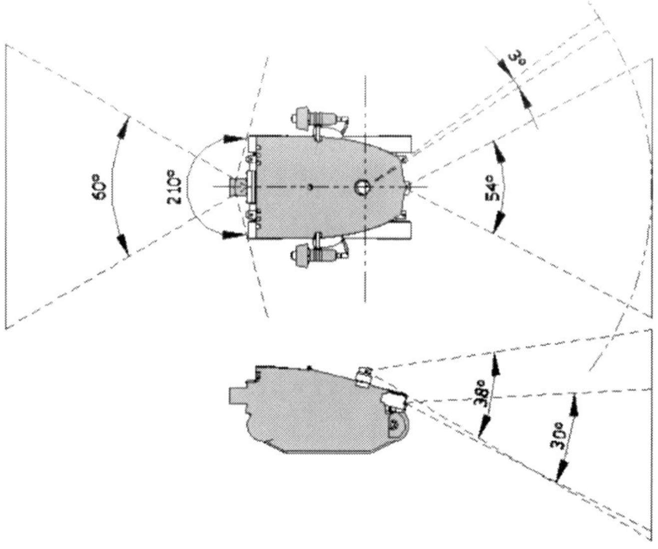

Figure 3. Field of view of acoustic modem, forward looking sonar and camera [15].

2.3. *Propulsion*

It is imperative that an ideal system should necessarily be designed under consideration of effective traction mechanism together with efficient drag reduction objective. Tracked wheel configuration makes the system capable of moving over fairly rough terrain when exploring over dry land, whereas, the thrusters come into action when it is underwater. Besides, chances of sinkage of the robotic system while operating on muddy or soft soil condition also needs to be evaluated.

2.4. *Buoyancy Management*

Wherever possible the heavy equipments were kept at lower cavity of the Hull (as depicted in figure 5) to achieve a considerable gap between the centre of gravity (CG) and centre of buoyancy (CB). More the separation the better is the stability and as a result the sway tends to stabilize faster. Figure 5 shows the restoring couple for roll stabilization generated due to such positing of CG and CB. Overall weight of the system is 41.5 Kg whereas buoyancy was measured 49.7 Kg. With dimensional parameters, the roll restoring couple-arm was measured to be 50 mm/30° tilt. The overall system was designed slightly over-buoyant. However, buoyancy packs and dead weights were provided to fine-tune overall buoyancy. The system can be deployed neutrally-buoyant, over-buoyant or under-buoyant depending upon the mission requirement. This enables to adjust the weight of the system to crawl on the mine bed of fully flooded mine galleries/tunnels or float on the surface of water in partially flooded tunnels

2.5. *Design Specification*

The system was designed to be remotely deployable and tether less. This enables the robot to explore through several bends in the mine gallery. However, this drastically reduces the mission time as the system solely relies on limited onboard battery power. Considering its scope of usage, the system was designed to operate up to a depth of 10 meter and was properly sealed and pressure-tested to eliminate any possibility of water leakage.

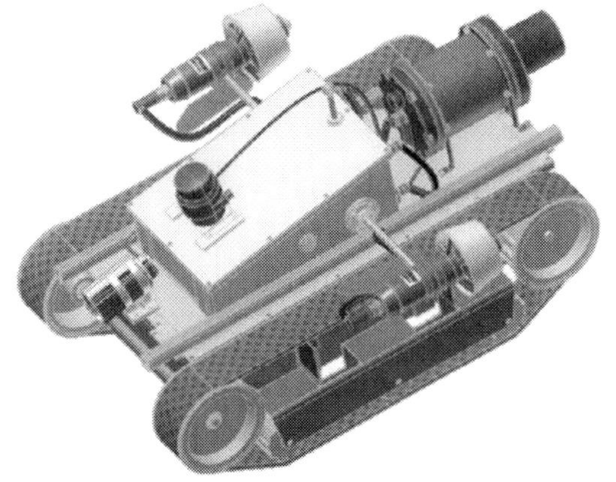

Figure 4. View of the SR with outer shell removed.

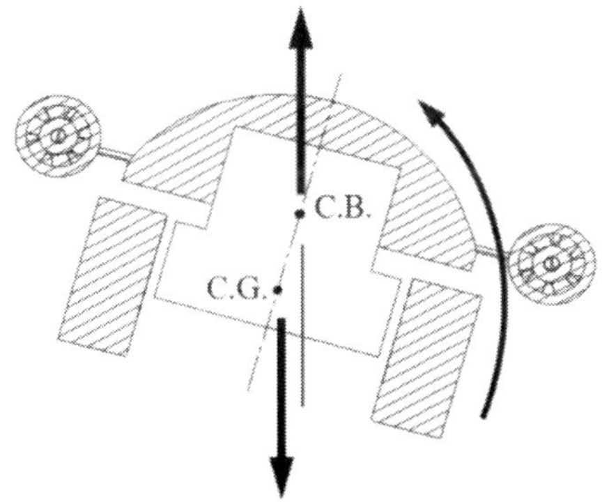

Figure 5. Judicious placement of equipments helps to achieve roll stabilization. [15]

The bearings mounted on these drive-shafts are adequately sealed to eliminate any possibility of water leakage. Also the system was provided with several moisture sensors to raise an alarm in the event of any water leakage into the cavity.

Table 2. Design Specification

Basic Dimension	550 x 270 x 250 mm
Weight	41.5 kg
Max. Depth	10 m
Speed	0.5 m/s
Drive	Separate differential drive with two motors and two thrusters Motor: Graphite brushed servo motors Thruster: DC brushless thrusters
Power Supply	Onboard Li-Ion battery banks
Processors	PC104+ based Single Board Computer (SBC)
Communication	Acoustic and RF based network
Architecture	Multiagent based Client-Server using TCP/IP
Sensors	Encoder, Sonar, Camera, Compass

3. Drag Analysis

To minimize propulsive power requirement, the shape of such a system should be as streamlined as possible. However, system requirement dictates that many equipments and sensors are to be packed within a small volume. For this reason it is not only difficult to achieve a good streamlined shape but also it is not an absolute necessity considering its slow operational speed. The hull of the robot was judiciously designed to achieve optimal shape. To estimate the hydrodynamic properties and to optimize drag, studies were carried out with the help of CFD tools. The 3D computational fluid dynamic (CFD) simulation of the robot was carried out to study hydrodynamic properties of the system. Based on these results, several improvements in design of the robot were done. Figure 6 shows the pressure distribution of flow past the SR explorer body for two different designs.

Formation of vortex vis-à-vis low pressure zone at the rear end of the SR contributes to drag. Thruster Power requirement is dictated by Total Drag (R_t) that an underwater body experiences while in motion and can be estimated as:

$$R_t = R_f + R_d \tag{1}$$

Where, R_f and R_d are *friction drag* and *pressure drag* respectively.
Again,

$$R_f = \frac{1}{2}\rho A_w V^2 C_f \tag{2}$$

$$R_d = \frac{1}{2}\rho A_f V^2 C_d \tag{3}$$

Where,

V = Velocity of the explorer
ρ = Density of water
A_w = Wetted surface area
A_f = Frontal projection area
C_f = Skin friction resistance co-efficient
C_d = Viscous pressure drag co-efficient

Flow separation causing higher pressure drag is a major phenomenon that hinders smooth propulsion when it is in motion under water. To minimize this, modifications in the shape of the SR were done as well as *through vents* were provided in the rear end of the casing. Vents allow smooth passage of the flow and alone helped to reduce drag by 30%. CFD analysis was done at different speed of the robotic vehicle and a typical C_d for the system was estimated to be 0.53, which is acceptable for such slow moving underwater system.

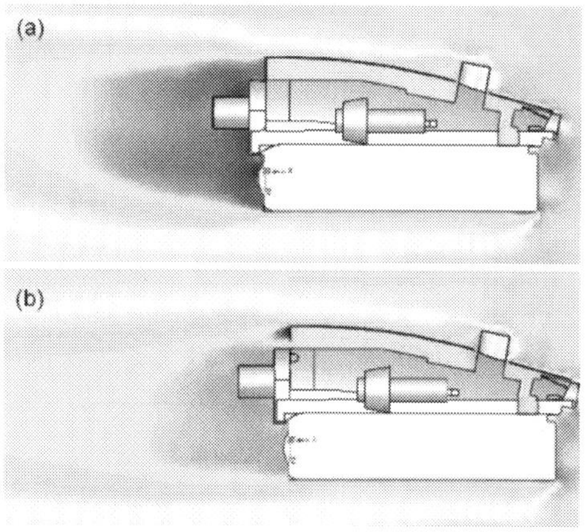

Figure 6. Pressure Plots are compared on the same scale. Vents were introduced on the body cover and the analysis result is shown in (b). The low pressure zone at trailing end is considerably reduced due to introduction of vents.

4. Manoeuvring

The two tracks are powered by two different motors. While moving over land, the steering action is achieved by differential drive of the track motors. Steering of the SR having *Track (d)* was achieved by differential drive of the individual track belts. If an equal differential speed (v) is applied on both tracks having initial speed (V),

$$V + v = \omega(R + \frac{d}{2}) \tag{4}$$

$$V - v = \omega(R - \frac{d}{2}) \tag{5}$$

The radius of curvature (R) of the robot path can be formulated as:

$$R = \frac{Vd}{2v} \tag{6}$$

From this equation, the following few inference can be drawn,
For $v = 0$, $R = \infty$, the robot moves on a straight line.
For $v \doteq V$, $R = d/2$, the robot turns centering on one track.
For $V = 0$, $R = 0$, the robot performs in-place turning.

Similarly Right and left thrusters are powered separately. While moving in water steering is achieved by differential action of the thrusters. Again the two thrusters are mounted on a common swivel rod which can be tilted. Pitching up or down is achieved by suitably selecting the thruster tilt angle.

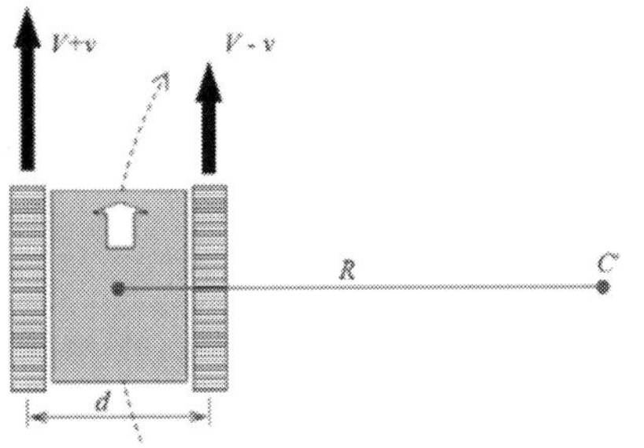

Figure 7. Differential drive of SR while moving over land.

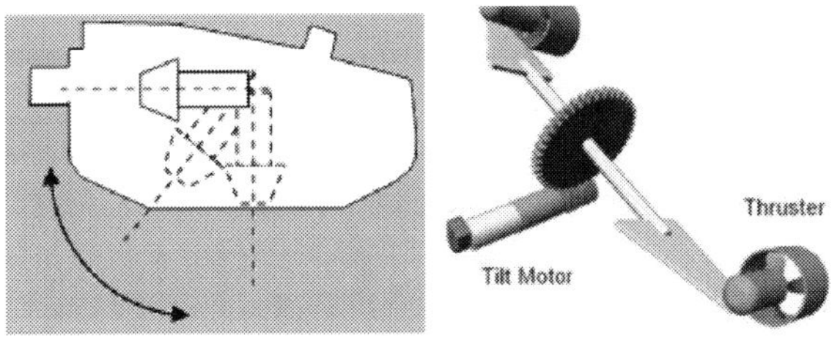

Figure 8. Thrusters are mounted on a common shaft and are tilted to pitch up or down while manoeuvring under water.

5. System Architecture

All the on-board equipments and accessories are connected to an on-board Single Board Computer (SBC) for control, communication and system operation. The sensors and other devices communicate with the SBC through multiple signal transmission and communication protocol ranging from pure analog to serial data transmission mode depending upon the specific sensor and device used. A block diagram depicting the system architecture is shown in figure 9.

6. Limited Field Trial and Deployment

Before actual deployment, several trial runs were carried out in simulated environment (figure 10) to ensure sealing and proper functioning of the equipments. Actual field testing at underground coal mine was carried out at Satgram Project under Eastern Coalfields Limited [10]. The mine shaft is approximately 211 m deep and a gassy-2 mine [11]. The location designated as 'Drift-3' was identified for field trial (figure 11). This drift is approximately 25 m long with slope of 1:6. The surface control station was set up above the mine shaft in order to demonstrate the capability of the robot to work in partially or fully flooded mine tunnels. Raw data obtained during the field trials were logged into the onboard SBC of the SR. Sonar and camera data (besides logging into the hard disk of SBC) was processed locally for obstacle avoidance and safe navigation.

Portability and remote deployment capability of the current system eliminates the need for expensive launching mechanism which is always an essential part of many underwater robotic systems such as ROVs and AUVs.

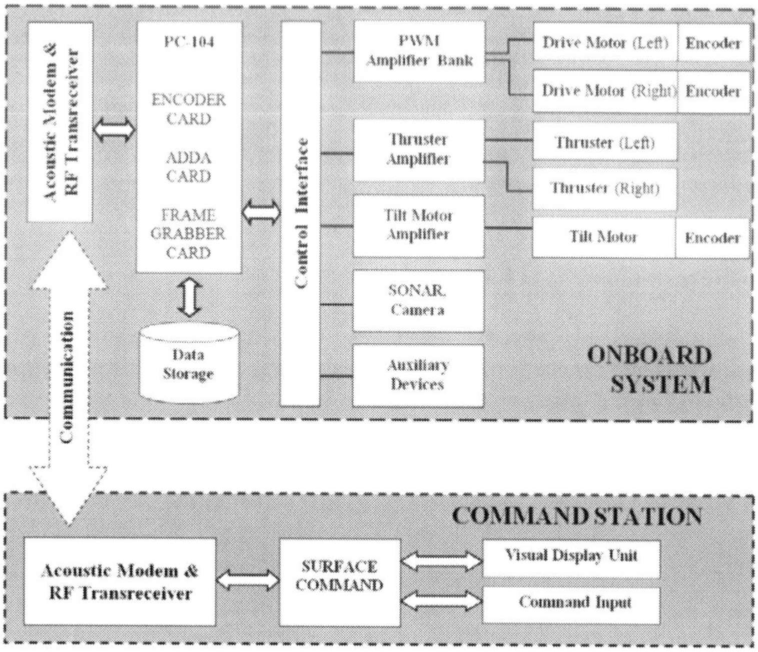

Figure 9. System Architecture of the robot together with the command station.

Although the main emphasis of this work was to develop a robotic system to locate entrapped miners, its potential was found to be far more extended. Apart from flooding, other disasters like subsidence, roof collapse, fire and explosion are a few common examples of mine accident. Work is also in progress to enhance the operational capability of the system to cater the need at time of such disasters.

Figure 10. Field testing of free swimming of the SR at CSIR-CMERI test facility.

Figure 11. The SR deployed in an inundated and submerged mine tunnel of Satgram Project.

The present system can be equipped with other mission specific sensors for disaster mitigation, surveillance, mapping of coal mines. Moreover, the amphibian subterranean robot can also be used for the purpose of autonomous mode of exploration in many shallow water bodies like pools, tunnels and lakes. In narrow mine tunnels, the performance of acoustic mode of communication reduces drastically due to noise caused by reflection even though suitable filter was used. However, the SR successfully achieved the objectives as envisaged.

Acknowledgment
This project was carried out with partial funding from Ministry of Coal, Govt. of India. Authors are grateful to CMPDI, Ranchi and Satgram Project, ECL for their valuable suggestions and assistance during field trial. Authors are thankful to the DG, CSIR and Director, CMERI, Durgapur, for their support for successful completion of the project.

References
[1] Ralston J C and Hainsworth D W 1998 *The Numbat: A Remotely Controlled Mine Emergency Response Vehicle, Field and Service Robotics* (London: Springer) pp 53-59
[2] Thrun S, Hähnel D, Ferguson D, Montemerlo M, Triebel R, Burgard W, Baker C, Omohundro Z, Thayer S and Whittaker W 2003 *A system for volumetric robotic mapping of abandoned mines, Proc. IEEE Int. Conf. Robot. Autom.* (Taiwan: Taipei) pp 4270–4275
[3] Baker C, Morris A, Ferguson D, Thayer S, Whittaker S, Omohundro Z, Reverte C, Whittaker W, Hähnel D and Thrun S *A Campaign in Autonomous Mine Mapping* (Pittsburgh: Carnegie Mellon University)
[4] Ferguson D, Morris A, Hähnel D, Baker C, Omohundro Z, Reverte C, Thayer S, Whittaker W, Burgard W and Thrun S *An Autonomous Robotic System for Mapping Abandoned Mines* (Pittsburgh: Carnegie Mellon University)
[5] Morris A, Silver D, Ferguson D, and Thayer S *Towards Topological Exploration of Abandoned Mines* (Pittsburgh: Carnegie Mellon University)
[6] Shaffer G and Stentz A 1992 *A Robotic System for Underground Coal Mining, IEEE International Conference on Robotics and Automation* Vol 1, pp 633-638
[7] Ibrahim M Y and Barfoot C 1997 *Robotization of Coal Harvesting in Open Cut Lignite Mines, Industrial Robot,* vol 24, No 5, pp 376-381
[8] Forster S and Schraft R D 1990 *First steps towards an autonomous mining robot for coal extraction, IEEE International Workshop on Intelligent Robots and Systems '90, Towards a New Frontier of Applications*, Vol 2, (New York) pp 487-492
[9] Ray D N, Dalui R, Maity A and Majumder S 2009 *Subterranean Robot: A Challenge for the Indian Coal Mines, The Online Journal on Electronics and Electrical Engineering (OJEEE),* Vol 2, No 2, pp 218-222
[10] Eastern Coalfielda Limited, *http://easterncoal.gov.in/ index.html,* Satgram Projects
[11] Indian Coal Mines Regulations, *The Coal Mines Regulations-1957 (Section 57 of the Mines Act, 1952)* Govt. of India
[12] Duff E S, Roberts J M and Corke P I 2002 *Automation of an Underground Mining Vehicle using Reactive Navigation and Opportunistic Localization*, (Auckland: Australasian Conference on Robotics and Automation)
[13] Singh T N and Singh B 1985 *Model simulation study of coal mining under river beds in India*, Vol 4(3), pp 1-10 (Madrid: International Journal of Mine Water)
[14] Vutukuri V S and Singh R N 1995 *Mine Inundation- Case Histories, Mine Water and the Environment*, Vol 14, Annual Issue, Paper 9, pp 107-130
[15] Atanu Maity, Somajyoti Majumder, Dip Narayan Ray; *Amphibian Subterranean Robot for Mine Exploration*, In proc. of 2013 International Conference on Robotics, Biomimetics, Intelligent Computational Systems (ROBIONETICS), Yogyakarta, Indonesia, November 25-27, 2013, pp. 242-246

Proposed Methodology for Application of Human-like gradual Multi-Agent Q-Learning (HuMAQ) for Multi-robot Exploration

Dip Narayan Ray, Somajyoti Majumder

SR Lab, CSIR-CMERI, Durgapur, West Bengal, India 713209
{dnray,sjm}@cmeri.res.in

Abstract. Several attempts have been made by the researchers around the world to develop a number of autonomous exploration techniques for robots. But it has been always an important issue for developing the algorithm for unstructured and unknown environments. Human-like gradual Multi-agent Q-learning (HuMAQ) is a technique developed for autonomous robotic exploration in unknown (and even unimaginable) environments. It has been successfully implemented in multi-agent single robotic system. HuMAQ uses the concept of Subsumption architecture, a well-known Behaviour-based architecture for prioritizing the agents of the multi-agent system and executes only the most common action out of all the different actions recommended by different agents. Instead of using new state-action table (Q-table) each time, HuMAQ uses the immediate past table for efficient and faster exploration. The proof of learning has also been established both theoretically and practically. HuMAQ has the potential to be used in different and difficult situations as well as applications. The same architecture has been modified to use for multi-robot exploration in an environment. Apart from all other existing agents used in the single robotic system, agents for inter-robot communication and co-ordination/ co-operation with the other similar robots have been introduced in the present research. Current work uses a series of indigenously developed identical autonomous robotic systems, communicating with each other through ZigBee protocol.

1. Introduction

The research on mobile robotics has now reached a stage, where it is being considered as an obvious choice for many applications, such as office/factory/hospital automation, personal care, domestic aids, exploration in land, water and other planets. Several mobile robots as well as navigation algorithms have been developed over the time by different researchers around the world. Some of them are autonomous and some of them are not only autonomous, but also intelligent in nature. Still there are several reasons where single robot cannot be preferred. The main tasks of these mobile robots for most of these cases are primarily autonomous explorations, general assistance, data collection of the unknown or slightly known environment. In such a situation predefined and preprogrammed approaches are inadequate to meet the challenges posed by the environmental dynamics. Another major constraint encountered by mobile robotics researchers is the robot's capability to process large amount of information in real time, obtained from various sensors in a dynamic and an unstructured environment. Also if the area is too large, it is difficult to explore the whole area by a robot in time due to limited on-board power. This time constraint becomes more important when a robot is deployed for security and surveillance purpose, such as landmine detection or bomb/ explosive detection in public places.

For the above reasons instead of using single robot, multiple robots can be used. This may effectively decrease the computational load due to handling large amount of data obtained from different sensors as well as drastically reduce overall exploration time for a large stretch of area. Still the challenge of exploration in completely unknown and unimaginable environments (that the programmer cannot think of) remains unanswered.

This has led towards the development of a generalized navigational (exploration) algorithm for safe navigation (towards the goal), that can be divided into multiple sensor specific subtasks. So that, in case of indoor exploration, these subtasks can be reordered in terms of priority as check battery power, seek goal, follow walls and corridors, detect door, detect obstacles and look for objects of interest, for example. For outdoor exploration, these can be re-ordered as check battery power, seek goal, detect obstacles and look for objects of interest. Capability of such approach is essential for the reason that in natural outdoor environment clearly lacks very common and easily identifiable geometric features such as corners or lines for example.

It is also a necessity to reduce the computational complexity and processing over load to achieve a reasonable response time for the system. This has been achieved by using the concept of multiagent system together with behaviour-based paradigm for certain states. The resultant performance has been found to be very satisfactory. Again for each of the sub-tasks, agent based approach has been implemented whose functional responsibility is to gather the state (s) from the sensors and suggest/ predict action (s). All the different actions from different agents are then modified through a coordination mechanism to produce a single definitive action. The interaction of the system with the environment makes the system to act rationally and effectively across various operating conditions starting from structured indoor environment to much unstructured outdoor environment under varying terrains and dynamic conditions.

Several works on multi robot exploration have been reported in literatures [1], [2], [3]. Different unsolved problems in this field emphasizing the theoretical issues have been presented in [4] along with a critical survey. Dudek et al. [5] has described the taxonomy for multi-agent robotics. The exploration strategies have been inspired from the fields of Artificial Intelligence, game theory, distributed computing, theoretical biology, animal ethology, artificial life etc. Initial inspiration as found in [6] was from the animal world. Arkin proposed a methodology for foraging and retrieving objects in a hostile environment by multiple mobile robots through co-operation without communication. Motor schemas, such as move-to-goal, avoid-robot, avoid-static-object and maintain-formation are used with the help of an arbiter to reach the goal. This work has been further extended for a mobile robot team by Balch & Arkin [7]. The work of Kube and Zhang [8] is based on the behaviours of different social insects, mostly ants and bees. They have implemented the box-pushing task using subsumption architecture [9], [10] as well as adaptive logic networks (ALN). Fukuda et al. [11] have also described similar studies using analogs to animal behaviour. A three layered control methodology for cooperation between heterogeneous mobile robots has been proposed in [12] for indoor environment. Those three layers are functional level, control level and high-level decision making planner level.

Recently, researchers are engaged to develop methodology for exploration by multiple mobile robots while making a group formation. Such line formations and general formations of non-holonomic robots have been studied by Yamaguchi [13] and Yoshida et al. [14]. Decentralised control laws using potential field approach has been implemented in [15] and [16] to guide the mobile robots away from obstacles. Similar approach for controlling a formation, guarding a perimeter and surrounding a facility has been described in [17] along with the stability analysis. Burgard et al. [18] have described an approach for the coordination of multiple robots for exploring an unknown environment taking into account the cost of reaching a target point and its utility. Similar approaches for exploration and mapping by multiple mobile robots have been presented in [19], [20], [21], [22]. Still the developments of different control laws/ architectures for autonomous navigation and exploration are going on as found in [23], [24]. HuMAQ is the new addition in this list. Initially the development of

HuMAQ was for Multi-agent systems, not necessary multiple systems, but finally it has been found that it has the potential for being used for multiple systems without any change.

2. Human-like-gradual Multi Agent Q-learning (HuMAQ)

Human-like Gradual Multi Agent Q-learning (HuMAQ) is a methodology already applied in [25] for single robotic system. It uses multiple agents as shown in Fig 1(a) for describing different behaviours/ tasks of the robot depending up on its type of applications.

Suppose a system has 'N' numbers of agents responsible for different states generated by the sensors and the internal states. In a simple system, two states of the robot can be handled by each agent. In case of a complex system, each agent can take care of 'n' states. Depending on the states, each agent selects different actions from the state-action table (Q-table) and interacts with the Co-ordinator. This co-ordinator arranges the agents to be invoked according to the predefined priority as shown in Fig 1(b). Additional details have been discussed with specific examples in [26].

Here the Co-ordinator plays an important role as action modifier, also referred as Modifier for simplicity. The modifier refines the actions proposed by different agents depending upon their priorities. For example, if the goal-seeking agent has directed the robot to move in forward direction, but the obstacle-avoidance agent has detected obstacle in the front. This clearly shows a conflict resolution required which is done by the modifier. It then searches for possible alternate position from where the goal is nearer and refines the action to move towards this alternate position. The modifier in essence tries to find a common area of interest where the states from most of the different agents are available and then refines the action from the descending list of states from that common space accordingly. It is not mandatory that all the states from all the agents should be present, because at any given instance of time all the agents may not be active. With the increase in number of agents, the complexity of the Co-ordinator and modifier also increases. Different rules for selection of the refined action by the modifier can be formed depending up on the type of sensors and applications. However, the basis of such rules should always be common space sharing.

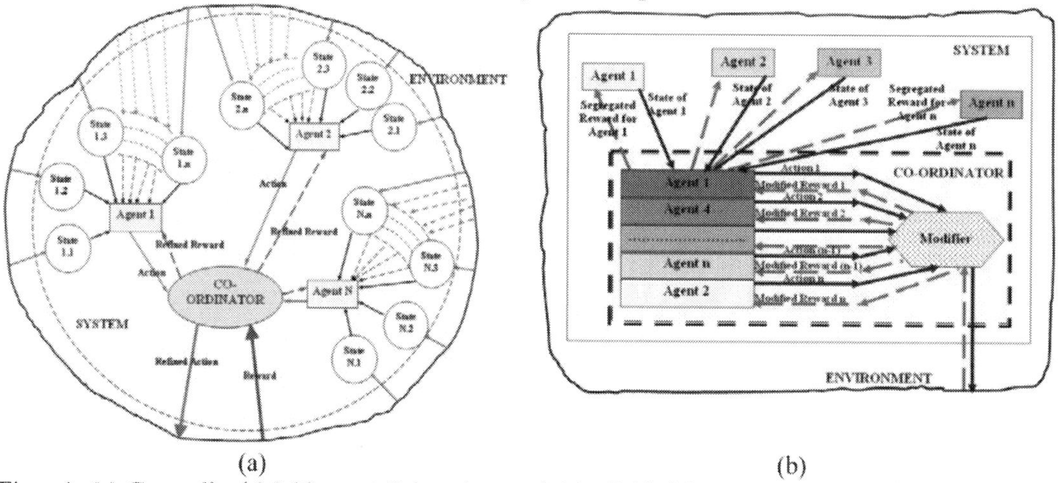

(a) (b)

Fig – 1: (a) Generalised Multi-agent Q-learning model is divided in two main parts: different Agents and the Co-ordinator (b) Generalized Co-ordinator for HuMAQ, also known as modifier, modifies the actions for different agents depending upon the priority and distributes the rewards accordingly

The refined action is now performed by the actuators and the obtained reward is again sent to the modifier to distribute them as per the priority of the agents only to those, which were active during determination of states. Therefore, instead of a single reward as in the case of Q-learning, a distributed and gradually reducing weightage system has been adopted here. A variable weightage system has been found to be the most suitable and therefore used to account the validity in priority of the agents.

IOP Conf. Series: Materials Science and Engineering **65** (2014) 012016 doi:10.1088/1757-899X/65/1/012016

For example, if there are four agents and the total reward obtained is R for any instance of time, when all the four agents were responsible for a refined action, it can be distributed as:

$$R = w_1 R_1 + w_2 R_2 + w_3 R_3 + w_4 R_4 = \sum_{i=1}^{4} w_i R_i \quad \cdots\cdots\cdots\cdots (1)$$

Where $w_1 > w_2 > w_3 > w_4$; w_1, w_2, w_3, w_4 are the distributed weightage for rewards for first agent (highest priority), second agent, third agent and fourth agent (low priority) respectively. R_1, R_2, R_3 and R_4 are the individual rewards for first agent, second agent, third agent and fourth agent respectively obtained directly due to system-environment interaction. With these separate rewards $w_i R_i (i \in 1,2,3,4)$, the separate Q-tables for different agents update themselves gradually. After a prolong trial run in any exploration task, if the system needs to be recharged or powered off for any reason, the final Q-table at that instance is stored and will be used as initial table for next run onwards.

3. Single Robot Vs. Multiple Robots

3.1. The Robotic Prototypes

As already mentioned, earlier works [24], [25] on HuMAQ used single robotic system. For this purpose the indigenously developed robot, called ARBIB (Autonomous Robot Based on Intelligent Behaviours) was used. The basic hardware module of ARBIB consists of a power supply, controlling and processing unit, sensor suite, communication module and the motors. The sensor suite includes three different pair of sensors i.e. light sensor (left and right), IR sensor (left and right) and battery-level indicator. The power supply is provided from an on-board 12 V Li-Ion battery bank. Only the battery-level indicator is connected directly to the power supply. Other sensors are obtaining power from the controlling/ processing board. Battery-level indicator constantly monitors the on-board battery voltage and informs the controlling unit when it goes below a pre-defined threshold level. The data from the sensors are fed to the controlling unit through one of their three terminals. The communication with the base station/ operator console has been established using commercially available Bluetooth module. The two motors are controlled via the H-bridge amplifier. The power supply for the motor is fed into the H-bridge and the control signal for the motor along with the change of directions is provided from the controlling/ processing unit to the H-bridge amplifier.

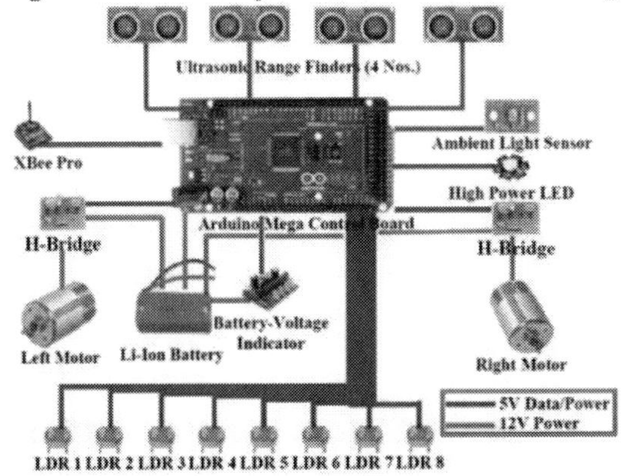

(a)

(b)

Fig – 2: (a) The hardware architecture (b) Prototype of a single robot showing major sensors

The present work uses ARBIB IV another upgraded version for testing the proposed multi-robot navigation methodology. The basic architecture of single robotic system is same throughout the ARBIB series of mobile robots. All the single robots are identical in all respect (in terms of mechanical structure, hardware and software architectures). All of them have on-board battery supply, controlling and processing unit, sensor suite, communication unit and two DC motors. Here on each single robot a high-power LED has been mounted for inter-robot communication. Sensor suite of the single robotic system includes four ultrasonic sensors (on each side), eight LDR (two fixed at the top along the side boundary of each side) and one ambient light sensor. The motors are controlled through H-bridge as done in earlier case. Fig 2 (a) and (b) depict the hardware architecture and the prototype of ARBIB IV respectively.

3.2. Communication

The overall communication has been divided in two different categories, inter-robot communication and the robot - base station/operator console communication as shown in Fig 3(a). For inter-robot communication the LED – LDR module has been employed. For sending the signal from the co-ordinator robot, LED of the co-ordinator robot is switched ON. The LDRs on the other robots are attracted towards the light emitting from the LED of co-ordinator and act accordingly. The communication between the robots and the base station/ operator console is established using ZigBee protocol. However, this communication is one way i.e. from the robot to the base station and only for sending data. Data from the robots have a definite header which separates the robots from each other. Initially for inter-robot communication the approach of distance measurement using signal strength was used. But due to carry out the experimentation in a small (a room of 10 Ft X 12 Ft dimension) most of the times, the signal strength was not significant enough for locating the robots as well as their directions.

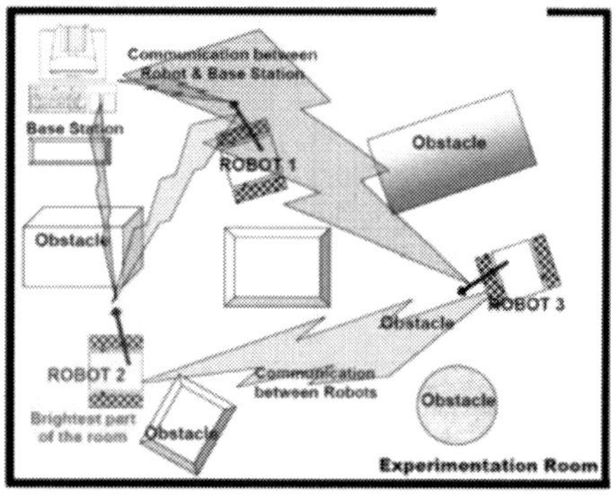

(a)

(b)

Fig – 3: (a) The inter-robot communication and the robot-base station communication strategy for multi-robot exploration (b) A view of the experimental area

4. Proposed Methodology

Here three homogeneous robots are deployed in a room as shown in Fig – 3(b) to find the brightest part of the room. As already mentioned four different sets of sensors and four related behaviours have been used. The four behaviours according to descending priority are Hunger, Obstacle-avoidance, Co-ordination and Goal-seeking. Hunger is related to battery level indicator and highest priority behaviour. This is referred by sub-states '0' (below threshold value) or '1'. Corresponding actions are 'shut down' or 'keep the system on' respectively. Obstacle-avoidance behaviour is associated with the Ultrasonic range finders and helps the robots to keep away from static and slow-moving obstacles. These four sensors are numbered 1, 2, 3, 4 clockwise when viewed from the top starting from the side having the castor wheel just at the bottom and the obstacles are defined at near [less than 20cm] (representing sub-state '3'), far [20 – 60cm] ('2') and farthest [above 60cm] ('1'). Accordingly the action is chosen in terms of direction and speed as shown in Table - II. Any of the three robots starts acting as co-ordinator when it reaches the predefined threshold value of the light read by the ambient light sensor. As soon as it reaches the area the eight LDRs mounted on the sides are made OFF and the LEDs are switched ON. Otherwise all the time the lowest priority behaviour, goal-seeking, is active. If there is no co-ordinator it uses the ambient light sensor to find out the maximum intense zone. It compares the current reading with the previous one. It may be either greater (sub-state '2') or lesser ('1') or same ('0'). If co-ordinator is present it will follow co-ordinator using LDRs to reach the goal. The LDRs are also numbered in similar way as done in case of the ultrasonic range finders. Here the sub-states are '1' (for output voltage of 0.1-1V), '2' (2.1-3V), '3' (3.1- 4V), '4' (4.1 -5V) and '0' (0V). The action for this agent is chosen in terms of direction as shown in Table – III.

Table – I: The actual representation of the state table as viewed by the robot

Sub-states for the eight LDRs for consecutive four sides								Sub-states for the URFs mounted on four sides				Battery -level	Current light value greater or not?
4	3	2	2	1	1	0	0	0	0	0	0	0	1
LDR [scale is 0, 1, 2, 3, 4 for eight LDRs]								Ultrasonic Range Finder [Scale is 0, 1, 2, 3 for four nos. of URFs]				Battery Level Detector [Low or high]	Ambient light sensor

Table – II: Action Table for Obstacle-avoidance Agent

Position/distance of obstacles	Near (less than 20cm)	Far (between 20 - 60 cm)	Farthest (more than 60cm)
RPM of DC Motors	15	30	60

Table – III: Action table for the Agent associated with goal-seeking behaviour

Serial of LDRs	0	1	2	3	4	5	6	7
Clockwise rotations	0 (0 × 45°)	45° (1 × 45°)	90° (2 × 45°)	135° (3 × 45°)	180° (4 × 45°)	225° (5 × 45°)	270° (6 × 45°)	315° (7 × 45°)

The above mentioned (sub) states and actions are denoted only in terms of numbers. Further details can be obtained from [25]. The most common action is chosen by the modifier as described earlier and the rewards obtained are updated in the corresponding state-action table (Q-table). The weightages for the most active agents (i.e. obstacle-avoidance and goal-seeking) are 0.6 and 0.4 respectively.

5. Experiments, Results & Discussion

A number of experiments have been carried out in a room of the laboratory as shown in Fig 3(b). A 100W lamp has been used as the light source in a slight dark environment. All the three robots are deployed at one place away from the light source facing each other in closed triangle. On reaching the predefined threshold value of the ambient light the robots stop at that position.

(a) (b) (c) (d)

Fig – 4: (a) At the starting (b) First robot has reached the goal and started acting as co-ordinator (c) Second robot has reached the goal (d) Third robot has also reached the goal

As shown in Fig 4(b) the first robot has reached the goal and started acting as co-ordinator. The second and third robots reached the goal afterwards following the signal from the co-ordinator. The data generated are directly transmitted to the base station through XBee. As soon as any robot reaches the goal it stops sending data. These data are further processed and represented in Table – IV.

Table – IV: Nos. of updates for reaching the goal for different trial runs

Nos. of updates	Run 1	Run 2	Run 3	Run 4	Run 5
Robot 1	298	239	216	232	182
Robot 2	283	291	263	274	237
Robot 3	254	281	248	220	256

It has been found that robot which acts as the co-ordinator takes less time than others. In the first run Robot 3 reached the goal first, thereafter Robot 1 and lastly Robot 2. Similar inferences can be made for other runs. If the numbers of updates for different runs are taken into consideration, it can be found the numbers of updates for most of the robots (except Robot 3) are lesser than the Run 1. This has been possible due to implementation of human-like gradual Q-learning. The numbers of updates for different robots versus the trial runs has been presented in Fig 5(a). The initial and final Q-table for different agents will clearly show refinement of state-action pairs which is a proof of learning. The initial and final Q-table for the obstacle-avoidance agent has been presented in Fig 5(b) and (c).

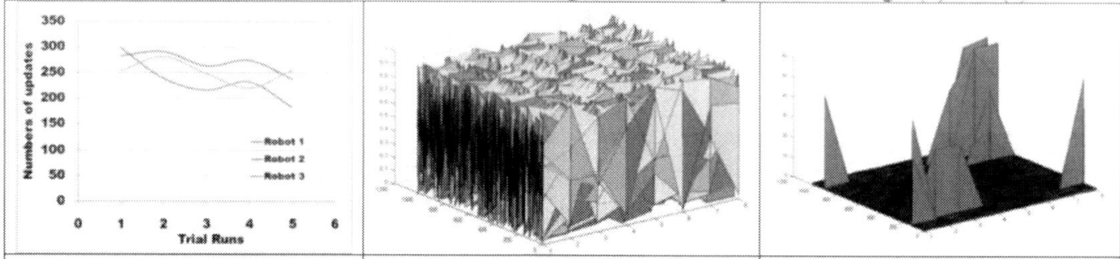

(a) (b) (c)

Fig – 5: (a) The number of updates for different robots versus trial runs (b) Initial Q-table for obstacle-avoidance agent (c) Final Q-table for obstacle-avoidance agent

6. Conclusion

Human-like Gradual Multi Agent Q-learning has already been proven for multiagent single robotic system. It has been tested in real time in both indoor and outdoor environments. Proof of learning has also been established. Present work aims to use the proposed HuMAQ for light exploration by a team of homogeneous robots. These indigenously developed robots use four different sensors and four different behaviours for safe exploration. Two different modes of one way communication have been implemented for successful operation. The proposed methodology has been successfully tested in an indoor environment in a limited space. The future scope of the work will include establishment of two way communications between the robots and base station, use of heterogeneous robotic systems and testing in rugged outdoor environments.

References

[1] Y. U. Cao, A. S. Fukunaga, and A. B. Kahng, "Cooperative mobile robotics: Antecedents and directions," in Proc. 1995 IEEE/RSJ IROS Conf., pp. 226–234.

[2] L. E. Parker, "Current state of the art in distributed autonomous mobile robotics," in Distributed Autonomous Robotic Systems 4, L. E. Parker,G. Bekey, and J. Barhen, Eds. New York: Springer-Verlag, 2000, pp. 3–12.

[3] J. Ota, "Multi-agent robot systems as distributed autonomous systems," Advanced engineering informatics (2006), pp. 59-70.

[4] Y. U. Cao, A. S. Fukunaga and A. B. Kahng, "Cooperative mobile robotics: Antecedents and Directions," Autonomous robots, MA, USA (1997), Vol. 4, pp. 7-27.

[5] G. Dudek, M. R. M. Jenkin, E. Milios and D. Wilkes, "A taxonomy for Multi-agent Robotics," Autonomous Robot (1996), Vol. 3, pp. 375-397.

[6] R. C. Arkin, "Cooperation without communication: Multiagent schema-based robot navigation,"J. Robot. Syst., vol. 9, no. 3, pp. 351–364, 1992.

[7] T. Balch and R. C. Arkin, "Behavior-based formation control for multi-robot teams," IEEE Trans. Robot. Automat., vol. 14, pp. 926–939, Dec.1998.

[8] R. C. Kube and H. Zhang, "Collective robotics: From social insects to robots," Adaptive Behavior, vol. 2, no. 2, pp. 189–218, Fall 1993.

[9] R. A. Brooks and A. M. Flynn, "Fast, cheap and out of control: A robot invasion of the solar system," J. Br. Interplanet. Soc., vol. 42, pp. 478–485, 1989.

[10] R. A. Brooks, "A robust layered control system for a mobile robot," IEEE J. Robot. Automat., vol. RA-2, pp. 14–23, Mar. 1986.
[11] T. Fukuda, H. Mizoguchi, K. Sekiyama, and F. Arai, "Group behaviour control for MARS (micro autonomous robotic system)," in Proc. Int. Conf. Robotics and Automation, Detroit, MI, May 1999, pp. 1550–1555.
[12] F. R. Noreils, "Toward a robot architecture integrating cooperation be-tween mobile robots: Application to indoor environment," Int. J. Robot. Res., vol. 12, no. 1, pp. 79–98, Feb. 1993.
[13] H. Yamaguchi and T. Arai, "Distributed and autonomous control method for generating shape of multiple mobile robot group," in Proc. IEEE Int. Conf. Intelligent Robots and Systems, vol. 2, 1994, pp. 800–807.
[14] E. Yoshida, T. Arai, J. Ota, and T. Miki, "Effect of grouping in local communication system of multiple mobile robots," in Proc. IEEE Int. Conf. Intelligent Robots and Systems, vol. 2, 1994, pp. 808–815.
[15] P. Molnar and J. Starke, "Communication fault tolerance in distributed robotic systems," in Distributed Autonomous Robotic Systems 4, L.E. Parker, G. Bekey, and J. Barhen, Eds. New York: Springer-Verlag, 2000, pp. 99–108.
[16] F. E. Schneider, D. Wildermuth, and H.-L. Wolf, "Motion coordination in formations of multiple robots using a potential field approach," in Distributed Autonomous Robotic Systems 4, L. E. Parker, G. Bekey, and J. Barhen, Eds. New York: Springer-Verlag, 2000, pp. 305–314.
[17] J. T. Feddema, C. Lewis and D. A. Schoenwald, "Decentralized control of cooperative robotic vehicles: Theory and application," IEEE transactions on Robotics and Automation, Vol. 18, Albuquerque, NM, USA (2002), pp. 852-864.
[18] W. Burgard, M. Moors, C. Stachniss and F. Schneider, "Coordinated multi robot exploration," IEEE traction on robotics, IEEE journals and magazines, Vol. 21 (2005), pp. 376-386.
[19] D. Fox, J. Ko, K. Konolige, B. Limketkai, D. Schulz and B. Steward, "Distributed multi-robot exploration and mapping"
[20] R. Simmons, D. Apfelbaum, W. Burgard, D. Fox, M. Moors, S. Thrun and H. Younes, "Coordination for multi-robot exploration and mapping "
[21] W. Burgard, M. Moors, D. Fox, R. Simmons and S. Thrun, "Collaborative multi-robot exploration," IEEE international conference on robotics and automation (2000), Vol. 1, pp. 476-481.
[22] Z. Yan, N. Jouandeau and A. A. Cherif, "Multi-robot decentralized exploration using a trade – based approach," Proceedings of the 8th international conference on informatics in control, automation and robotics(2) (2011).
[23] L. E. Parker, "Designing control laws for cooperative agent teams", Robotics and automation (1993), Vol. 3, pp. 582-587.
[24] T. Laengle, T. C. Lueth, U. Rembold, H. Woern, "A distributed control architecture for autonomous mobile robots – implementation of the Karlsruhe multi-agent robot architecture kamara," Advanced robotics (1997), pp. 411-431.
[25] D. N. Ray, Amit K. Mondal, S. Mukhopadhyay, S. Majumder, "A Proposed Methodology for Behaviour-based Multi-Agent Q-learning for Autonomous Exploration", 26th International Conference on CAD/CAM, Robotics and Factories of Future 2011, Kuala Lumpur, Malaysia, July 26 -28, 2011, Vol. 2, Page – 583 – 593, Eds. Dr. M. Khurshid Khan & Dr. Ni Lar Win
[26] D. N. Ray, "Autonomous Navigation of Mobile Robots: A Fusion of Behaviour-based Robotics and Reinforcement Learning", Ph D Thesis, National Institute of Technology, Durgapur, India, 2012

IOP Conf. Series: Materials Science and Engineering **65** (2014) 012017 doi:10.1088/1757-899X/65/1/012017

Dynamics and Control of Constrained Multibody Systems modeled with Maggi's equation: Application to Differential Mobile Robots Part I

Yawo H. Amengonu[1], **Yogendra P. Kakad**[2]

[1,2]Electrical and Computer Engineering Department
University of North Carolina at Charlotte
Charlotte, NC 28223

E-mail: [1]yhameng1@uncc.edu, [2]kakad@unnc.edu

Abstract. Quasivelocity techniques such as Maggi's and Boltzmann-Hamel's equations eliminate Lagrange multipliers from the beginning as opposed to the Euler-Lagrange method where one has to solve for the n configuration variables and the multipliers as functions of time when there are m nonholonomic constraints. Maggi's equation produces n second-order differential equations of which (n-m) are derived using (n-m) independent quasivelocities and the time derivative of the m kinematic constraints which add the remaining m second order differential equations. This technique is applied to derive the dynamics of a differential mobile robot and a controller which takes into account these dynamics is developed.

1. Introduction

Differential Wheeled Mobile Robots (DWMR) are perhaps one of the most widely used systems to demonstrate modeling techniques and control of mechanical systems subject to nonholonomic constraints [1] [2]. Often, kinematic constraints cannot be written as time derivatives of some functions of the generalized coordinates. The control of a mechanical system with nonholonomic constraints has been studied extensively and quite often, kinematic control is designed ignoring the dynamics of the system. It has been demonstrated in [3] that a mechanical system with nonholonomic constraints can be controlled regardless of the structure of the nonholonomic constraints. It was demonstrated also that a nonholonomic system cannot be stabilized to a single equilibrium point by a smooth feedback [4]. One approach to obtaining the differential equations of motion of a mechanical system described by n generalized coordinates and m nonholonomic constraints is the use of Euler-Lagrange equation. In order to enforce the m nonholonimic constraints, m Lagrange multipliers λ_s are introduced which are functions of time. Overall one has a total of $(n+m)$ equations to solve. Often, however, one is not interested in the Lagrange functions λ_s. Maggi's equation eliminates the λ_s from the beginning. In this paper, the dynamics of a DWMR will be modeled using Maggi's equation and a controller will be designed using the Dynamic Extension Algorithm to track an arbitrary path. The control law design is shown in the companion paper [6].

2. Kinematics of a differential mobile robot

Consider the diagram of the differential mobile robot depicted in figure 1 there are three bodies involved; the platform of the robot and the wheels. There are four degrees-of-freedom associated with the DWMR. The coordinates (x_p, y_p) of the midpoint of the axle, the heading angle ϕ, the mobile robot's right wheel's rotation through θ_r, and the robot's left wheel's rotation through angle θ_l. A vector of generalized coordinates for the mobile robot is thus given by $\underline{\gamma} = [x_p, y_p, \phi, \theta_r, \theta_l]^T$, where the subscript T stands for transpose. Using *Koenig's theorem*, the kinetic energy is written as in (1). Here, it is convenient to select the midpoint P of the axle of the mobile robot as the reference point.

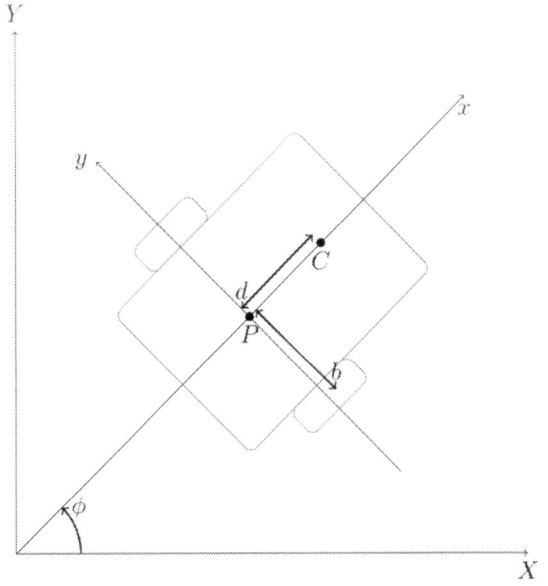

Figure 1. Diagram of a differential mobile robot

$$T = \frac{1}{2}\sum_{i=1}^{3} m_i v_i^2 + \frac{1}{2}\sum_{i=1}^{3} \omega_i I_i \omega_i + \sum_{i=1}^{3} m_i v_i.\dot{\rho}_i. \tag{1}$$

where v_i is the velocity of the reference point of the ith body, ρ_i is the position of the center of mass of the ith body relative to P expressed in a translating frame placed at P, and ω_i is the angular velocity of the body measured in the body's frame. The rotation matrix ${}^{I}T_B$ which transforms body coordinates to inertial coordinates is given by

$${}^{I}T_B = \begin{bmatrix} \cos(\phi) & -\sin(\phi) & 0 \\ \sin(\phi) & \cos(\phi) & 0 \\ 0 & 0 & 1 \end{bmatrix}$$

. The coordinates of the left wheel W_L and right wheel W_R expressed in the inertial frame are given respectively by

$${}^{I}W_L = \begin{bmatrix} \cos(\phi) & -\sin(\phi) & x_p \\ \sin(\phi) & \cos(\phi) & y_p \\ 0 & 0 & 1 \end{bmatrix} \begin{bmatrix} 0 \\ b \\ 1 \end{bmatrix} = \begin{bmatrix} x_p - b\sin(\phi) \\ y_p + b\cos(\phi) \\ 1 \end{bmatrix}. \tag{2}$$

$$ {}^{I}W_R = \begin{bmatrix} \cos(\phi) & -\sin(\phi) & x_p \\ \sin(\phi) & \cos(\phi) & y_p \\ 0 & 0 & 1 \end{bmatrix} \begin{bmatrix} 0 \\ -b \\ 1 \end{bmatrix} = \begin{bmatrix} x_p + b\sin(\phi) \\ y_p - b\cos(\phi) \\ 1 \end{bmatrix}. \tag{3} $$

The velocity of the left and right wheels are computed respectively and are given by

$$ {}^{I}\dot{W}_L = \begin{bmatrix} \dot{x}_p - b\dot{\phi}\cos(\phi) \\ \dot{y}_p - b\dot{\phi}\sin(\phi) \\ 0 \end{bmatrix}. \tag{4} $$

$$ {}^{I}\dot{W}_R = \begin{bmatrix} \dot{x}_p + b\dot{\phi}\cos(\phi) \\ \dot{y}_p + b\dot{\phi}\sin(\phi) \\ 0 \end{bmatrix}. \tag{5} $$

To find the kinematic constraint equations, the velocities in equations (4) and (5)are transformed back to the body coordinate frame by multiplying equations (4) and (5) by the transpose of ${}^{I}T_B$. They are given by

$$ {}^{B}\dot{W}_R = \begin{bmatrix} \dot{x}_p\cos(\phi) + \dot{y}_p\sin(\phi) + b\dot{\phi} \\ -\dot{x}_p\sin(\phi) + \dot{y}_p\cos(\phi) \\ 0 \end{bmatrix}. \tag{6} $$

$$ {}^{B}\dot{W}_L = \begin{bmatrix} \dot{x}_p\cos(\phi) + \dot{y}_p\sin(\phi) - b\dot{\phi} \\ -\dot{x}_p\sin(\phi) + \dot{y}_p\cos(\phi) \\ 0 \end{bmatrix}. \tag{7} $$

The constraints of rolling without slipping implies that the velocity of a wheel is equal to the angular velocity of the wheel about its axis multiplied by the radius of the wheel and the side velocities of the wheels must equal zero. In other words the velocity of the wheels expressed in the body coordinates frame must be given by

$$ \begin{aligned} \dot{x}_p\cos(\phi) + \dot{y}_p\sin(\phi) - b\dot{\phi} &= r\dot{\theta}_l \\ \dot{x}_p\cos(\phi) + \dot{y}_p\sin(\phi) + b\dot{\phi} &= r\dot{\theta}_r \\ -\dot{x}_p\sin(\phi) + \dot{y}_p\cos(\phi) &= 0. \end{aligned} \tag{8} $$

The kinetic energy of the mobile robot using eq.(1) is given by

$$ T = \frac{1}{2}(m_c + 2m_w)v_p^2 + \frac{1}{2}I\dot{\phi}^2 + \frac{1}{2}I_w\dot{\theta}_r^2 + \frac{1}{2}I_w\dot{\theta}_l^2 - m_c d\dot{\phi}(-\dot{x}_p\sin(\phi) + \dot{y}_p\cos(\phi)). \tag{9} $$

where $I = I_c + 2I_m + 2m_w b^2 + m_c d^2$, and m_c, m_w are the masses of the platform and the mass of the wheels respectively.

3. Dynamics of the differential mobile robot

In general, the dynamical equations of motion of a mobile robot with n generalized coordinates $\underline{\gamma} = [\gamma_1 \ldots \gamma_n]^T$ subject to m constraints can be written as shown in (10) by applying Euler-Lagrange's equation

$$ \underline{H}(\underline{\gamma})\ddot{\underline{\gamma}} + \underline{D}(\underline{\gamma}, \dot{\underline{\gamma}}) + \underline{G}(\underline{\gamma}) = \underline{B}(\underline{\gamma})\underline{\tau} - A^T(\underline{\gamma})\underline{\lambda} \tag{10} $$

The description of each component can be found in the appropriate literature. In the case considered here the gravitational vectors are equal to zero. The equation of motion of the DWMR can be written as

$$\underline{H}\left(\underline{\gamma}\right)\ddot{\underline{\gamma}}+\underline{D}\left(\underline{\gamma},\dot{\underline{\gamma}}\right)+\underline{G}\left(\underline{\gamma}\right)=\underline{\tau}. \tag{11}$$

$$\frac{d}{dt}\left(\frac{\partial T}{\partial\dot{\gamma}_i}\right)-\left(\frac{\partial T}{\partial\gamma_i}\right)=Q_i+\sum_{j=n-m+1}^{n}\lambda_i\psi_{ji}\quad(i=1\dots n). \tag{12}$$

Maggi's equation is a quasivelocity method. In general, there is no set procedure on selecting quasivelocities. For a mechanical system of n generalized coordinates with m nonholonomic constraints, n quasivelocities are selected such that m of them span the constraint space, and (n-m) are independent quasivelocities. During the selection, one keeps in mind that the matrix Ψ obtained as in (13) is invertible with an inverse Φ.

$$\underline{u}=\Psi\dot{\underline{\gamma}} \tag{13}$$

where $\underline{u}=[u_1\dots u_n]^T$, is a quasivelicity vector. Equation (13) can be rewritten as

$$u_j = \sum_{j=1}^{n}\psi_{ji}(\underline{\gamma})\dot{\gamma}_i\quad(j=1\dots m) \tag{14}$$

$$u_j = \sum_{j=1}^{n}\psi_{ji}(\underline{\gamma})\dot{\gamma}_i=0\quad(j=n-m+1\dots n), \tag{15}$$

and

$$\dot{\gamma}_i=\sum_{i=1}^{n-m}\phi_{ij}(\underline{\gamma})u_j\quad(i=1\dots n). \tag{16}$$

Note that these equations do not explicitly involve time otherwise there must be additional terms.
Using equation(15), virtual displacements are given as

$$\delta\theta_j = \sum_{i=1}^{n}\psi_{ji}(\underline{\gamma})\delta\gamma_i=0\quad(j=n-m+1\dots n)\quad\text{or}$$

$$\delta\gamma_i = \sum_{i=1}^{n-m}\phi_{ij}(\underline{\gamma})\delta\theta_j\quad(i=1\dots n). \tag{17}$$

The first (n-m) δ_j are selected independently. Using Lagrange's principle of virtual displacement we have

$$\sum_{i=1}^{n}\left[\frac{d}{dt}\left(\frac{\partial T}{\partial\dot{\gamma}_i}\right)-\left(\frac{\partial T}{\partial\gamma_i}\right)-Q_i\right]\delta\gamma_i=0 \tag{18}$$

with $T(\underline{\gamma},\dot{\underline{\gamma}})$, the unconstrained kinetic energy defined above. After substituting for the virtual displacement as defined in (17), we have

$$\sum_{j=1}^{n-m}\sum_{i=1}^{n}\left[\frac{d}{dt}\left(\frac{\partial T}{\partial\dot{\gamma}_i}\right)-\left(\frac{\partial T}{\partial\gamma_i}\right)-Q_i\right]\phi_{ij}(\underline{\gamma})\delta\theta_j=0. \tag{19}$$

Using the fact that the first $(n\text{-}m)$ δ_j are independent, the coefficients of (19) must equal zero and we have

$$\sum_{i=1}^{n}\left[\frac{d}{dt}\left(\frac{\partial T}{\partial \dot{\gamma}_i}\right)-\left(\frac{\partial T}{\partial \gamma_i}\right)\right]=\sum_{i=1}^{n} Q_i \phi_{ij}(\underline{\gamma})\delta\theta_j \tag{20}$$

The resulting equation (20) is Maggi's equation. Let us explicitly define quasivelocities as

$$\begin{aligned}
u_1 &= \dot{x}_p\cos(\phi)+\dot{y}_p\sin(\phi)\\
u_2 &= \dot{\phi}\\
u_3 &= -\dot{x}_p\sin(\phi)+\dot{y}_p\cos(\phi)\\
u_4 &= \dot{x}_p\cos(\phi)+\dot{y}_p\sin(\phi)+b\dot{\phi}-r\dot{\theta}_r\\
u_5 &= \dot{x}_p\cos(\phi)+\dot{y}_p\sin(\phi)-b\dot{\phi}-r\dot{\theta}_l.
\end{aligned} \tag{21}$$

$$\Phi=\begin{bmatrix}\cos(\phi) & 0 & -\sin(\phi) & 0 & 0\\ \sin(\phi) & 0 & \cos(\phi) & 0 & 0\\ 0 & 1 & 0 & 0 & 0\\ \frac{1}{r} & \frac{b}{r} & 0 & -\frac{1}{r} & 0\\ \frac{1}{r} & -\frac{1}{r} & 0 & 0 & -\frac{1}{r}\end{bmatrix}. \tag{22}$$

Applying (20) to the unconstrained kinetic energy and using the first and second columns of Φ, we have

$$\begin{aligned}
m\cos(\phi)\ddot{x}_p+m\sin(\phi)\ddot{y}_p+m_c d\dot{\phi}^2+\frac{I_w}{r}\ddot{\theta}_r+\frac{I_w}{r}\ddot{\theta}_l &= \frac{\tau_r}{r}+\frac{\tau_l}{r}\\
m_c d\cos(\phi)\ddot{x}_p-m_c d\sin(\phi)\ddot{x}_p+\frac{I_w b}{r}\ddot{\theta}_r-\frac{I_w b}{r}\ddot{\theta}_l &= \frac{b\tau_r}{r}-\frac{b\tau_l}{r}.
\end{aligned} \tag{23}$$

The time derivative of the constraint equations together with (23) form the second-order differential equations of motion of the differential mobile robot and is given by

$$\underline{H}\left(\underline{\gamma}\right)\ddot{\underline{\gamma}}+\underline{D}\left(\underline{\gamma},\dot{\underline{\gamma}}\right)=B\underline{\tau}. \tag{24}$$

where

$$\underline{H}\left(\underline{\gamma}\right)=\begin{bmatrix}m\cos(\phi) & m\sin(\phi) & 0 & I_w/r & I_w/r\\ m_c d\sin(\phi) & -m_c d\cos(\phi) & I & I_w b/r & -I_w b/r\\ -\sin(\phi) & \cos(\phi) & 0 & 0 & 0\\ \cos(\phi) & \sin(\phi) & b & -r & 0\\ \cos(\phi) & \sin(\phi) & -b & 0 & -r\end{bmatrix}, \tag{25}$$

$$\underline{D}\left(\underline{\gamma},\dot{\underline{\gamma}}\right)=-\begin{bmatrix}-m_c d\dot{\phi}^2\\ 0\\ \dot{x}_p\dot{\phi}\cos(\phi)+\dot{y}_p\dot{\phi}\sin(\phi)\\ \dot{x}_p\dot{\phi}\sin(\phi)-\dot{y}_p\dot{\phi}\cos(\phi)\\ \dot{x}_p\dot{\phi}\sin(\phi)-\dot{y}_p\dot{\phi}\cos(\phi)\end{bmatrix}. \tag{26}$$

and

$$\underline{B}=\begin{bmatrix}1/r & 1/r\\ b/r & -b/r\\ 0 & 0\\ 0 & 0\\ 0 & 0\end{bmatrix}. \tag{27}$$

Let us define the state vector $q = [x_p, y_p, \phi, \theta_r, \theta_l, \dot{x}_p, \dot{y}_p, \dot{\phi}, \dot{\theta}_r, \dot{\theta}_l]^T$ by choosing the state space variables $q_i, 1 \leq i \leq 10$ as

$$\begin{cases} q_i = \gamma_i & \text{for } 1 \leq i \leq 5 \\ q_i = \dot{\gamma}_{i-5} & \text{for } 6 \leq i \leq 10 \end{cases}$$

The system of equations can be expressed in the form

$$\underline{\dot{q}} = f(\underline{q}) + \underline{g_1}(\underline{q})\tau_1 + \underline{g_2}(\underline{q})\tau_2. \tag{28}$$

More explicitly

$$\underline{\dot{q}} = \begin{bmatrix} q_6 \\ q_7 \\ q_8 \\ q_9 \\ q_{10} \\ f_6(q_3, q_6, q_7, q_8) \\ f_7(q_3, q_6, q_7, q_8) \\ f_8(q_3, q_6, q_7, q_8) \\ f_9(q_3, q_6, q_7, q_8) \\ f_{10}(q_3, q_6, q_7, q_8) \end{bmatrix} + \begin{bmatrix} 0 & 0 \\ 0 & 0 \\ 0 & 0 \\ 0 & 0 \\ 0 & 0 \\ g_{61}(q_3) & g_{61}(q_3) \\ g_{71}(q_3) & g_{71}(q_3) \\ g_{81}(q_3) & -g_{81}(q_3) \\ A & B \\ B & A \end{bmatrix} \begin{bmatrix} \tau_1 \\ \tau_2 \end{bmatrix}. \tag{29}$$

where the entries of (29) are explicitly given in the companion paper.

4. Conclusion

Maggi's equation is a quasivelocity method used to derive the equations of motion of dynamical systems subject to linear nonholonomic constraints. It results from the application of the concept of Lagrange's virtual displacements *i.e.*, the Lagrange's principle. Maggi's method eliminates the Lagrange multipliers which are used to enforce the nonholonomic constraints from the start. This technique is used to derive the equation of motion of a differential wheeled mobile robot. Because of limited space, the controller developed and the method used are shown in a companion paper. The system has two inputs, however, it can be transformed into a "single input" system first by noticing that by taking the difference between the first and second equations of (8), and by taking time derivative of the resulting equation, together with row 8th of (29) one can transform the system into "single input system". Thereafter one can use single input-single output feedback linearization techniques to simultaneously compute the inputs torques.

References

[1] Fierro R and Lewis F 1995 *Control of a nonholonomic mobile robot: backstepping kinematics into dynamics* Decision and Control, Proceedings of the 34th IEEE Conference on. Vol. 4. IEEE
[2] Yu Ti, Sidek N and Sarkar N 2009 *Modeling and control of a nonholonomic wheeled mobile robot with wheel slip dynamics.* Computational Intelligence in Control and Automation IEEE
[3] Campion G, d'Andrea-Novel B and Bastin G 1991 *Controllability and state feedback stabilization of non holonomic mechanical systems.* Lecture Notes in Control and Information Science (Berlin:Springer-Verlag) p 106-24
[4] Bloch A and McClamroch N 1989 *Control of mechanical systems with classical nonholonomic constraints* Proc. of 28th IEEE Conference on Decision and Control (Florida) p 201-05
[5] Sarkar N, Yun X and Kumar R 1992 *Control of Mechanical Systems with Rolling Contraints: Application To Dynamic Control of Mibile Robots* (Pennsylvania:University of Pennsylvania, Technical Report (CIS)).
[6] Amengonu Y and Kakad Y 2014 *Dynamics and control for Constrained Multibody Systems modeled with Maggi's equation: Application to Differential Mobile Robots Part II Preprint* , 27th International Conference on CADCAM, Robotics and Factories of the Future 2014

IOP Conf. Series: Materials Science and Engineering **65** (2014) 012018 doi:10.1088/1757-899X/65/1/012018

Dynamics and control for Constrained Multibody Systems modeled with Maggi's equation: Application to Differential Mobile Robots PartII

Yawo H. Amengonu[1] **and Yogendra P. Kakad**[2]

Electrical and Computer Engineering Department
University of North Carolina at Charlotte
Charlotte, NC 28223

E-mail: [1]yhameng1@uncc.edu, [2]kakad@unnc.edu

Abstract. Quasivelocity techniques were applied to derive the dynamics of a Differential Wheeled Mobile Robot (DWMR) in the companion paper. The present paper formulates a control system design for trajectory tracking of this class of robots. The method develops a feedback linearization technique for the nonlinear system using dynamic extension algorithm. The effectiveness of the nonlinear controller is illustrated with simulation example.

1. Introduction

Maggi's quasivelocity method was used to derive the dynamical equations of motion of a Differential Wheeled Mobile Robot (DWMR) and the derivation is shown in the companion paper [1]. The system has two inputs, however, it can be transformed into a "single input" system by noticing that by taking the difference between the first and second equations of (8) in [1], duplicated here for completeness, and by taking time derivative of the resulting equation, together with row 8th of (2) one can transform the system into "single input system". Thereafter one can use single-input single-output techniques developed for example in [2] to simultaneously compute the input torques. The equations resulting from the transformation are given in the first part of the work presented here. In the second part of the controller design, a multi-input multi-output feedback linearization technique is considered. The first difficulty encountered is that there is no relative degree with respect to the output vector considered. Dynamic extension algorithm is used to obtain relative degree with respect to each output in the output vector.

2. Dynamic transformation

The kinematic constraints of the mobile robot derived in the companion paper is shown below for completeness

$$\begin{aligned} \dot{x}_p \cos(\phi) + \dot{y}_p \sin(\phi) - b\dot{\phi} &= r\dot{\theta}_l \\ \dot{x}_p \cos(\phi) + \dot{y}_p \sin(\phi) + b\dot{\phi} &= r\dot{\theta}_r \\ -\dot{x}_p \sin(\phi) + \dot{y}_p \cos(\phi) &= 0. \end{aligned} \quad (1)$$

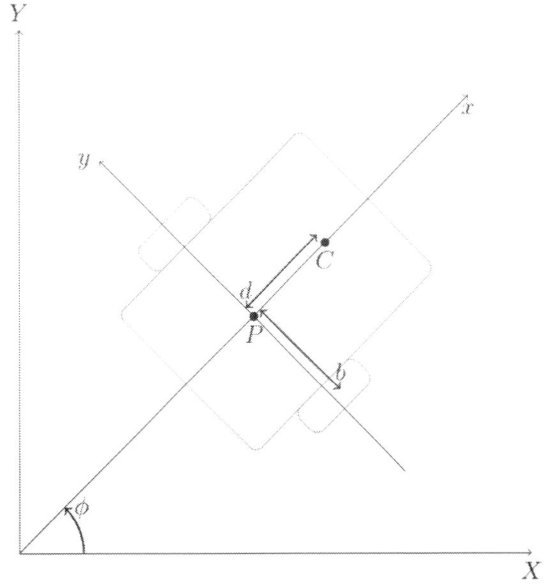

Figure 1. Diagram of a differential mobile robot

The dynamical equation of motion of the mobile robot derived in [1] is given by

$$\underline{\dot{q}} = \begin{bmatrix} q_6 \\ q_7 \\ q_8 \\ q_9 \\ q_{10} \\ f_6(q_3, q_6, q_7, q_8) \\ f_7(q_3, q_6, q_7, q_8) \\ f_8(q_3, q_6, q_7, q_8) \\ f_9(q_3, q_6, q_7, q_8) \\ f_{10}(q_3, q_6, q_7, q_8) \end{bmatrix} + \begin{bmatrix} 0 & 0 \\ 0 & 0 \\ 0 & 0 \\ 0 & 0 \\ 0 & 0 \\ g_{61}(q_3) & g_{61}(q_3) \\ g_{71}(q_3) & g_{71}(q_3) \\ g_{81}(q_3) & -g_{81}(q_3) \\ g_{91} & g_{92} \\ g_{10} & g_{11} \end{bmatrix} \begin{bmatrix} \tau_1 \\ \tau_2 \end{bmatrix} \tag{2}$$

By taking the difference between the first and second equations of (1) in [1], we have

$$\dot{\phi} = \frac{r}{2b}\left(\dot{\theta}_r - \dot{\theta}_l\right). \tag{3}$$

Taking time derivative of (3), the resulting equation is equal to row 8 of (2). From this , we can solve for τ_1 which is given by

$$\tau_1 = \frac{r}{2bg_{81}}\left(\dot{\theta}_r - \dot{\theta}_l\right). \tag{4}$$

Upon substitution in (2) and noticing that the substitution does not affect the upper half equation in (2), the lower half of (2) is given by

$$\begin{bmatrix} 1 & 0 & 0 & -g_6K & g_6K \\ 0 & 1 & 0 & -g_7K & g_7K \\ 0 & 0 & 1 & -\frac{r}{2b} & \frac{r}{2b} \\ 0 & 0 & 0 & 1-g_{91}K & g_{91}K \\ 0 & 0 & 0 & -g_{10}K & 1+g_{10}K \end{bmatrix} \begin{bmatrix} \dot{q}_6 \\ \dot{q}_7 \\ \dot{q}_8 \\ \dot{q}_9 \\ \dot{q}_{10} \end{bmatrix} = \begin{bmatrix} f_6 - \frac{g_{61}f_8}{g_{81}} \\ f_7 - \frac{g_{71}f_8}{g_{81}} \\ 0 \\ f_9 - \frac{g_{91}f_8}{g_{81}} \\ f_{10} - \frac{g_{10}f_8}{g_{81}} \end{bmatrix} + \begin{bmatrix} 2g_{61} \\ g_{71} \\ 0 \\ g_{91}+g_{92} \\ g_{10}+g_{11} \end{bmatrix} \tau_2 , \quad (5)$$

where $K = \frac{r}{2bg_{81}}$, g_{81} is a constant so the division by it is well defined. The first half of (2) together with (5) give a system of "single input". Upon computing the control law for τ_2 one can compute τ_1 and $\ddot{\phi}$ explicitly. Because of limited space we do not pursue the design of this control law.
The control law derived in the rest of the paper involves the two inputs system shown in (2) using dynamic extension.

3. Tracking Control Design of the two inputs system

The dynamics of the mobile robot are formulated in the state space as dynamics of a standard nonlinear system as

$$\dot{\underline{q}} = \underline{f}(\underline{q}) + G(\underline{q})\underline{u} \quad (6)$$

Immediate calculations show that $rank\{\underline{g_1}, \underline{g_2}\} = 2$ *i.e.* the vectors $\underline{g_1}, \underline{g_2}$ are linearly independent. The distribution G=$\{\underline{g_1}, \underline{g_2}\}$ is also involution because simple calculations show that $rank[\underline{g_1}, \underline{g_2}, [\underline{g_1}, \underline{g_2}]]$=2 for all q. where $[\underline{g_1}, \underline{g_2}]$ is the Lie bracket of $\underline{g_1}$ and $\underline{g_2}$ defined by

$$[\underline{g_1}, \underline{g_2}] = \frac{\partial g_2}{\partial \underline{q}}\underline{g_1} - \frac{\partial g_1}{\partial \underline{q}}\underline{g_2} \quad (7)$$

Thus by Frobenius' Theorem, we deduce the existence of *8 real-valued functions* that span the space orthogonal to the space spanned by $\{\underline{g_1}, \underline{g_2}\}$. However, it is not always very easy to find these functions because it involves solving partial differentials equations. It is well known that systems with at least one nonholonomic constraints are not input-state linearizable [3] and since the system cannot be made asymptotically stable by a smooth feedback [4], we seek feedback control which achieves input-output stability. A natural choice of outputs are the coordinates of the reference point P figure 1 [1] which defines a position of the mobile robot in the plan. The output vector is defined as

$$\underline{y} = \begin{bmatrix} h_1 \\ h_2 \end{bmatrix} = \begin{bmatrix} q_1 \\ q_2 \end{bmatrix} \quad (8)$$

With respect to these outputs, the system does not have a relative degree because immediate calculations show that

$$L_g h = \begin{bmatrix} 0 & 0 \\ 0 & 0 \end{bmatrix} \quad (9)$$

and

$$A(\underline{q}) = L_g L_f h(\underline{q}) = \begin{bmatrix} g_{61}(q_3) & g_{61}(q_3) \\ g_{71}(q_3) & g_{71}(q_3) \end{bmatrix} \quad (10)$$

which has rank 1 for all q. We apply the dynamic extension algorithm. Relative degree $r=\{4,3\}$ is achieved after two iterations of Dynamic Extension Algorithm when the following compensator is cascaded with the system. (Refer for example to [2][5] for Dynamic Extension).

$$\begin{aligned} \tau_1 &= \frac{1}{g_{61}(q_3)}\Big(-f_6(\underline{q}) + \eta\Big) - v_2 \\ \dot{\eta} &= \zeta \\ \dot{\zeta} &= v_1 \\ \tau_2 &= v_2 \end{aligned} \tag{11}$$

4. Simulation

A simulation is developed to demonstrate the effectiveness of the control law designed above. As figures 2 and 3 demonstrate, the tracking control which takes the dynamics of the mobile robot into account perform good tracking. One should also note that the control law is valid for $g_{16}(q_3) \neq 0$, i.e.,$q_3 \neq \frac{\pi}{2}(2k+1)$, where k is an integer.

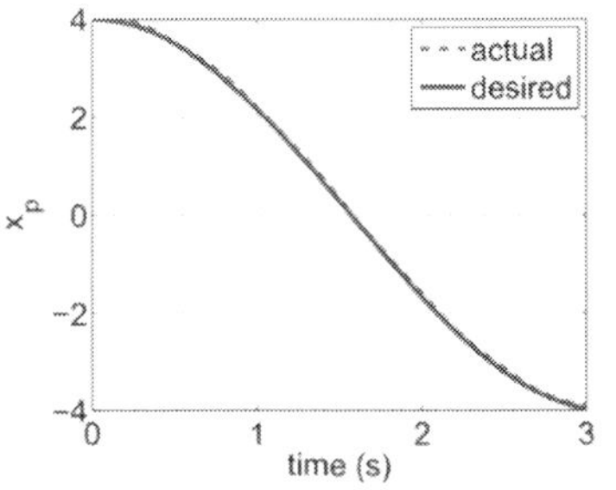

Figure 2. x_p and the desired trajectory.

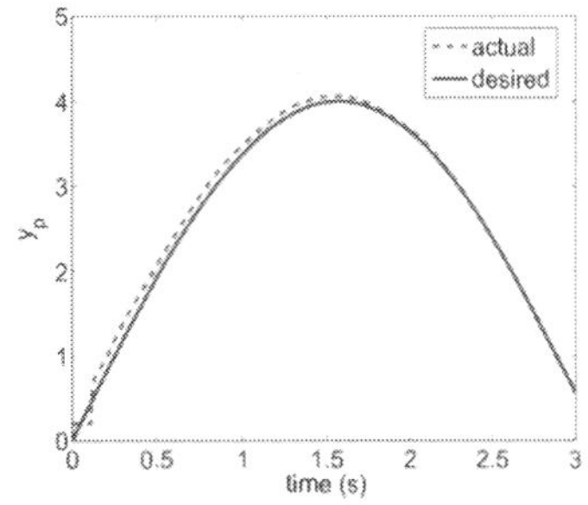

Figure 3. y_p and the desired trajectory.

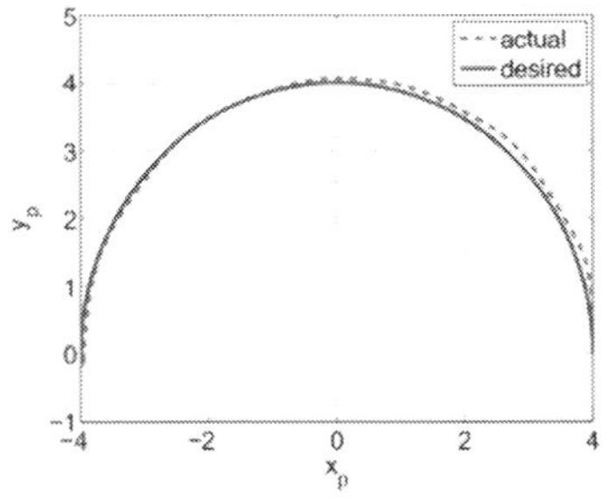

Figure 4. Circular path.

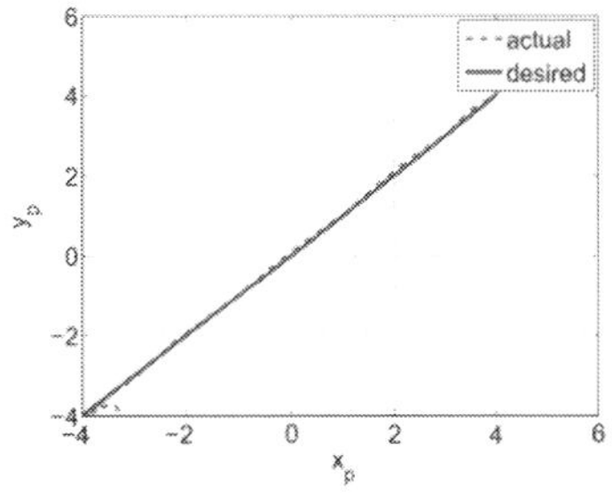

Figure 5. Straight line path.

5. Conclusion

We consider the modeling of the dynamics of a differential mobile robot and the derivation of a nonlinear control law to track a predefined trajectory. Maggi's method, a quasivelocity technique was used to derive the dynamic. This method eliminates the Lagrange multipliers used to enforce the nonholonomic constraints from the start as most of the time one is not interested in the Lagrange multipliers. It thus reduce the number of the variables and the number of equations to solve. Feedback linearization is the technique employed in this paper to derive a trajectory control law for the wheeled mobile robot. It is well known that one of the key point to start feedback linearization is the selection of the output vector. Often, some state variable are not available for measurement and the choice is limited. With respect to the output vector selected in this paper, there is no relative degree. Dynamic extension is then used to obtain a relative degree vector relative to each component of the output vector. Computer simulations are then added to demonstrate the effectiveness of the control law implemented for this class of wheeled mobile robots.

References

[1] Amengonu Y. and Kakad Y 2014 *Dynamics and control for Constrained Multibody Systems modeled with Maggi's equation: Application to Differential Mobile Robots Part I*. 27th International Conference on CADCAM, Robotics and Factories of the Future 2014

[2] Isidori A 1999 *Nonlinear control systems* Vol. 2. (Berlin: Springer)

[3] Campion G, d'Andrea-Novel B and Bastin G 1991 *Controllability and state feedback stabilization of non holonomic mechanical systems*. Lecture Notes in Control and Information Science (Berlin:Springer-Verlag) p 106-24

[4] Bloch A and McClamroch N 1989 *Control of mechanical systems with classical nonholonomic constraints* Proc. of 28th IEEE Conference on Decision and Control (Florida) p 201-05

[5] Khalil H 2002 *Nonlinear systems* Vol. 3. Upper Saddle River: Prentice hall.

27th International Conference on CADCAM, Robotics and Factories of the Future 2014 IOP Publishing
IOP Conf. Series: Materials Science and Engineering **65** (2014) 012019 doi:10.1088/1757-899X/65/1/012019

Synthesis of system for program signals formation for electric drives of multilink manipulators

V F Filaretov[1,2] **and A S Gubankov**[1,2]

[1] Department of Automation and Control, Far Eastern Federal University, Vladivostok, Russia
[2] Robotics Lab, Institute of Automation and Conrtol Processes FEB RAS, Vladivostok, Russia

E-mail: `filaretov.vf@dvfu.ru`, `gubankov.as@dvfu.ru`

Abstract. The paper is devoted to the problem of synthesis of system for formation of input signals for actuators of multilink manipulators. This method provide movement of working tools with the maximum speed depending on the location of processed objects in the working area of manipulator without decrease of the set dynamic accuracy. In addition method provides an approach to the objects without overshooting.

1. Introduction

Robotics equipment used in modern industry has to perform working operations as quickly as possible and moving of the working tool (WT) from one detail or item to another should be smooth with exact stop near working objects without overshooting.

Currently at WT trajectory forming points of the working area is interpolated by splines not less than third order. And if necessary to ensure smooth movement the order of splines can be increased up to the fifth (see [1]). Because of arbitrary and constantly changing coordinates of details in the working area of manipulators computational load on the control computer increases significantly. In this case to maintain the required accuracy of the WT movement in all parts of the trajectory the average speed of their movement is often especially slows down, tuning on the most negative case in order to avoid reject in some parts of these trajectories even when it can be significantly increased. As a result performance of the manipulator is decreases.

In [2] a method of planning the trajectories of the manipulator WT using the four order B-splines is considered. To set these splines multi-parametric optimization algorithm is used. But the algorithm has a very high computational complexity which requires the use powerful processors to implement the control in real time.

In [3] the control system (CS) synthesis method for movement mode of the underwater vehicle which automatically generates a maximum current speed of its movement along spatial path with a given dynamic accuracy is considered. However, for effective use of the CS for the multilink manipulator one have to consider kinematic and dynamic features and then continuously determine the deviation of the working tool from the given trajectory. And this is very complicated during the manipulator movement.

In [4] and [5] a method of turning of program signals parameters of one degree of freedom manipulator movement is offered. It is based on the amplitude-frequency characteristics of manipulator actuators and provides maximum possible speed without reducing given dynamic accuracy. This method can be applied for the synthesis of system for formation of program signals of multilink manipulators motion, but one have to take into account the kinematic and dynamic features of the considered dynamic control object and continuous change its configuration during the execution of the required process steps.

2. Task Settings

The main goal of this paper is designing of the synthesis method of easy realizable control systems for multilink manipulators. These systems depending on the current location of processing objects have to form such input signals for all actuators which will provide movement of WT as quickly as possible and for given dynamic accuracy. The approach to the objects must be performed smoothly and without overshooting.

3. Description of the control object

Synthesis of the proposed system let us consider on two degree of freedom manipulator 1 (see fig. 1) making technological operations with objects 2 located on a pallet 3 which is located in the working area of the manipulator. A new pallet is moved in the working area of the manipulator each working cycle and location of objects is differing from the previous one. Working operations with all objects are performed only when the WT speed with respect to them is zero. Camera 4 is over the pallet and allows to determine the current location of all objects in it (coordinates x_j and y_j in the absolute coordinate system Oxy, where j - the number of objects on a pallet).

Actuators of the manipulator (see fig. 1) are DC drives with separate excitation or permanent magnet. Considering differential equation of the DC drive anchor circuit and its torque equations (see [6]) and interactions between all degrees of freedom (see [7]) the i-th actuator can be described by the equation (where the index i is omitted for convenience):

$$\begin{aligned} &L(J_\Sigma + H^*)\dddot{\alpha} + (L\dot{H}^* + Lh^* + RJ_\Sigma + RH^*)\ddot{\alpha} + (Rh^* + L\dot{h}^* + K_\omega K_M)\dot{\alpha} + \\ &+ R(M_C + M_{ex}^*) + L(\dot{M}_C + \dot{M}_{ex}^*) = K_M K_y u, \end{aligned} \quad (1)$$

where R, L - resistance and inductance of an anchor chain of actuator respectively; K_M, K_ω - torque and emf coefficients respectively; K_y - gain coefficient; u - input signal; $H^* = H(q)/i_p$; $h^* = h(q, \dot{q})/i_p$; $M_{ex}^* = M_{ex}(q, \dot{q}, \ddot{q})/i_p$; $H(q)$ - component characterizing the inertial properties of the corresponding degrees of freedom of manipulator; $h(q, \dot{q})$ - component of the Coriolis force; $M_{ex}(q, \dot{q}, \ddot{q})$ - torque influence, taking into account the gravitational forces and the effects of interaction between all the degrees of freedom of manipulator during its motion; i_p - gear ratio; $q_i, \dot{q}_i, \ddot{q}_i$ - generalized coordinates, velocities and accelerations of manipulator links respectively ($i = 1, 2$ - number of degrees of freedom of manipulator); J_Σ - moment of inertia of the rotating parts of the motor and gear unit; $\dot{\alpha}$, $\ddot{\alpha}$ - velocity and acceleration of the DC motor shaft respectively; M_C - Coulomb friction.

To provide the invariance of the quality of each actuator to the effects of interaction between all the degrees of freedom of moving manipulator (to stabilize their parameters and hence the coefficients of (1) at a nominal level) in each CS were added self-turning corrective devices described in [7].

To provide the required quality of work to both direct chains of actuators with already stabilize parameters were added standard corrective devices with transfer functions (see [6]):

$$W_{ki}(s) = (T_{1i}s + 1)/(T_{2i}s + 1), \quad (2)$$

where T_{1i} and $T_{1i} > T_{2i}$ - time constants that depend on the particular characteristics of DC drives. Their choice will be explained below.

4. Program signals forming for actuators of manipulator

The algorithm of work of the proposed system is described below.

After determining the coordinates (x_j, y_j) of j-th item in the working area a two-link manipulator 1 by means of camera 4 (see fig. 1) and solving the inverse kinematics problem by method described, for example in [8], it is easy to find generalized coordinates $\tilde{q}^*_{1j}(x_j, y_j)$ and $\tilde{q}^*_{2j}(x_j, y_j)$. In trivial case for transfer of the manipulator WT from one item to another one can form the actuators input signal as a sequence of steps 1 (see fig. 2) corresponding to calculated generalized coordinates.

However, as shown in [4] and [5], than input signal is the sequence of steps it is necessary to significantly increase performance time of each stages of a running cycle specially limiting essentially increasing current in windings of electric drives in a start of motion and to provide the acceptable value of overshooting.

But the high quality of the transients and the required accuracy in the system can be achieved using simple regulators (2), forming the program signals as a sections of smooth curves (parts of harmonic signals). As these signals in paper is offered to use the sections harmonic curves of different frequencies (see curve 2 in fig. 2):

$$q^*_{ij} = \begin{cases} A_{pij}(1 - \cos(\omega_{pij}\tilde{t}_j)) + \tilde{q}_{i,j-1}, \text{if } \tilde{t}_j \leq \pi/\omega_{pij}; \\ 2A_{pij}, \text{if } \dot{q}^*_{ij} = 0, \dot{q}^*_{kj} \neq 0, i \neq k, \end{cases} \tag{3}$$

where A_{pij}, ω_{pij} - amplitudes and frequencies of program signals q^*_{ij}; $\tilde{t}_j$ - performance time of j-th working operation, which is cleared at the beginning of everyone operation; $\tilde{q}_{i,j-1}$ - generalized coordinates of manipulator links, corresponding to location of its WT at j-1-th item. It is assumed that the transient time in all manipulator actuators less than WT moving time from one place to another.

In the beginig of the working cicle the manipulator is in initial position in which $q^*_{i0} = 0$. The amplitudes of input signals for actuators are calculated according to $A_{pij} = 0.5(\tilde{q}_{ij} - \tilde{q}_{i,j-1})$. In [4] and [5] are shown that without reducing the accuracy of the dynamic control the frequency ω_{pij} can be changed depending on the current parameters of manipulators and amplitudes of harmonic input, essentially increasing working speed and accordingly the performance of the technological equipment without changing the structure and parameters of used regulators. The value ω_{pij} for the next move of the WT can be calculated as follows

$$\omega_{pij} = \frac{\sqrt{-1 + \sqrt{1 + 4T_i^2 K_i^2 \varepsilon_{di}^2 A_{pij}^{-2}}}}{T_i\sqrt{2}}, \tag{4}$$

where $T_i = 0.5T_{3i} + \sqrt{0.5T_{3i}^2 - T_{4i}}$; $T_{3i} = R_i J_{\Sigma i}/K_{Mi}K_{\omega i}$; $T_{4i} = L_i J_{\Sigma i}/K_{Mi}K_{\omega i}$; $K_i = K_{yi}/K_{\omega i}i_{pi}$; ε_{di} - given dynamic error of WT moving.

In this case at movement of WT from one object to another during one cycle the i-th actuator of the manipulator according to (3) will move along part of the harmonic path at maximum frequency (speed) (4). Thus each of the two motors have to work out an angle $2A_{pij}$, when turn is finished one of them stops expecting a stop of another.

5. Results of simulation

The installation process of some details on six objects placed on a pallet (see fig. 1) by the manipulator was investigated during the simulation. These details were located in the manipulator part magazine. The manipulator links with the same lengths $l_1 = l_2 = 0.4$ m were moved by identical DC motor with the following parameters K_y=800, K_M=0.02 N · m/A, K_ω=0.02 V · s, R=0.4 Ω, L=0.004 H, T_1=0.07 s, T_2=0.0035 s, $i_p = 100$.

In the initial position of the manipulator $q_{i0}^{*} = 0$, that is all its link located along axis y. Its WT (part magazine) consistently approaches to items that are placed on a pallet at the following points: (-0.7;0.1), (-0.15;0.65), (-0.6;0.3), (0.1;0.7), (0.2;0.5), (0.1;0.4). J_{Σ} remains unchanged and equal to 0.0005 kg $\cdot$ m^2 due to using self-turning CS [7] and $\varepsilon_d = 0.0015$ rad.

On figure 3 input program signals for manipulator actuators are shown. They formed by the proposed system according to (3) and the considered technique. On figure 4 the laws of change of ω_p in above input signals are shown, they are calculated using (4).

Fig. 3 shows that when moving to the first item the first link of the manipulator needs to work out almost twice larger angle than its second link. That is the amplitude of the input signal to the first actuator is twice more than amplitude of the signal applied to the second actuator. Therefore to save the given accuracy of the WT approach to the item (see fig. 5) CS automatically reduces the frequency (speed) of rotation of the first link using (4). As a result the second link without exceeding set value ε_d turns on a given angle with greater speed, stops and waits for the approach of the first link. Figure 3 also shows that the first link rotates faster and expect the second when moving to the fifth item.

It should be noted that the greatest increase (4-9 times) ω_p (see fig. 4) is observed when links turn on angles less than 30 degrees. The study showed if program signals is forming according to (3) and ω_p according to (4) then total run time with six items is 19.7 s. If ω_p is selected as constant taking into account the required accuracy then total time increases to 30.5 seconds. And finally if the inputs of all actuators are step signals then run time of this work increases to 55 s, taking into account that the next step on the corresponding actuator will be formed only when the steady-state error will not exceed ε_d=0.0015 rad. That is by means of turning only ω_p one can improve the performance of the whole system up to 1.55 times and by means of harmonic signals this performance can be improved up to 2.79 times. The above confirms the high efficiency of the developed system allowing to form program signals for actuators of multilink manipulators as segments of harmonics with variable amplitude and turning frequency.

6. Conclusion

The paper proposes a synthesis method of systems for forming program signals for actuators of multilink manipulators. It allows to provide movement of manipulator links with the maximum speed (frequency) and given level of dynamic accuracy. This provide a smooth approach of the manipulator working tool to the object (item) without overshooting. The implementation of the proposed system is not difficult.

Acknowledgments

This paper is supported by Govenment of Russia (02.G25.31.0025), RFBR grants and grants of Far Eastern Federal University

References

[1] Fu K S , Gonzalez R C and Lee C S G 1987 *Robotics: Control, Sensing, Vision and Intelligence* (McGraw-Hill, Inc)

[2] Koray Ayten K, Iravani P and Necip Sahinkaya M 2011 Optimum trajectory planning for industrial robots through inverse dynamics *Proc. of 8th International conference on Informatics in Control, Automation and Robotics* (Noordwijkerhout, The Netherlands) **1** p 105–110

[3] Filaretov V F and Yukhimets D A 2010 Synthesis of a system of automatic calculation of program signals for controlling motion of an underwater vehicle along a complex spatial trajectory *Journal of Computer and Systems Sciences International* **49** 1 p 96–104

[4] Filaretov V F and Gubankov A S 2009 Synthesis of the Adaptive Control Systems with Turning of the Input Signal Parameters *Proc. of the 20th International DAAAM Symposium "Intelligent Manufacturing & Automation: Theory, Practice & Education"* (Vienna, Austria) p 187-188

[5] Filaretov V F and Gubankov A S 2010 Synthesis of program control signals for mechatronic systems motion speed *Proc. of The First Joint International Conference on Multibody System Dynamics - IMSD 2010* (Lappeenranta, Finland) CD-ROM p 1–8

[6] Andreev V D Ivkin A M Kuleshov V S et al 1978 *Principles of tracking systems design* ed Lakota N A (Mashinostroenie, Moscow)(*in Russian*)

[7] Filaretov V F 2000 *Self turning control systems of manipulators actuators* (FENTU Press, Vladivostok) (*in Russian*)

[8] Zenkevich S L and Yushenko A S 2004 *The basis of manipulator robots control* (BMSTU Press, Moscow) (*in Russian*)

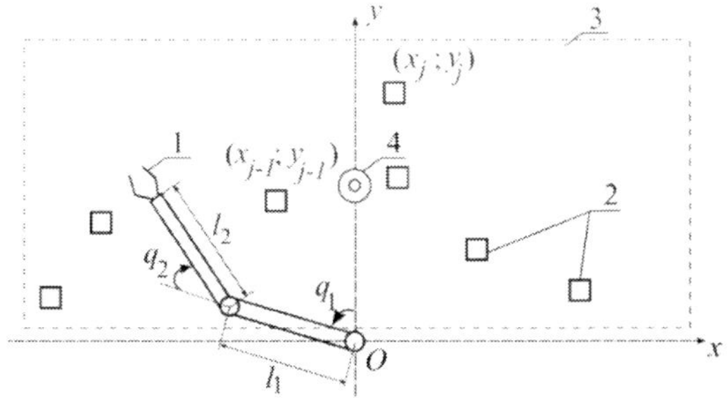

Figure 1. Scheme of technological process

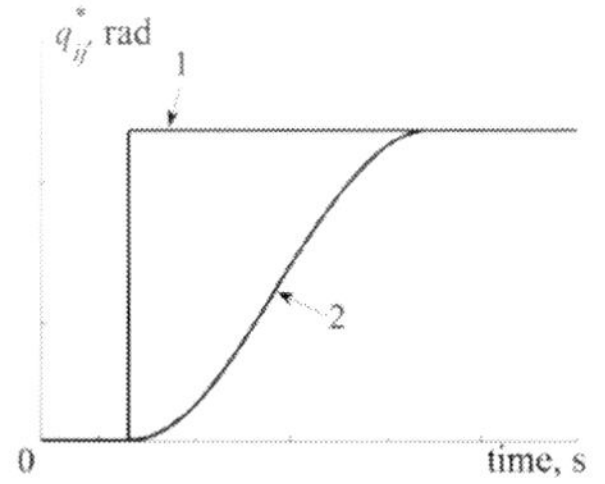

Figure 2. Forms of the input signals

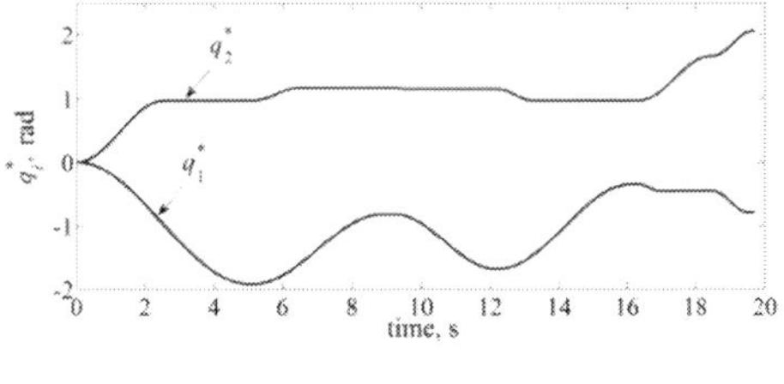

Figure 3. Input program signals for manipulator actuators

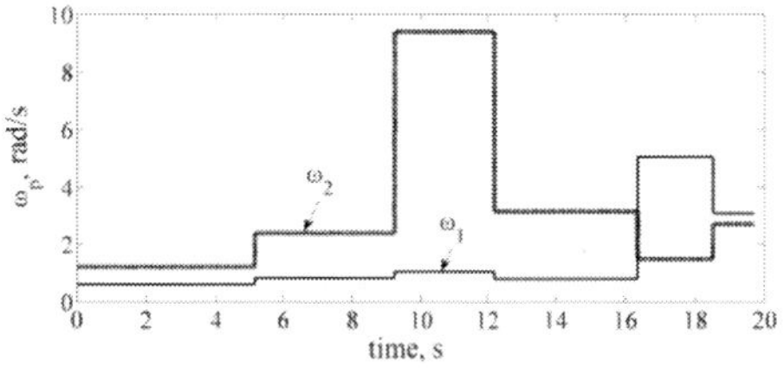

Figure 4. Frequencies of input signals

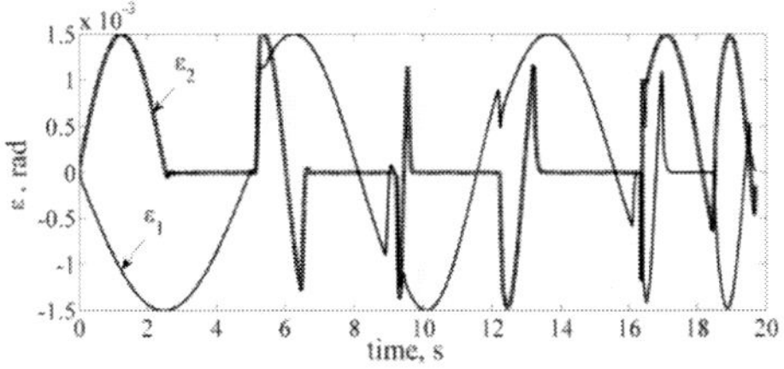

Figure 5. Dynamic errors

Investigating the feasibility of a BCI-driven robot-based writing agent for handicapped individuals

Chanan S. Syan[1], **Randy E. S. Harnarinesingh and Rishi Beharry**
Department of Mechanical and Manufacturing Engineering
The University of the West Indies
St. Augustine, Trinidad

E-mail: chanan.syan@sta.uwi.edu

Abstract. Brain-Computer Interfaces (BCIs) predominantly employ output actuators such as virtual keyboards and wheelchair controllers to enable handicapped individuals to interact and communicate with their environment. However, BCI-based assistive technologies are limited in their application. There is minimal research geared towards granting disabled individuals the ability to communicate using written words. This is a drawback because involving a human attendant in writing tasks can entail a breach of personal privacy where the task entails sensitive and private information such as banking matters. BCI-driven robot-based writing however can provide a safeguard for user privacy where it is required.

This study investigated the feasibility of a BCI-driven writing agent using the 3 degree-of-freedom Phantom Omnibot. A full alphanumerical English character set was developed and validated using a teach pendant program in MATLAB. The Omnibot was subsequently interfaced to a P300-based BCI. Three subjects utilised the BCI in the online context to communicate words to the writing robot over a Local Area Network (LAN). The average online letter-wise classification accuracy was 91.43%. The writing agent legibly constructed the communicated letters with minor errors in trajectory execution. The developed system therefore provided a feasible platform for BCI-based writing.

1. Introduction

A Brain-Computer Interface (BCI) is a system that allows users to control external devices using only their inherent brain activity. BCI users are required to perform mental or physical tasks that embody device commands. For example, users can focus on directional stimuli on a computer screen in order to issue motive commands to a mouse cursor. BCIs are responsible for recording brain data and identifying the resulting modulation of brain features that signify user intention. BCIs are an attractive option for disabled individuals because they rely on cognitive ability without necessarily requiring sensorimotor functions [1].

BCIs have been used to control virtual keyboards thereby allowing disabled individuals an avenue for communication [2,3]. However, the current work on BCI-based physical writing systems is limited. Human attendants can assist in writing tasks for disabled individuals. However, this can entail a breach of privacy where the task entails sensitive information such as personal letters and legal contracts. BCI-based writing therefore provides a safeguard for the user's privacy where it is required.

[1] To whom any correspondence should be addressed.

A previous study utilized the 5-axis Mitsubishi RV-2AJ robot [4] for BCI-based writing. The robot was used to write letters in real time over the internet using a BCI-based speller. This paper investigates the feasibility of a BCI-based writing agent using the lower cost 3-axis Quanser Phantom Omnibot for a full alphanumerical English character set.

The BCI is based on the P300 signal. The P300 is positive deflection in the Electroencephalogram (EEG) that is evoked approximately 300-400ms following the presentation of a rare, deviant or target stimulus [5]. The P300 wave can be evoked due to visual, auditory or somatosensory stimulation. However, this study uses the visually evoked P300.

2. BCI Experimental Study

2.1. Subject demographics
Three subjects hereinafter referred to as S1, S2 and S3 each participated in 1 training session and 1 testing session using the P300 Speller. All of the subjects were male and had normal or corrected-to-normal vision. Subjects S2 and S3 were naïve to BCI experiments and Subject S1 was non-naïve.

2.2. Data Recording Circumstances
The g.tec g.MOBILAB was used to record EEG at 8 sites according to the 10/20 electrode placement guideline at a sampling frequency of 256 Hz. Subject EEG cap sizes were chosen based on head circumference and the CZ position was centred along the Nasion to Inion points and the left to right pre-auricular areas. The 8 EEG sites used for data collection were P7, P3, P4, P8, FZ, CZ, PZ and OZ. Electrical reference was taken at the left mastoid and ground was taken at the Inion.

2.3. Stimulus Presentation Paradigm
The Stimulus Presentation Paradigms (SPPs) used for stimulus delivery were programmed as Graphical User Interfaces (GUIs) in MATLAB® 7.8.0.347. Subjects were seated approximately 80cm from a 22-inch computer monitor in a dimly lit room for the experiments.

2.3.1. Training session. Training sessions were used in order to obtain features to train classifiers to recognise subject-specific P300 responses. The GUI used for training sessions is presented in Figure 1. The matrix consisted of 36 grey alphanumerical elements against a white background. Target characters were highlighted in red for 2 seconds and subjects were instructed to mentally count the number of times the target characters were evoked in the subsequent flashing sequence.

Each character of the matrix was highlighted in black for 100ms followed by a 75ms Inter-Stimulus Interval in which no character was highlighted. The order of character presentation was random and the presentation of all 36 characters is referred to as a trial. Five trials were used to communicate each target character. Each character appeared once as the target. This generated a total of 36 5-trial segments.

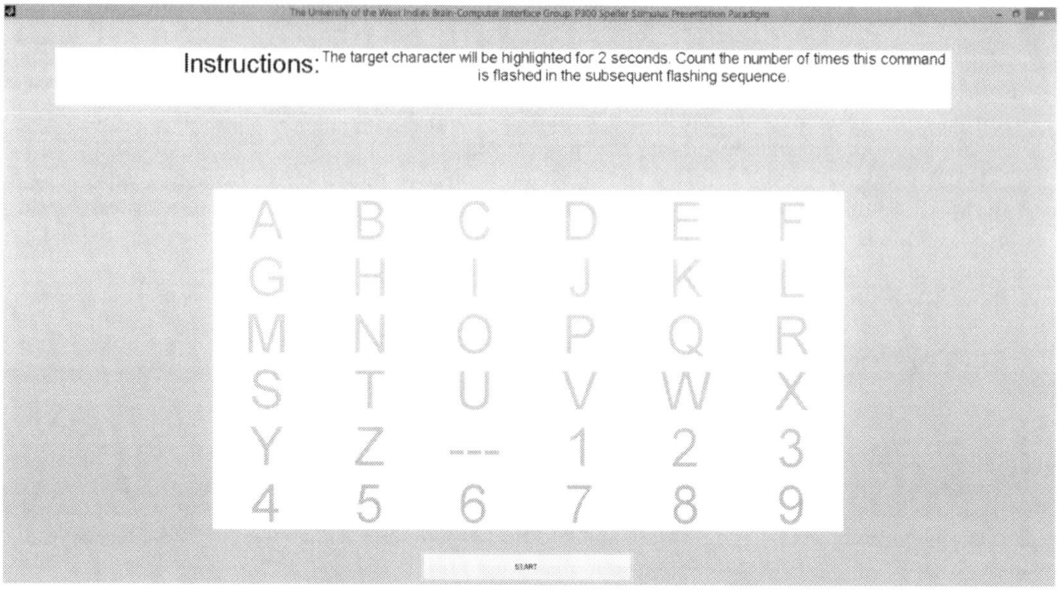

Figure 1. Stimulus Presentation Paradigm GUI for Training session

2.3.2. Testing sessions. The trained classifiers were implemented in online testing sessions in which the subjects selected characters on their own volition. In these testing sessions, the BCI had no prior knowledge of subject intention. This provided a basis for the unbiased determination of classifier performance. The GUI used for online testing session is presented in Figure 2.

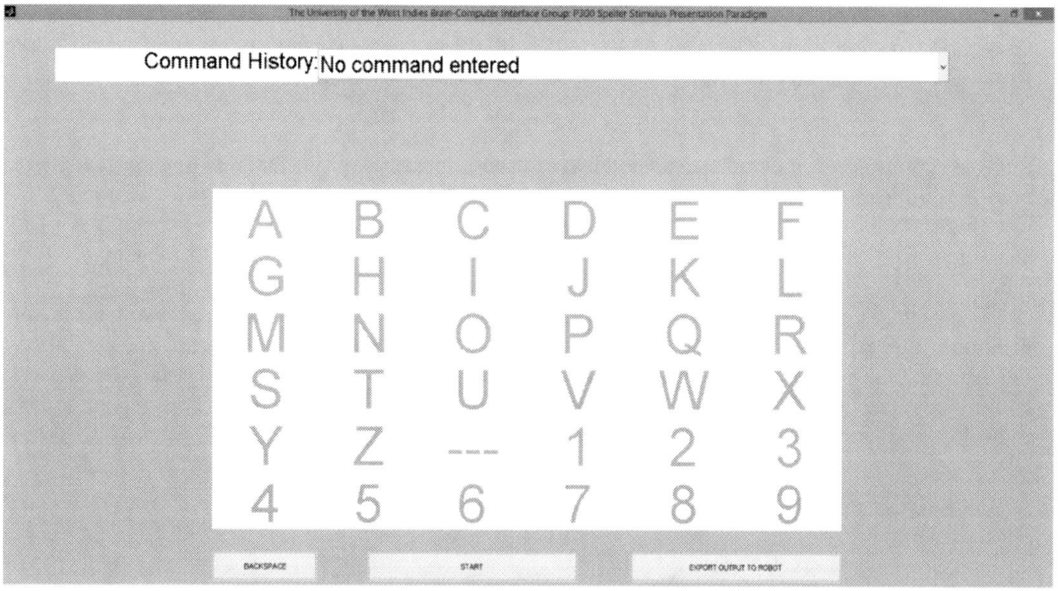

Figure 2. Stimulus Presentation Paradigm GUI for Testing session

Preliminary simulations were performed to select the appropriate number of trials to utilise for online testing for each subject. Three trials of online presentation were used to communicate characters for Subject S1. However, 5 trials of online presentation were used for Subjects S2 and S3. The classified letters were sequentially ordered in a text box at the top of the GUI. Eight letters were communicated in total to send to the writing robot. A "backspace" functionality was programmed into the GUI to allow for deletion of misclassified letters to ensure that only the intended letters were written by the robot.

2.4. Pre-processing, Feature Extraction and Pattern Classification

The EEG was collected raw with no pre-processing applied for the offline sessions. N-fold Cross-Validation (NFCV) was used to grid search pre-processing and feature extraction parameters for Fishers Linear Discriminant Analysis (FLDA) classifiers for Subjects S1-S3. The NFCV parameter search is described further.

Each EEG channel was forward-reverse filtered using a digital Butterworth filter of order n with 3DB cut-off frequencies at 1Hz and 12Hz respectively. The 500ms segments subsequent to stimulus delivery were down-sampled by a factor of K to create a set of temporal features for each channel. The temporal features were combined across the 8 channels to obtain a single spatiotemporal feature vector for each post-stimulus segment. The values of n and K were determined using a NFCV search on the offline data with N=3 for single trial P300 classification using FLDA.

3. Robot Forward Kinematics

Forward kinematic modelling was utilised to determine the end-effector position of the Omnibot I terms of the individual joint angles. The coordinate frame assignment for the Phantom Omnibot is presented in Figure 3.

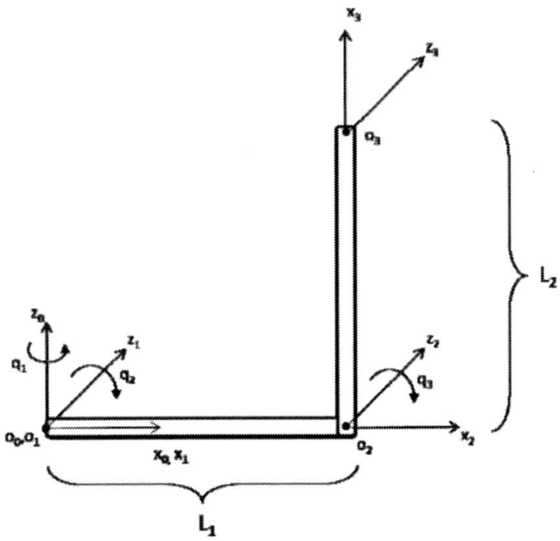

Figure 3. Phantom Omnibot coordinate frame assignments

The Denavit-Hartenberg (DH) [6] parameters for the Omnibot joints are presented in Table 1.

Table 1. DH joint parameters for Omnibot

Link	**a** (m)	**d** (m)	α	Θ
1	0	0	-π/2	q1
2	0.132	0	0	q2
3	0.132	0	0	q3- π/2

The DH matrix relating joint 'i-1' to joint 'i' takes the form shown in (1).

$$H_{i-1}^{i} = \begin{pmatrix} cos\theta & -sin\theta cos\alpha & sin\theta sin\alpha & acos\theta \\ sin\theta & cos\theta cos\alpha & -cos\theta sin\alpha & asin\theta \\ 0 & sin\alpha & cos\alpha & d \\ 0 & 0 & 0 & 1 \end{pmatrix} \tag{1}$$

In this study, a whiteboard marker was fixed to the final joint of the Omnibot. The marker tip can be referenced to the base frame of the robot using the Forward Kinematic model as shown in (2). The coordinates of the marker tip referenced to joint 3 are given as tip_x, tip_y and tip_z.

$$pos = H_0^1 H_1^2 H_2^3 \begin{pmatrix} tip_x \\ tip_y \\ tip_z \\ 1 \end{pmatrix} \tag{2}$$

4. Robot Inverse Kinematics and Controller Design

A flat whiteboard surface was used for character writing. The vertical z-coordinate of the whiteboard was determined by measurement. This allowed for the specification of the xy coordinates of character points whilst leaving the z-coordinate fixed. Finding the joint angles that satisfy the forward kinematics matrix multiplication for a given end-effector position solves the inverse kinematics problem.

A Proportional-Derivative (PD) controller was used to obtain the joints angles identified by Inverse Kinematics. The PD position controller for a joint is provided in Figure 4 where 'J' and 'B' are the inertial and frictional terms respectively. The placement of the derivative block in the inner loop averts the generation of impulses due to step changes in the set-point.

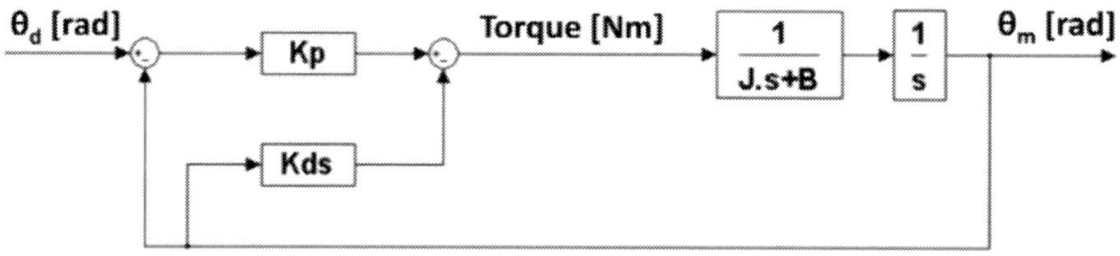

Figure 4. Feedback control block diagram for joint positioning

The controllers for 3 Omnibot joints were designed to meet the specifications in (3) and (4).

Percentage Overshoot (PO) = 25% (3)
Peak time (tp) = 0.14s (4)

The closed loop transfer function for the feedback loop in Figure 4 is given in (5).

$$\frac{\theta_m}{\theta_d} = \frac{K_p/J}{s^2+\left(\frac{B+K_d}{J}\right)s+K_p/J} \tag{5}$$

This is recognized as a reference 2nd order transfer function with natural frequency ω_n and damping ratio ζ whose relationships with the closed loop parameters are given in (6) and (7).

$$\omega_n{}^2 = \frac{K_p}{J} \tag{6}$$

$$2\zeta\omega_n = \frac{B+K_d}{J} \tag{7}$$

The peak overshoot, M_p and peak time, t_p for a second order system are given in (8) and (9).

$$M_p = e^{\frac{-\pi\zeta}{\sqrt{1-\zeta^2}}} \tag{8}$$

$$t_p = \frac{\pi}{\omega_n\sqrt{1-\zeta^2}} \tag{9}$$

This permits the simultaneous solution of ζ and ω_n using the design criteria specified in (3) and (4)

$$\zeta = \frac{\ln 4}{\sqrt{\pi^2+(\ln 4)^2}} = 0.404 \tag{10}$$

$$\omega_n = \frac{\pi}{0.14\sqrt{1-\zeta^2}} \tag{11}$$

The PD parameters were obtained using Equations (6)-(11) are presented in Table 2.

Table 2. PD controller parameters for Omnibot joints

Link	J (kgm^2)	B (Nmsrad^{-1})	Kp (Nmrad^{-1})	Kd (Nmsrad^{-1})
1	0.0031	0.0089	1.86	0.0525
2	0.0022	0.0170	1.32	0.0266
3	0.0009	0.0058	0.54	0.012

5. Character generation

The alphanumerical characters were defined using a set of xy coordinates on the whiteboard that were spatially interpolated to traced the desired character. The characters points were referenced to the base frame of the robot. The Simulink program determined the intermediate points between the user-defined points at equally spaced intervals for the purpose of interpolation. These points were input to the inverse kinematics system for the determination of joint angles and the PD controller was used to realize the required joint angles. The representative plots for the 'E' and 'J' characters are shown in Figure 5.

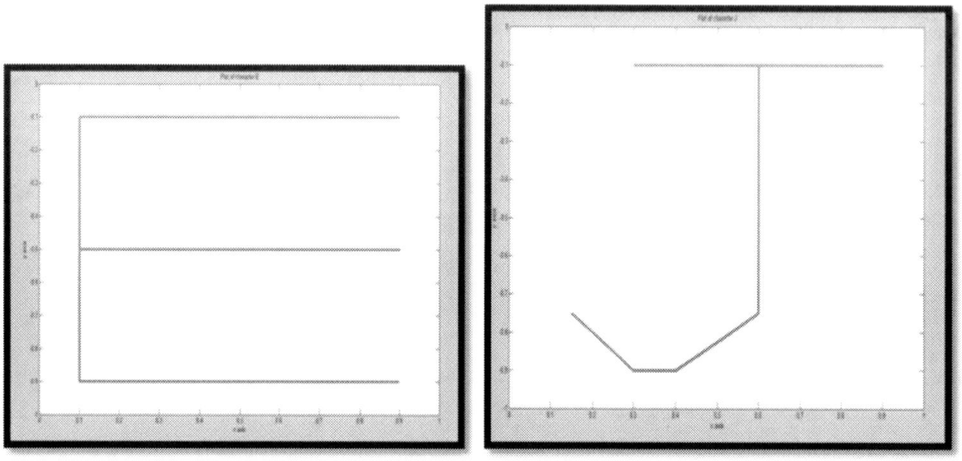

Figure 5. MATLAB Character generation of 'E' and 'J'

6. Results

The NFCV searched parameters for Subjects S1-S3 are presented in Table 3.

Table 3. NFCV Parameters for Subjects S1-S3

Subject	K	n
S1	10	1
S2	12	1
S3	16	2

The online BCI accuracies obtained for letter classification are presented in Table 4.

Table 4. Online BCI accuracies for subjects S1-S3

Subject	Number of letters correctly communicated	Number of misclassified letters	Total number of letters communicated	Percentage classification accuracy
S1	24	2	26	92.31%
S2	16	3	19	84.21%
S3	24	1	25	96.00%
TOTAL	64	6	70	91.43%

Words were sent to the writing robot over the LAN to another computer connected to the Omnibot after the successful classification of 8 letters for Subjects S1-S3. Figures 6-8 present the Omnibot realisation of 8 letter words and phrases communicated by Subjects S1-S3.

Figure 6. Omnibot generation of 'POSTPONE' communicated by Subject S1

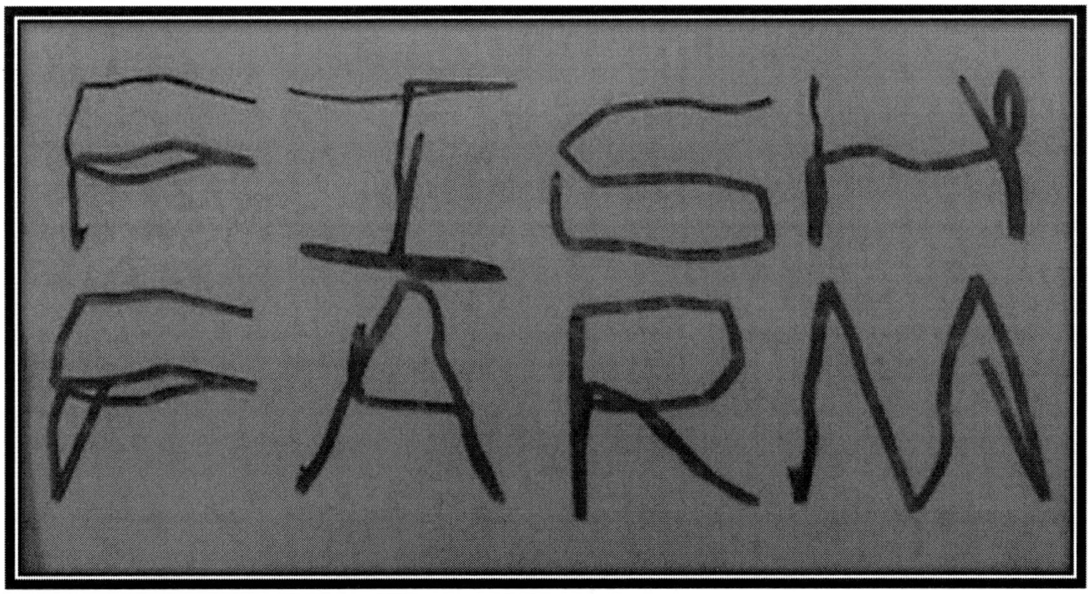

Figure 7. Omnibot generation of 'FISHFARM' communicated by Subject S2

Figure 8. Omnibot generation of 'BONERAGE' communicated by Subject S3

7. Discussion

The average online BCI classification accuracy for Subjects S1-S3 was 91.43%. Subject S3 attained an online classification accuracy of 96% which was significant for a naïve BCI subject. In contrast, Subject S1 was non-naïve and attained an accuracy of 92.31%. This seemed like a counter-intuitive result since naïve subjects in general perform worse at BCI usage than non-naïve subject. However, Subject S1 used fewer trials for online communication than Subject S3. Subject S1 used 3 trials for online communication where Subject S3 used 5 trials. Trial averaging has been shown to reduce misclassification errors in BCI [7] and provides an explanation for the observed online accuracies.

The 3-axis Omnibot was programmed to write 36 English characters consisting of 25 letters, numbers 0-9 and a hyphen. The robot-written characters were legible for the developed character set. However, curved realization of linear characters segments were observed for Omnibot written characters. This can be seen in Figures 6-8 and can be attributed to multiple reasons.

First, the transience inherent to second order controller response manifests itself as errors in trajectory execution. Second, the controllers were designed in the continuous-time domain however the MATLAB/Simulink program was discrete-time. The approximations in controller design can also manifest themselves as trajectory errors. Finally, the whiteboard surface used was not perfectly 2-dimensional. Ridges and sinks on the surfaces would have appeared as output perturbations for the control system which can degrade the response at those regions.

8. Conclusion

This paper investigated the feasibility of an online BCI-based robot writing agent using the 3-axis Phantom Omnibot. A full English alphanumerical character set was developed in MATLAB. The writing robot was interfaced to the BCI on another PC over a Local Area Network (LAN). Three subjects utilised the developed system to communicate words and phrases to the writing robot. The generated characters were legible with observed deviations in programmed trajectory due to controller transience, approximation in controller design and uneven writing surfaces. These issues however did not significantly demerit the legibility of the written characters can be addressed for future implementation.

References

[1] Itturate I, Antelis J, Kubler A and Minguez J 2009 A Non-Invasive Brain-Actuated Wheelchair Based on a P300 Neurophysiological Protocol and Automated Navigation *IEEE Transactions on Robotics* **25(3)** 617-627

[2] Andrews S, Palaniappan R, Teoh A and Kiong L C 2008 Enhancing P300 Component by Spectral Power Ratio Principal Components for a Single Trial Brain-Computer Interface *American Journal of Applied Science* **5(6)** 639-644

[3] Citi L, Poli R, Cineland C and Supulveda F 2008 P300-Based BCI Mouse With Genetically-Optimized Analogue Control *IEEE transactions on Neural Systems and Rehabilitation Engineering* **16(1)** 51-61

[4] Perez-Marcos D, Buitrage J A and Velasquez F D 2011 Writing through a robot: a proof of concept for a brain-machine interface *Med. Eng. Phys.* **33(10)** 1314-1320

[5] Edlinger G, Wach P and Pfurtscheller G 1998 On the Realization of an Analytic High-Resolution EEG *IEEE Transactions on Biomedical Engineering* **45(6)** 736-745

[6] Denavit J and Hartenberg R S 1955 A kinematic notation for lower-pair mechanisms based on matrices *Trans ASME J. Appl. Mech.* **23** 215-221

[7] Pires G, Nunes U and Castelo-Branco M 2008 Brain Computer Interface Approaches to Control Mobile Robotic Devices *Proceedings of the 11th International Conference on Climbing and Walking Robots and the Support Technologies for Mobile Machines (Coimbra, Portugal, 8-10 September 2008)* pp 489-496

IOP Conf. Series: Materials Science and Engineering 65 (2014) 012021 doi:10.1088/1757-899X/65/1/012021

Asynchronous vehicle pose correction using visual detection of ground features

Randy E S Harnarinesingh and Chanan S Syan[1]
Department of Mechanical and Manufacturing Engineering
The University of the West Indies
St. Augustine, Trinidad

E-mail: chanan.syan@sta.uwi.edu

Abstract. The inherent noise associated with odometry manifests itself as errors in localization for autonomous vehicles. Visual odometry has been previously used in order to supplement classical vehicle odometry. However, visual odometry is limited in its ability to reduce errors in localization for large travel distances that entail the cumulative summing of individual frame-to-frame image errors. In this paper, a novel machine vision approach for tiled surfaces is proposed to address this problem. Tile edges in a laboratory environment are used to define a travel trajectory for the Quansar Qbot (autonomous vehicle) built on the iRobot iRoomba platform with a forward facing camera. Tile intersections are used to enable asynchronous error recovery for vehicle position and orientation. The proposed approach employs real-time image classification and is feasible for error mitigation for large travel distances.

The average position error for an 8m travel distance using classical odometry was measured to be 0.28m. However, implementation of the proposed approach resulted in an error of 0.028m. The proposed approach therefore significantly reduces pose estimation error and could be used to supplement existing modalities such as GPS and Laser-based range sensors.

1. Introduction

Mobile robots are utilized for a wide range of applications including interplanetary exploration [1], bomb disposal [2] and motorized wheelchair platforms for disabled individuals [3]. The success of a mobile robot in a particular task however requires the robots to have accurate knowledge of its position and orientation (pose). This is referred to as localization and is therefore a fundamental requirement for mobile robots. Localization can be realized using a host of technologies and approaches. The common approaches to mobile robot localization are discussed further.

Shaft encoders are electromechanical devices that can be used to measure angular position and velocity of rotating shafts. Multiple shaft encoders can be utilized to provide estimates of travel velocity and heading for mobile robots which can be used to iteratively update the position of a mobile robot with reference to a previous position. This is termed relative localization.

Shaft encoder readings however are typically noisy and contribute to errors in the determination of robot pose. These errors are cumulative and become significant for large travel distances [4]. Kalman filtering and particle filtering have been applied to treat with encoder based noise for robot localization problems [5]. These approaches however only mitigate errors and cannot eliminate them. The implementation of algorithms such as Kalman filtering also increases the computational complexity of the localization program.

[1] To whom any correspondence should be addressed.

In addition to sensor noise, encoder-based localization is hampered by wheel slippage. Slippage is caused by momentary reductions in wheel to surface traction. In the event of slippage, the wheel spins but vehicular motion is not realized. The rotation of the wheel however is measured by the encoders and contributes to pose update. Slippage therefore makes it difficult for a localization systems based on shaft encoders to perform reliably on low traction surfaces such as sand or gravel paths. Sonar, infrared and laser-based technologies are however robust to wheel slippage and have been applied to the problem of mobile robot localization.

Sonar sensors are used in [6] to determine the pose of a mobile robot. The implemented algorithm detected straight segments from the obtained ranges and compared it to a known reference model of the room. The algorithm was able to discriminate between straight line room segments such as walls and smaller elements such as furniture and persons. This allowed for model building to exclude furniture such as tables and desks.

A network of coded infrared beacons are used in [4] whose coordinates were known *a priori* to the robot. Infrared signals were detected using an upward facing CCD camera and an infrared receiver. Infrared signals are visible to digital CCD cameras due to the phenomena of baseband replication that arises from signal sampling. Detection of a coded infrared transmission identified the position of the transmitter and allowed for triangulation-based absolute localization of the vehicle.

Laser-based Range Finding (LRF), also known as Laser Radar (LIDAR) is the process of illuminating an object with laser and using the reflected signal to determine distance. LRF allows for high accuracy range detection and has been applied to the problem of mobile robot localization [7]. LRF approaches are advantageous because they do not necessarily require the addition of artificial beacons to an environment like those used in [4]. This allows for rapid deployment of robots in unknown surroundings without the need for environmental modifications.

The high accuracy of LRF allows makes it possible to localize a robot without knowledge of the robot kinematic model [8]. Successful localization without the need for robot kinematic modelling reduces the conceptual and computational complexity of odometry. LRF also permits the localization of mobile robots without prior knowledge of an environmental map [9]. This is useful in disaster recovery scenarios. For example, earthquake damaged buildings do not lend themselves to rapid mapping prior to ad hoc exploratory and recovery missions.

The inherent advantages of LRF are however outweighed by the costly nature of laser transmitters and receivers. This renders LRF inappropriate for low cost mobile platforms. LRF also entails high computational expense when map building algorithms are implemented. In addition, active sensing technologies such as laser, sonar and infrared possess common intrinsic disadvantages. Robots can misinterpret a signal transmitted from another device as a reflected signal originating from itself. This can corrupt the localization ability of active sensors. Passive sensors such as cameras address this problem. Cameras are an advantageous sensor for localization because they provide a large amount of information and are not susceptible to the noise inherent to light and sound based sensors [10].

Cameras are used in [11] to implement a Simultaneous Localization and Mapping (SLAM) algorithm using naturally occurring scale-invariant landmarks in an environment. The invariance of environmental landmarks to image translation, scaling and rotation enhanced their suitability for mobile robot localization and map building since it allowed for feature construction from various viewpoints. However, the algorithm required 2-3 seconds for the processing of acquired stereo images. Localization would therefore be achieved for stationary robots however, a travelling robot would realize a localization error because of the distance covered during the attendant image processing.

The use of cameras for localization in general requires the use of image databases [12], filtering algorithms [13] or geometric image models which increase the computational expense of the localization system. A hybrid approach to mobile robot localization using shaft encoders and a monocular camera is proposed in this paper. The hybrid approach averts the need for image databases, geometric image models and filtering algorithms and relies on the inherent visual features of a tiled laboratory environment to supplement encoder-based localization.

2. Hybrid approach to robot localization

The hybrid approach to mobile robot localization is illustrated in Figure 2. The main components of this approach are described further.

2.1. Colored line following

The colored line following approach is designed for tiled environments in which the tile edges are color-wise distinguishable to the inner tile portions. Colored line following is implemented to follow an externally defined trajectory for the mobile robot and to allow for the acquisition of intersections for localization. This requires the use of a forward facing monocular camera. The Quanser Qbot is a mobile differential drive robot built on the iRobot iRoomba platform. The Qbot features a forward facing monocular camera and is used for this study. The approach however may be modified to operate with ceiling tiles that satisfy the color-wise discrimination requirement using an upward facing monocular camera. In general, any floor visual features that are repetitive and overlay the floor surface can be utilized. The tiled environment used for testing the hybrid approach is pictured in Figure 1 using a smartphone camera.

Figure 1. Tiled Laboratory surface pictured using a smartphone camera

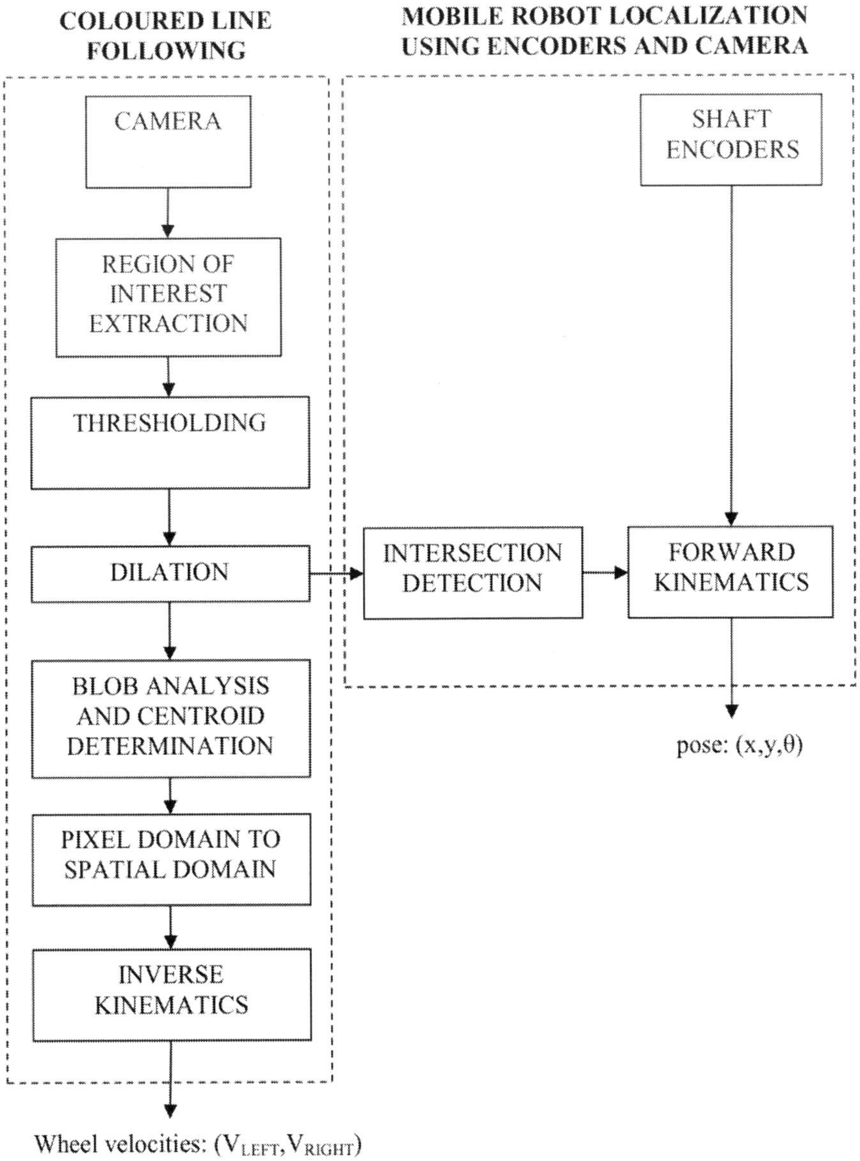

Figure 2. Hybrid approach to mobile robot localization

2.1.1. Region of Interest Extraction. The black line as viewed from the forward facing camera on the Qbot is pictured in Figure 3. The entire image however does not have to be utilised for the coloured line following approach. The important image data resides at the bottom centre of the image where the line is immediately in front of the robot. This area is outlined in red in Figure 3. In this regard, a Region-of-Interest-Extraction (ROIE) is implemented to acquire this image segment. This has the benefit of reducing the computational expense of the later processing stages.

Figure 3. Black line tile edge pictured from forward facing Qbot camera with outline Region of Interest

The ROIE image is illustrated in Figure 4.

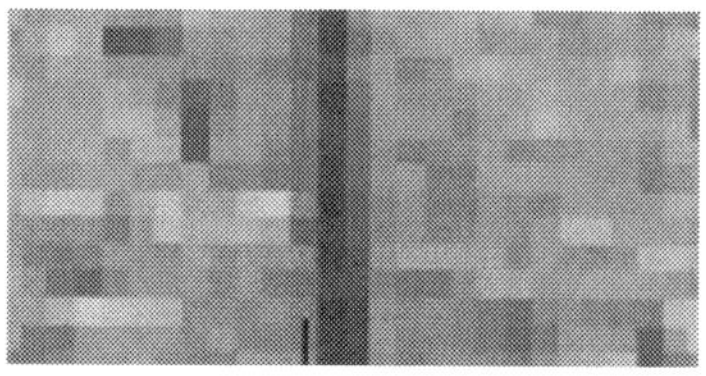

Figure 4. Region of Interest Extracted Image of Black line tile edge

2.1.2. Thresholding. Thresholding was implemented in order to discriminate the black line tile edges from the surrounding tile interior. The mathematical description of thresholding is given in (1).

$$h(x,y) = \begin{cases} 255, & if\ th1 \le f(x,y) \le th1 \\ 0, & otherwise \end{cases} \quad (1)$$

The thresholding ranges were determined experimentally using grayscale histograms. The thresholded image is presented in Figure 5.

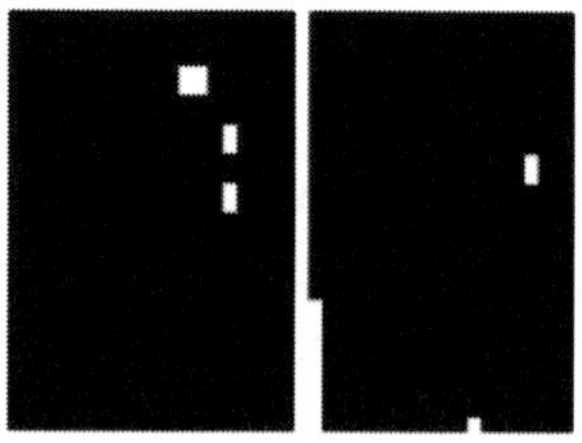

Figure 5. Thresholded image of black line tile edge

2.1.3. Dilation. Discontinuities in the thresholded image are possible due to glare and dirt on the tile edges. This is responsible for the variable width of the detected line in Figure 5. This presents a challenge to intersection detection which is the basis for the localization approach. Dilation is a morphological operation that can be used to remove discontinuities in images due to noise. The mathematical description of dilation is presented in (2).

$$A, B \in Z^2$$
$$dilation(A, B) = \left\{z|\left(\hat{B}\right)_z \cap A \neq \emptyset\right\} \text{ where}$$
$$\hat{B} = \{w|w = -b, for\ b \in B\}, and$$
$$(\hat{B})_z = \{c|c = w + z, for\ w \in \hat{B}\} \quad (2)$$

Figure 6 and Figure 7 present the case of thresholded images of the same tile intersection before and after dilation respectively.

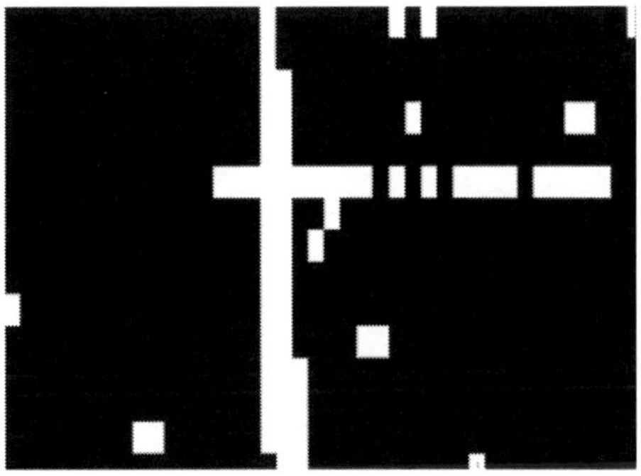

Figure 6. Thresholded image before dilation operation

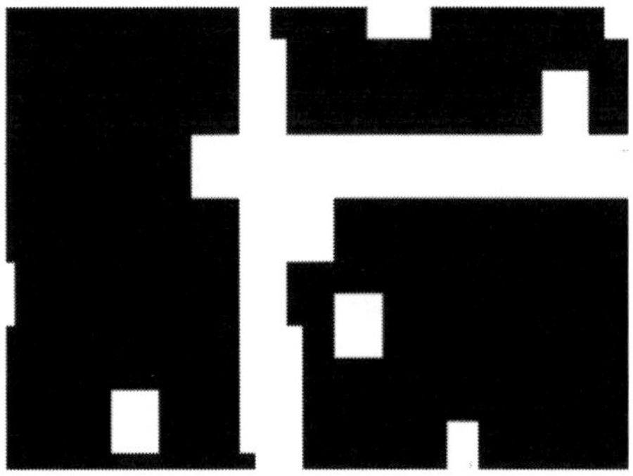

Figure 7. Thresholded image after dilation

A blob is define as a contiguous set of pixels of the same color. Analysis was performed on the dilated image in order to determine the centroid and width of the largest blob. The centroid of the largest pixel was mapped to the spatial domain and input to the Inverse Kinematic model of the vehicle in order to obtain the wheel velocities for motion. The width of the blob was used to detect intersections for localization. The derivation of the robot kinematic model is presented in Section 3.

3. Differential Drive Kinematic Modelling

The Quanser Qbot is a differential drive mobile vehicle consisting of two wheels mounted on a common axis. Differential drive vehicles possess zero-minimum turning radius which is advantageous since it allows rotation in tight spaces. Each wheel can be controlled independently

IOP Conf. Series: Materials Science and Engineering **65** (2014) 012021 doi:10.1088/1757-899X/65/1/012021

in order to navigate the vehicle. However, the vehicle must rotate about a point that lies along the common left and right wheel axes. This point is known as the Instantaneous Centre of Curvature (ICC) and is depicted in Figure 8 (Dedek and Jenkin, 2000).

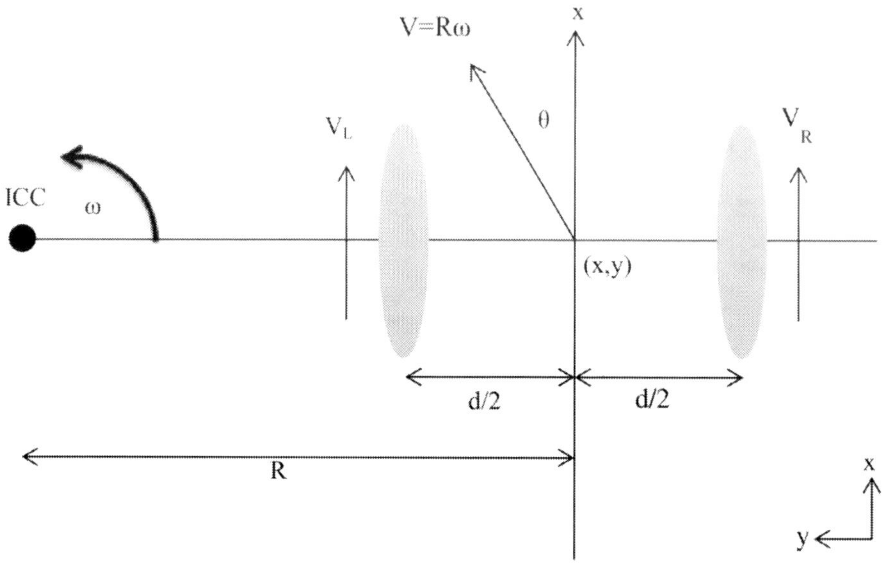

Figure 8. Differential drive kinematic model of Quanser Qbot

The rate of angular rotation about the ICC is the same for both wheels. The following equations therefore establish a relationship between the parameters of motion for a differential drive vehicle:

$$V_R = \omega(R + \frac{d}{2}) \quad (3)$$

$$V_L = \omega(R - \frac{d}{2}) \quad (4)$$

Equations (3) and (4) are independent equations in R and w, the Radius of Curvature (ROC) and the angular velocity, and can therefore be uniquely solved to give the following expressions:

$$\omega = \frac{(V_R - V_L)}{d} \quad (5)$$

$$R = \frac{d}{2} \times \left(\frac{V_R + V_L}{V_R - V_L}\right) \quad (6)$$

Equations (5) and (6) can be used to determine the instantaneous tangential velocity V of the vehicle.

$$V(t) = R(t)w(t) = \frac{d}{2} \times \left[\frac{V_R(t) + V_L(t)}{V_R(t) - V_L(t)}\right] \times \frac{[V_R(t) - V_L(t)]}{d} = \frac{[V_R(t) + V_L(t)]}{2} \quad (7)$$

IOP Conf. Series: Materials Science and Engineering **65** (2014) 012021 doi:10.1088/1757-899X/65/1/012021

3.1.1. Forward Kinematics. The forward kinematics problem deals with finding the vehicle position and orientation (x,y,θ) at time t using the initial position (x_0,y_0,θ_0) and the wheel velocities. The forward kinematic solution in discrete-time form is presented in (8) for the case of encoder-based odometry only. Discrete-time form is necessary since the Qbot Application Programming Interfaces (APIs) for accessing the camera output and assigning the wheel velocities are MATLAB/Simulink.

$$\begin{bmatrix} x[k+1]) \\ y[k+1] \\ \theta[k+1] \end{bmatrix} = \begin{bmatrix} x[k] + \frac{1}{2}(V_R + V_L)\cos(\theta)T_S \\ y[k] + \frac{1}{2}(V_R + V_L)\sin(\theta)T_S \\ \theta[k] + \frac{1}{d}(V_R - V_L)T_S \end{bmatrix} \quad (8)$$

The hybrid approach however relies on the detection of tile edge intersections for localization. Tile edge intersection are equally spaced at 1.23m. The approach constrains the robot to start at an intersection facing the black tile edge. The starting intersection is taken as the origin regardless of which intersection is used. The Forward Kinematic update rule for intersection detection is given in (9).

$$\begin{bmatrix} x[k+1]) \\ y[k+1] \\ \theta[k+1] \end{bmatrix} = \begin{bmatrix} x_{previous\_intersection} + 1.23 \\ 0 \\ 0 \end{bmatrix} \quad (9)$$

3.1.2. Inverse Kinematics. The inverse kinematics problem deals with finding the required wheel velocities to obtain a specified vehicle pose (x,y,θ) at time t. This problem is modified in the discrete-time case as finding the required wheel velocities to get the vehicle to pose (x_{k+1},y_{k+1},θ_{k+1}) from (x_k,y_k,θ_k) where *k* is the discrete-time index.

The vehicle pose at discrete-time indices *k* and *k+1* allows for the determination of the distance and the angular heading *α* between the two poses.

$$distance = \sqrt{(x_{k+1} - x_k)^2 + (y_{k+1} - y_k)^2} \quad (10)$$

$$\alpha = \theta_{k+1} - \theta_k \quad (11)$$

The distance and heading between the two poses can be used to determine *V* and *ω*.

$$V = {distance}/{T_s} \quad (12)$$

$$\omega = {\alpha}/{T_s} \quad (13)$$

The left and right wheel velocities are then obtained from *V* and *ω*.

$$V_R = \frac{2V + d\omega}{2} \quad (14)$$

$$V_L = 2V - V_R \quad (15)$$

In this approach, (x_k,y_k,θ_k) is assumed to be (0,0,0) and $(x_{k+1},y_{k+1},\theta_{k+1})$ are the coordinates of the largest blob in the dilated image mapped to the spatial domain.

4. Results

4.1. Colored line following

The coloured line following approach was used to drive the vehicle for 100 executions of 8m travel distances. The vehicle remained on the black line without deviation for all executions.

4.2. Hybrid approach to mobile robot localization

The vehicle was driven for 30 executions of 8m of black line following. Figure 9 presents the average trajectory plot for forward kinematics using shaft encoders only. Figure 10 present the average trajectory plots for forward kinematics using the hybrid approach.

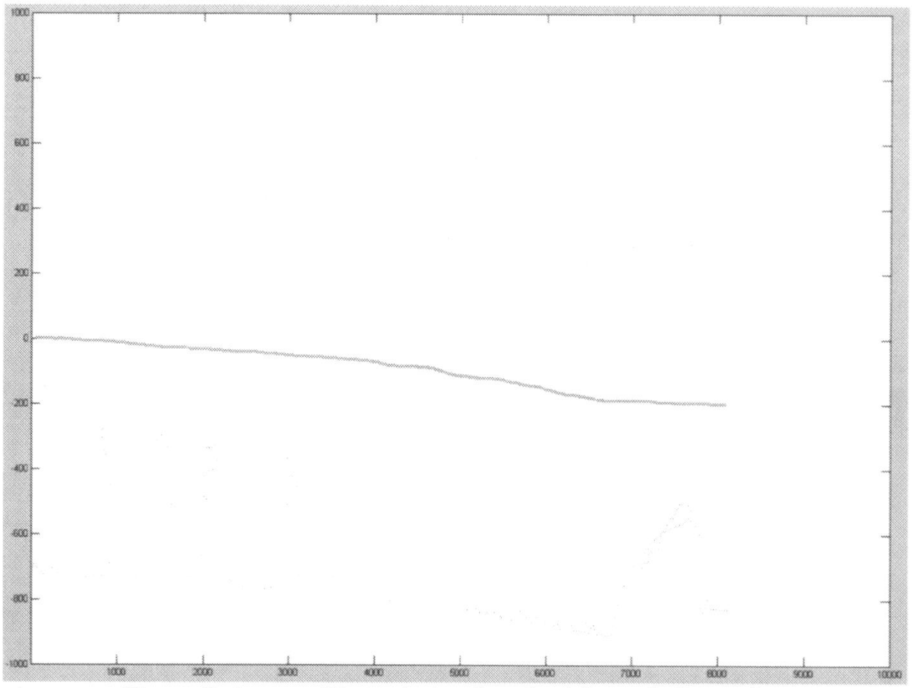

Figure 9. Average Qbot trajectory based solely on shaft encoders

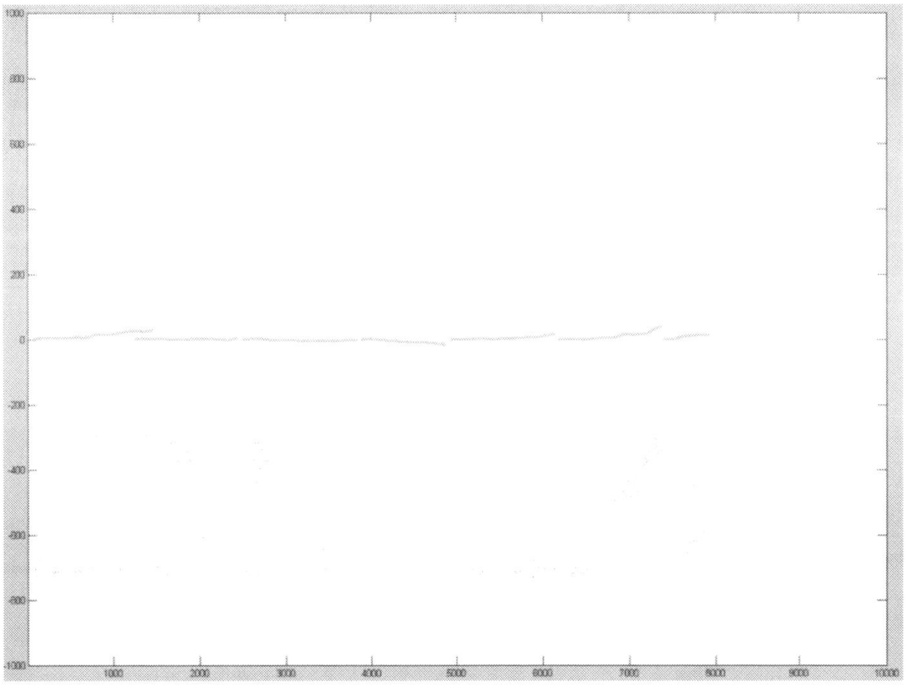

Figure 10. Average Qbot trajectory based on hybrid approach

5. Discussion

Colored line following was implemented for following the black line tile edges. The colored line following approach worked reliably for 100 executions. This validated the development of the colored line following subsystems.

Figure 9 presents the average trajectory for the Qbot using encoder-based Forward Kinematics. Localization drift is evident in the trajectory plot and grows as the travel distance is increased as suggested in [4]. The absolute position error at the end of travel was 0.28m.

Figure 10 presents the average trajectory using the hybrid approach to localization. Discontinuities are evident in these trajectory plots. These are due to the auto-corrective feature of the approach. For example, if the vehicle has travelled the length of 1 intersection which is 1.23m but encoder-based odometry measures a distance not equal to 1.23m, the position correction feature is manifested as a finite jump in the trajectory plot or a discontinuity. This asynchronous adjustment mechanism ensures that the error for any possible travel distance is arrested to a finite band. The absolute error at the end of travel was 0.028m.

The proposed approach therefore resulted in a 10-fold reduction in position error over the traversed space. This is a significant reduction in error and validates the proposed approach for tiled environments.

6. Conclusion and future work

The proposed hybrid approach to localization resulted in a ten-fold reduction in position error from 0.28m to 0.028m by supplementing the encoder-based forward kinematic system with the visual detection of ground tile intersections. This ensures that the autonomous vehicle has an accurate report of its position which is essential for successful task completion. The proposed approach is therefore feasible and could be adopted to supplement existing modalities such as GPS and LRF.

Future work will investigate the feasibility of grid based navigation using the Qbot. In the current implementation, the Qbot detects intersection approximately 450mm ahead of travel but travels past the intersection. Grid based navigation would require the vehicle to stop at intersections and allow turning to face other intersections before resuming travel. This presents a challenge due to the error in the measurement of rotation angle by shaft encoders. However the results of this paper motivate a strategy for intersection turning based solely on visual feedback.

References

[1] Volpe R, Balaram J, Ohm T and Ivlev R 1997 The Rocky 7 Mars Rover Prototype *Proc. IEEE/RSJ International Conference on Intelligent Robots and Systems (Osaka, Japan, 4-8 November 1996)*

[2] Costo S and Molfino R 2004 A New Robotic Unit Ror Onboard Airplanes Bomb Disposal *Prof. 35th International Symposium on Robotics (Paris, France, 23-26 March 2004)*

[3] Iturrate I, Antelis J, Kubler A and Minguez J 2009 A Non-Invasive Brain-Actuated Wheelchair Based on a P300 Neurophysiological Protocol and Automated Navigation *IEEE Transactions on Robotics* **25(3)** 614-627

[4] Brassart E, Pegard C and Mouaddib M 2000 Localization using infrared beacons *Robotica* **18** 153-161

[5] Von Der Hardt H, Arnould P, Wolf D and Dufaut M 1994 A method of mobile robot localisation by fusion of odometric and magnetometric data *The International Journal of Advanced Manufacturing Technology* **9(1)** 65-69

[6] Drumheller M 1987 Mobile Robot Localisation Using Sonar *IEEE Transactions on Pattern Analysis and Machine Intelligence* **9** 325-332

[7] Ribeiro M I and Goncalves 1996 Natural landmark based localisation of mobile robots using laser range data *Proc. of the First Euromicro Workshop on Advanced Mobile Robot (Kaiserslautern, Germany, 9-11 Oct 1996)* pp 186-193

[8] Spero D J and Jarvis R A 2004 Toward Exteroceptive Based Localisation *Proc. Of the 2004 IEEE International Conference on Robotics, Automation and Mechatronics (Singapore, 1-3 December 2004)* pp 822-827

[9] Zhang D, Kurazume R, Iwashita Y and Hasegawa T 2011 Robot localization under perceptual aliasing conditions based on laser reflectivity using particle filter *Proc of the 2011 IEEE/SICE International Symposium on System Integration (Kyoto, Japan, 20-22 December 2011)* pp 90-95

[10] Wolf J, Burgard W and Burkhardt H 2002 Using an Image Retrieval System for Vision-Based Mobile Robot Localization *Lecture Notes in Computer Science* **2383** 108-119

[11] Se S, Lowe D and Little J 2002 Mobile Robot Localization and Mapping with Uncertainty using Scale-Invariant Visual Landmarks *International Journal of Robotics Research* **21** 735-758

[12] Lamon P, Nourbakhsh I, Jensen B and Siegwart R 2001 Deriving and matching image fingerprint sequences for Mobile Robot Localization *Proc. of the 2011 IEEE International Conference on Robotics and Automation* **2** pp 1609-1614

[13] Georgiev A and Allen P K 2004 Localization methods for a mobile robot in urban environment *IEEE Transactions on Robotics* **20(5)** 851-864

IOP Conf. Series: Materials Science and Engineering **65** (2014) 012022 doi:10.1088/1757-899X/65/1/012022

Robosapien Robot used to Model Humanoid Interaction to Perform tasks in Dangerous Manufacturing Environments

R. Stopforth[1] **and G. Bright**[2]
[1,2]Mechatronics and Robotics Research Group (MR2G), University of KwaZulu-Natal, Durban, South Africa

E-mail: [1] stopforth@ukzn.ac.za and [2] brightg@ukzn.ac.za

Abstract. Humans are involved with accidents in manufacturing environments. A possibility to prevent humans from these scenarios is, to introduce humanoid robots within these industrial areas. This paper investigates the control scenario and environments required at a small scale level, with the use of the Robosapien robot. The Robosapien robot is modified to control it with a task of removing a cylinder and inserting it into a hole. Analysis is performed on the performance of the Robosapien robot and relating it with that of a humanoid robot. A discussion with suggestions is concluded with the efficiency and profitability that would need to be considered, for having a humanoid robot within the manufacturing environment.

1. Introduction

Studies in advanced manufacturing environments have shown that 20 % (of 103 reported cases) were maintenance and repair personnel that have been injured in accidents [1]. Dangerous environments could be considered as areas with hazardous chemicals, gases, voltage and currents, and machinery that have moving parts. Service, maintenance and repair robots have been limited with the tasks that they perform and were designed to perform set specific tasks. The maintenance robots are many times only located and used for a specific section of a manufacturing environment.

Humanoid robots such as the Honda Asimo [2] robot have been developed to walk in a stabilized manner over different obstacles. General Motors have developed for NASA the Robonaut [3], which was designed to perform operations and maintenance that were too dangerous for astronauts.

The question that arises is whether humanoid robots will be efficient in the manufacturing environment, considering the complexity of their design and development, therefore having a high purchase and maintenance price themselves. As to whether it would be more beneficial for a company to have one humanoid robot to perform multiple tasks, or a number of cheaper service robots able to perform each only one task, will depend on a company's needs and requirements.

Before investment in humanoid robots for manufacturing companies are pursued, it is required to evaluate the performance at a small scale and to determine the technologies that will be incorporated into the manufacturing environment so that the humanoid robot will be able to perform desired tasks.

This paper views some of the requirements that will need to be considered by investigating the performance of a small-scale humanoid robot. The Robosapien robot is used with a personal digital assistance device in order to control it remotely from a host computer. The task that was assigned to the robot was to go to a destination, pick up an object and drop it off at another destination. This was a simple task that was used to determine if the humanoid was capable of performing a user defined function of replacing a part. As parts within a manufacturing environment have become modular for quick repairs and a decreased downtime of the plant, or as for recently research reconfigurable manufacturing systems (RMS) [4], the task assigned to the Robosapien robot is ideal, as it represents the removal of a faulty module.

2. Communication and Software Interaction

The HP RX3715 Personal Digital Assistant (PDA) or pocket PC was used to relay commands from a computer to the Robosapien robot. The purpose for the use of the PDA as a relay station was that it allowed for Wireless network communication for the PC to interact with the designed software and having the ability to communicate with the Robosapian robot via infrared transmission. Further descriptions of the configuration is described later on in this section.

There were two main types of pocket PC software used for the Robosapien interaction. They were Nevo and dotPocket. There is a lot of remote control software available that enables one to control a device that has infrared capabilities. The remote control software that has been chosen to be used in this research and analysis is Nevo. There are others similar to Nevo such as Ultramote and VitoRemote. They all have the basic capabilities of controlling infrared devices.

Nevo offers complete audio visual, home and digital media control and is only available for users of mobile devices. It allows the user to seamlessly connect, control and interact with digital media and electronic devices in a network. Some interesting features that are available with Nevo are:

- The ability to control automation devices.
- It has an intuitive touch screen visual interface.
- It allows the user to customise the graphical interface.

The Nevo software was programmed to map with all of the existing functions of the Robosapien robot and additional tasks that enable the robot to perform a specific routine. This function is very useful as it allows the user to control the Robosapien and multiple devices as well.

Programs can also be written and executed on the PDA to control the robot. The software used needs to be specifically for embedded devices because the operating system of the PC is different from that of the PDA and some programs that are written for a PC do not work on the PDA.

The dotPocket software allows the user to control the pocket PC from a host computer. The screen of the pocket PC appears on the monitor of the host computer (as seen in figure 1) and the user is able to have full control of the pocket PC. A wireless connection is requires for a link to be formed between the PDA and the host computer. This tool is very useful as it makes it possible to control the pocket PC while it is in a different location within a set proximity.

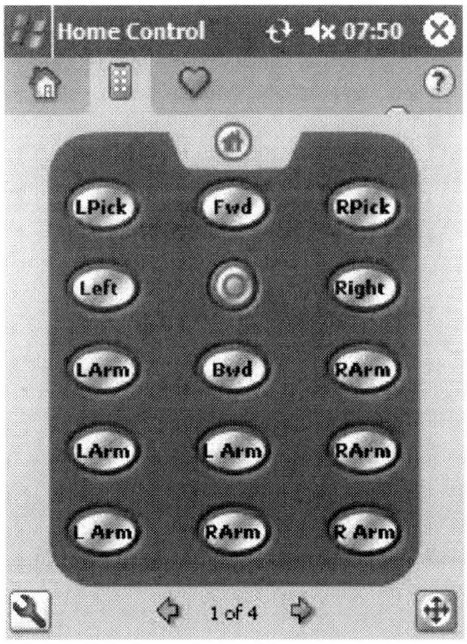

Figure 1. Picture of the PDA screen that is displayed on the host computer

A pocket PC can also be used to control the robot using Infrared. In this instance, the PDA acts as a remote but it can also be programmed to do a specific task, not limiting it to the pre-programmed functions. The PDA that is used supports Infrared and this feature was used to communicate with the Robosapien. The PDA supports slow Infrared which has a maximum transmission rate of 1.15 Mbps. With tests performed it indicated that the infrared could be used for short range transmission of 0.2 m.

The use of infrared requires low power therefore it was beneficial for use in the PDA. The use of the PDA was required it to be on the robot (Figure 2). This means that the PDA will be as mobile as the robot and will continually be in line of sight of the Infrared receiver that is situated on the robot. Therefore, commands that are sent to the robot from the PDA cannot be accidentally sent to other devices instead.

Although Infrared operates on the line of sight technique and it is usually troublesome to use in some controls, this was not a problem between the PDA and the robot as the PDA was attached to the back of the robot.

Figure 2. Robosapien with the PDA on its back

The wireless LAN (or WiFi) was used to communicate between the host computer and the PDA. The host computer has 802.11g wireless capabilities but is also compatible to 802.11b standards. The PDA has built in WiFi, and thus a wireless network can be set up between the PDA and the host computer. This wireless network enables the host computer to connect to the PDA and even share the same internet connection. This was very useful and was used to view the PDA screen on the host computer and give commands to be send to the Robosapien robot. An overview of the communication system between the Robosapien robot and the host computer is shown in figure 3.

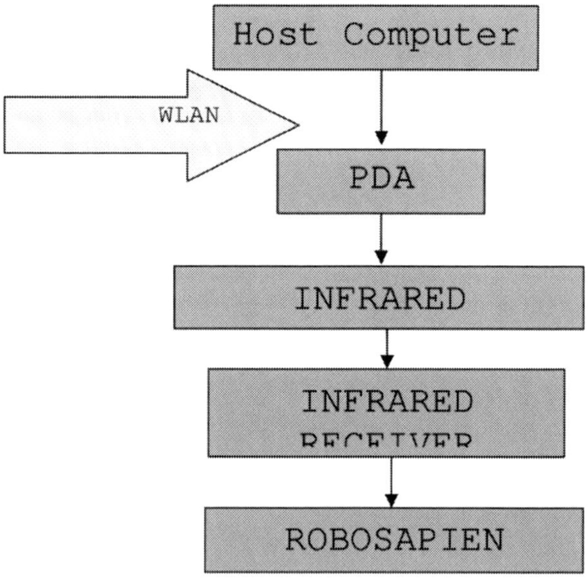

Figure 3. Flow diagram showing communication methods between the PC and Robot using the PDA

3. Communication and Software Interaction

The task that the Robospaien robot is to perform, involves picking up pipes from a table and inserting them in a subassembly, which could represent any part that is serviced or replaced in a manufacturing environment. The tasks are accomplished by two Robosapien robots. There are three pipes to be picked and inserted into a block with three holes. The lead robot will insert the first pipe in the first hole of the block, and then the slave robot will insert the second pipe in the second hole of the block. In order to model this task, it is important to determine the work envelope and all robot limitations relating to the task.

The robot's work envelope was computed with relation to the model. Figure 4 shows a rough sketch of robosapien with numbered joints indicated by asterisks.

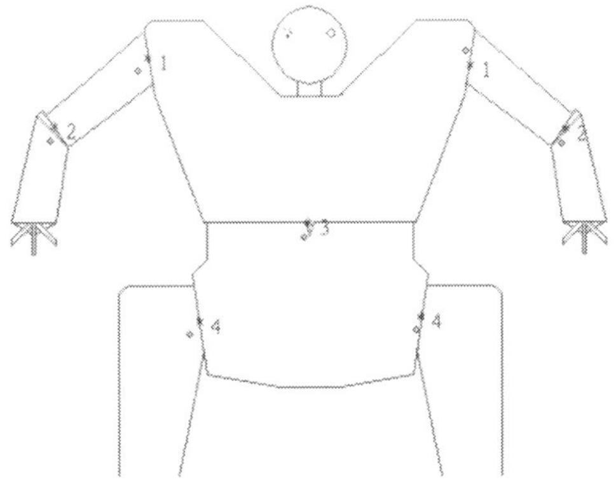

Figure 4. Rough 2D sketch of the *Robosapien robot and its joints*

3.1 Joints

Robosapien has rotational joints only; there are no direct translations that can be performed by the robot. The joints rotational angles numbered one to four in figure 4, were obtained by executing commands on the remote control and can be represented as follows in figure 5:

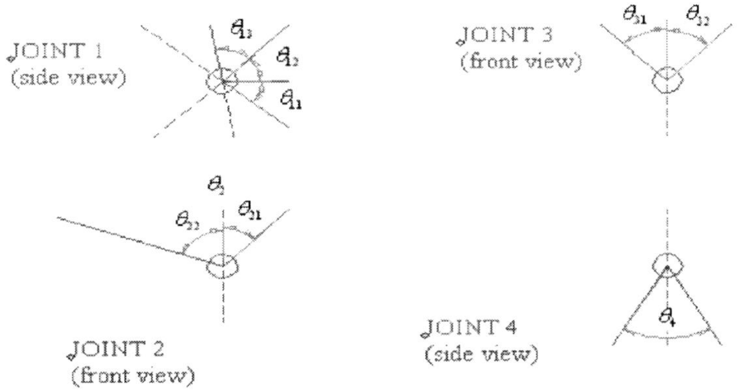

Figure 5. *Joints*

3.2 Object description

The pipes 30mm in diameter and 80 mm in length will be on top of a table 140 mm in height. The pipe is described by a coordinate system bearing a fixed relationship to it (see figure 6). Joint 3 is taken as a reference coordinate system. The center line of the pipe must be described by a vector 100**i**+210**j**, from the origin of the fixed coordinate system. The transforms are described in terms of direction cosines from the coordinate system axes.

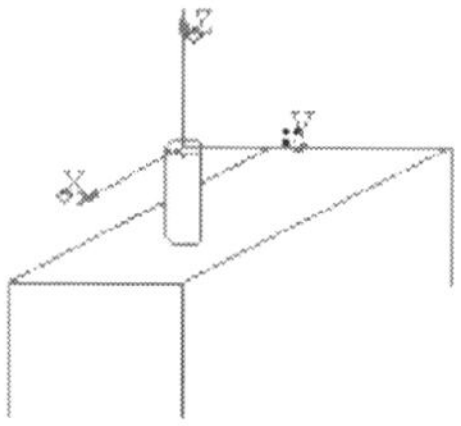

Figure 6. *Pipe definition*

A graphical representation of the pipe and its coordinate axis is as shown in figure 6. A specification of the position and orientation of the coordinate system is sufficient to reconstruct the object in any other position and orientation.

3.3 Task description

Transformations are used to describe a task. The task consists of the two robosapien robots picking up pipes from a table and inserting them into a subassembly. Figure 7 illustrates the concept of one robosapien robot ready to perform the task.

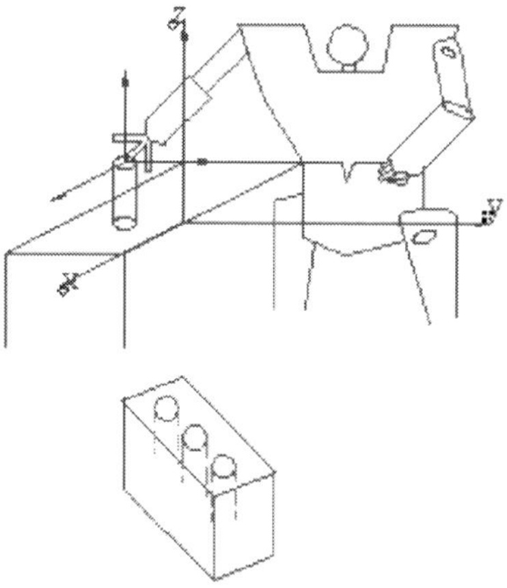

Figure 7. *Task*

By defining a series of the arm and end effector positions with subscript pn, the task can be described as a sequence of arm moves and actions. Referring to these positions, the task positions can be described as:

MOVE	p1	Approach the pipe
MOVE	p2	Move over the pipe
GRASP		Grasp the pipe
MOVE	p3	Lift it vertically

MOVE	p4	Approach hole at an angle
MOVE	p5	Stop on contact with the hole
MOVE	p6	Stand the pipe up
MOVE	p7	Insert the pipe
RELEASE		Let go of the pipe
MOVE	p8	Move away

The above description is very rigid, yet similar actions are performed with the removing and insertion of modules or parts within a manufacturing environment.

3.1.1. Strusture of the Task

The structure of the task can be firstly defined by defining the structure of the arm. The arm is defined by the product of three transformations such that the positions in the task description are replaced by

$$\text{MOVE}_{pn} = \text{MOVE } Z\, T_6\, E$$

where:

Z represents the position of the arm with respect to the base coordinate system;
T_6 represents the end of the arm with respect to joint 3;
E represents the end effector at the end of the arm.

With such a description, the calibration of the arm to the work station is represented by Z. The structure of the task is now represented in terms of the following transforms:

P	position of the pipe in base coordinate;
H	the position of the block with three holes;
$^{H}HR_i$	the position of the ith hole in the block with respect to the H coordinate system;
^{P}PG	the position of the gripper holding the pipe with respect to the pipe;
^{P}PA	the gripper approaching the pipe;
^{P}PD	the gripper departing with the pipe;
^{HR}PHA	the pipe approaching the ith hole;
^{HR}PCH	the pipe in contact with the hole;
^{HR}PAL	the pipe at the beginning of insertion;
^{HR}PN	the pipe inserted.

The task can now be represented as a series of transform equations solvable for T_6, the control arm input, as shown in figure 8.

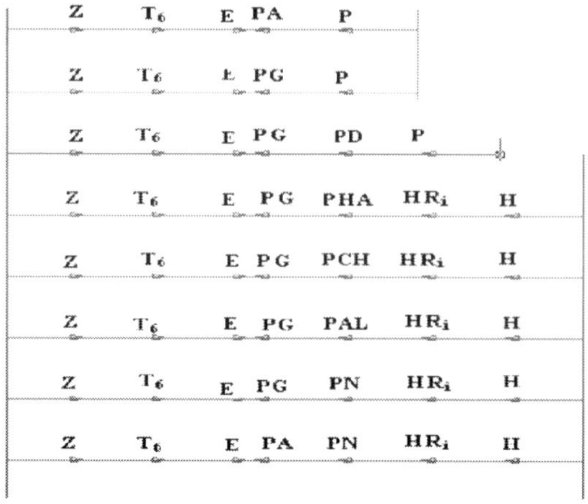

Figure 8. Task Position Transform Graph

The above represents the essential structure of the task, and each transformation represents a separate piece of information. In order to define the transforms a combination of definition and teaching-by-doing will be used. Some of the transforms are logically defined symbolically, such as HR_i, which may be obtained from engineering drawings.

It is important that transformations representing pure translation or rotation are defined as such, for this information can be used to speed up and simplify matrix multiplication during task execution. Other transforms are logically defined by moving the arm from position to position, obtaining T_6, and solving for required transforms. The coordinate frames corresponding to P, H, and Z are shown in figure 9. While the interrelationship of these three coordinate frames is fixed by the physical object, the location of the base coordinate system is arbitrary, and can be specified in terms of any of the three frames.

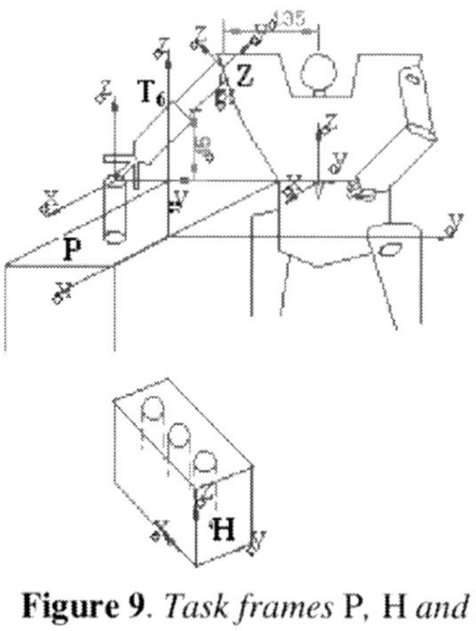

Figure 9. *Task frames* P, H *and* Z

The base frame will be specified in terms of Z. While the arm coordinate frame is located in the shoulder, it is appropriate for the base coordinate frame to be at joint 3 or at ${}^{T6}p_z = -45$. The arm cannot reach joint 3, so it will be set behind the origin of base coordinate such that ${}^{T6}p_x = 0$ and ${}^{T6}p_y = -135$.

$$\text{So } Z\begin{bmatrix} 1 & 0 & 0 & 0 \\ 0 & 1 & 0 & -135 \\ 0 & 0 & 1 & -45 \\ 0 & 0 & 0 & 1 \end{bmatrix} = I \tag{1}$$

where I is the identity transform, and thus

$$Z = \begin{bmatrix} 1 & 0 & 0 & 0 \\ 0 & 1 & 0 & 135 \\ 0 & 0 & 1 & 45 \\ 0 & 0 & 0 & 1 \end{bmatrix} \tag{2}$$

The end effector transformation is shown in figure 10.

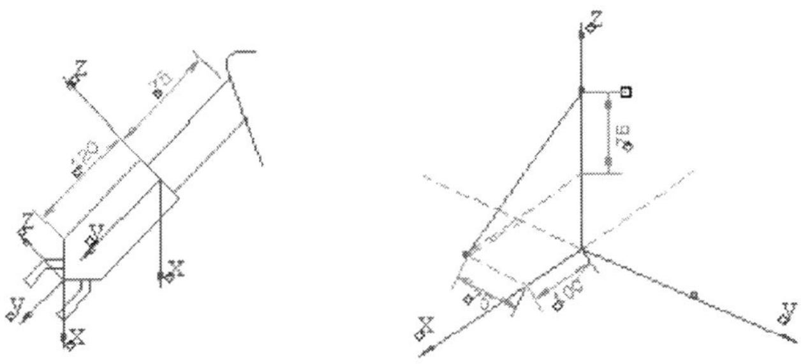

Figure 10. End effector transformation

The end effector is defined by a transform with respect to the end of the arm. Using the convention that the z axis of the end effector is in its direction of approach to a task and the y axis describes orientation, the end effector shown in figure 10 is described as:

$$^{T6}E = \begin{bmatrix} 1 & 0 & 0 & 100 \\ 0 & 1 & 0 & 75 \\ 0 & 0 & 1 & 75 \\ 0 & 0 & 0 & 1 \end{bmatrix} \tag{3}$$

While the pipe have been described in figure 7, a part with three holes is now looked at, H. At top elevation its features are defined by means of a transform array HR (see figure 11).

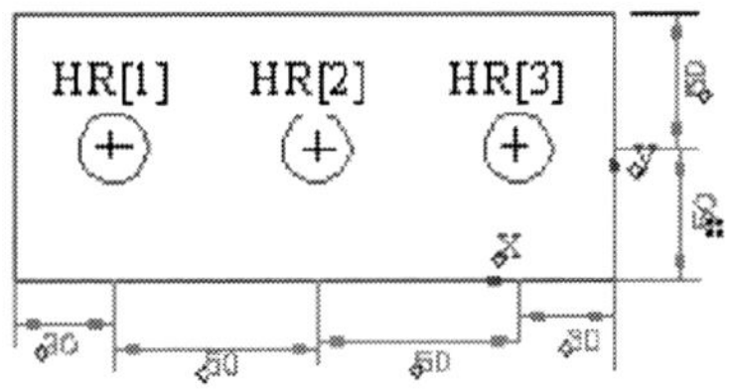

IOP Conf. Series: Materials Science and Engineering **65** (2014) 012022 doi:10.1088/1757-899X/65/1/012022

Figure 11. Block with three holes

$$HR_1 = \begin{bmatrix} 1 & 0 & 0 & 130 \\ 0 & 1 & 0 & 50 \\ 0 & 0 & 1 & 0 \\ 0 & 0 & 0 & 1 \end{bmatrix} \quad (4)$$

$$HR_2 = \begin{bmatrix} 1 & 0 & 0 & 80 \\ 0 & 1 & 0 & 50 \\ 0 & 0 & 1 & 0 \\ 0 & 0 & 0 & 1 \end{bmatrix} \quad (5)$$

$$HR_3 = \begin{bmatrix} 1 & 0 & 0 & 30 \\ 0 & 1 & 0 & 50 \\ 0 & 0 & 1 & 0 \\ 0 & 0 & 0 & 1 \end{bmatrix} \quad (6)$$

The most important transform is the pipe inserted in a hole (see figure 12). The z axis of the pipe must agree with the axis of the hole. The direction of x and y is arbitrary, as the pipe has cylindrical symmetry.

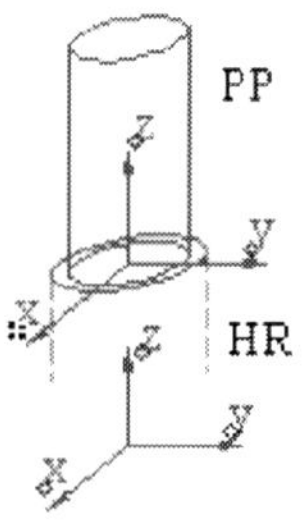

Figure 12. *Pipe inserted in Hole*

The gripper approaches and grips the pipe from its initial position by rotating position T_6 by θ_2 to the right about the y axis. The transform results:

$$^{P}PA = {}^{P}PG = \begin{bmatrix} \cos\theta_2 & 0 & \sin\theta_2 & 0 \\ 0 & 1 & 0 & 0 \\ -\sin\theta_2 & 0 & \cos\theta_2 & 0 \\ 0 & 0 & 0 & 1 \end{bmatrix} \quad (7)$$

The departure position of the gripper holding the pipe with respect to the pipe will be a result of rotating position T_6 by $-\theta_{22}$ about the y axis from the approach transform. The transform results:

$$^{P}PD = R(y,-\theta_{22})R(y,\theta_2) = \begin{bmatrix} f & 0 & -g & 0 \\ 0 & 1 & 0 & 0 \\ g & 0 & f & 0 \\ 0 & 0 & 0 & 1 \end{bmatrix} \tag{8}$$

To insert the pipe the arm is again rotated by θ_{11} upwards about the y axis of Z, and is again rotated by θ_{22} to the right about the y axis of position T_6. The transform with the pipe in the hole is given by

$$^{HR}PN = Rot(y,\theta_{22})Rot(y,\theta_{11})\,^{P}PD = \begin{bmatrix} m & 0 & -n & 0 \\ 0 & 1 & 0 & 0 \\ n & 0 & m & 0 \\ 0 & 0 & 0 & 1 \end{bmatrix} \tag{9}$$

The remaining transforms can now be determined by using the arm. The end effector is placed on the pipe at its pick up position and the following transform equation is true:

$$Z\,T_6\,E = P\,PG \tag{10}$$

which defines P:

$$P = Z\,T_6\,E\,PG^{-1} \tag{11}$$

It will be assumed that the approach position is the same as when the pipe is contact with the grippers during pick, so:

$$Z\,T_6\,E = P\,PA \tag{12}$$

A departure point relative to P is now defined by rotating the pipe in the gripper to the departure position PD, resulting:

$$Z\,T_6\,E = P\,PD\,PG \tag{13}$$

which defines PD

$$PD = P^{-1}\,Z\,T_6\,E\,PG^{-1} \tag{14}$$

The position of the block H is defined by:

$$Z\,T_6\,E = H\,HR_1\;PN\,PG \tag{15}$$

and solving for H

$$H = Z\,T_6\,E\,(HR_1\,PN\,PG)^{-1} \tag{16}$$

The pipe at the beginning of insertion is defined by:

$$Z\ T_6\ E = H\ HR_1\ PAL\ PG \tag{17}$$

and

$$PAL = (H\ HR_1)^{-1}\ Z\ T_6\ E\ PG^{-1} \tag{18}$$

The pipe on first contact with the PCH and an approach point PHA to the contact point are defined by

$$PCH = (H\ HR_1)^{-1}\ Z\ T_6\ E\ PG^{-1} \tag{19}$$

$$PHA = (H\ HR_1)^{-1}\ Z\ T_6\ E\ PG^{-1} \tag{20}$$

which completes the definition of all transforms.

For insertion of the second and third pipe the same program should be executed, the only thing that is going to change is the position of the hole from HR_1 to HR_2 and HR_3 respectively.

4. Results and Discussion

The Robosapien robot was programmed to perform a specific task of picking up a cylinder and inserting it into a hole. Due to the lack of motion (minimal degrees of freedom (DOF) with the different joints), the task were seldom performed accurately, even if it was possible to control Robosapien from the host computer to guide its motion.

The right end effector could only open to approximately 45 mm in diameter. So the object that had to be picked have to be less than 45 mm. The left end effector can only open to about 25 mm in diameter, so the object that had to be picked by the left hand had to be less than 25 mm.

The Robosapien robot can only pick objects from the side due to the limited DOF and can therefore not pick up objects from its front and from its back. The grippers could not close at the exact pick up point due to the limitation of friction on the actuators. The robot is limited to pick up objects that are very light in weight and which are not fragile, because it releases the objects and does not put them down gently. The weight of the object affects the stability or the balance of the robot.

Even the above mentioned limitations of the Robosapien robot, it could be assumed that these limitations would not occur with well designed and developed humanoid robots. To reflect back to the question that was indicated in the introduction: Would it be feasible for a company to consider a humanoid robot and what other modifications and costs might be associated with the new technology?

Humanoid robots for manufacturing environments will be pre-programmed to control the manipulators. Feedback vision system will also be onboard of the humanoid robot for guidance and navigation to certain areas, but more importantly to identify objects and units that are to be maintained and serviced. These units could be unique for each manufacturing environment, as each company could prefer to use specific makes and models which could have different configurations. Unless the manufacture of the humanoid incorporates all models and makes within the onboard database, each company will need to train the humanoid robot to identify the units and the modules that will be required to be replaced. The training could consist of uploading of photos of different angles the units onto the humanoid robot. The above description will allow for generic humanoid robots to be built and be adapted to different manufacturing environments.

As per the research performed on the Robosapien robot, there was obviously an onboard communication system and a wireless network communication system that was required. The onboard communication system would be the responsibility of the mechatronics integration performed by the manufactures of the humanoid robot. The wireless network communication system would be the company's responsibility and access points will be required to be installed at short distances from each other, so that their ranges overlap and allow coverage throughout the manufacturing environment. Similar protocols as the Robotics Communication Protocol (RCP) [5] could be used to transfer instructions and commands between the access points, machines, robots and humanoid robot. Cellular networks could be an option o communication, but there is a prone for the loss of signal due to metallic structures and building material.

A vision system will also need to be placed within the manufacturing environment, to allow the server to use image processing techniques to identify the humanoid robot's location. Global Positioning System (GPS) would not be a possible location method, as the satellite signal will not be received through the metal and building material. The accuracy of GPS is also decreased if the satellite signals are bounced about the building structure.

The standard localization method used within the plant will also need to be installed for other robotic platforms and machinery. The localization system incorporated with the wireless system and any other sensory system located within the manufacturing environment will be required to be controlled by the same server, or at least a network of servers that would be able to communicate with each other different types of information. The server would be able to, in turn, send instructions to the humanoid robot of different tasks that have to be executed, similar to those performed form the host computer to the Robosapien robot. The instruction set would be dependent on the level of autonomy and intelligence of the humanoid robot to perform tasks. The instruction set could be as complex as: LOCATE unit XYZ, MOVE to unit XYZ, OPEN panel, DETECT module xyz, REMOVE module xyz, THROW removed module xyz into trash, INSERT new module xyz into unit, CLOSE panel, AWAIT for new instruction. Alternatively, for more advanced humanoid systems, the instruction set could be as simple as: LOCATE unit XYZ, REPLACE module xyz, AWAIT for new instruction.

5. Conclusion

A humanoid robot will only be useful in a Smart Manufacturing and Industrial environment. Companies that have already been established as a Smart Environments would benefit from the humanoid robot, as the only cost to the company would be that of the humanoid robot. Unless a company has budgeted to change into a Smart Environment for other robotic and sensory systems, the company will have to consider carefully whether it is beneficial and efficient to introduce a humanoid robot to their manufacturing environment. The reason for this statement is that the company will have to install a sensory system and a wireless network within the manufacturing environment before the humanoid robot can be operated. The cost of the systems that have to be installed to make the plant a Smart Environment, could be unexpected and high.

Saving human lives and the prevention of people involved in accidents are of high importance within a manufacturing environment. Due to this, the question is therefore raised: How much money is too much money to replace people with the involvement of dangerous environments and accidents? The answer to this problem is that it might be cheaper for the company to purchase multiple cheaper maintenance robots that is only able to perform a single task than a humanoid robot. In this way humans are saved from being less involved in manufacturing accidents.

As manufacturing environments evolve and improve, companies will be forced to start implementing Smart Environments. It will take time before these Smart Environments are established to introduce humanoid robots. Once these Smart Environments are established, it will definitely be beneficial for companies to purchase humanoid robots that will assist with dangerous tasks and maintenance within the manufacturing environment.

[1] Aderhold J, Davydov V Yu, Fedler F, Klausing H, Mistele D, Rotter T, Semchinova O, Stemmer J and Graul J 2001 *J. Cryst. Growth* **222** 701

6. References

[1] Järvinen, J. & Karwowski, W. 1995 Analysis of self-reported accidents attributed to advanced manufacturing systems, *International Journal of Human Factors in Manufacturing*, Wiley Publishers **5** Issue 3, 251 – 266.

[2] Asimo by Honda, The World's most Advanced Humanoid Robot, http://asimo.honda.com [accessed 18/01/2013].

[3] NASA – Robonaut 2, http://www.nasa.gov/mission_pages/station/main/robonaut.html [accessed 18/01/2013].

[4] Padayachee, J., Bright, G. & Masekamela, I. 2009 Modular Reconfigurable Machine Tools: Design, Control and Evaluation, *South African Journal of Industrial Engineering* **20**, No 2, 127 – 143

[5] Stopforth, R., Bright, G., Davrajh, S., Walker, A. 2011 Improved Communication between Manufacturing robots, *South African Journal of Industrial Engineers* **22**, Issue 1, 99 - 107

IOP Conf. Series: Materials Science and Engineering **65** (2014) 012023 doi:10.1088/1757-899X/65/1/012023

Design and Development of an Automatic Tool Changer for an Articulated Robot Arm

H Ambrosio[1] M Karamanoglu[2]

[1]Designer in Residence, Department of Design Engineering and Mathematics, School of Science and Technology, Middlesex University, London, UK
[2] Head of the Deparment Design Engineering and Mathematics, School of Science and Technology, Middlesex University, London, UK

Email: h.ambrosio@mdx.ac.uk

Abstract. In the creative industries, the length of time between the ideation stage and the making of physical objects is decreasing due to the use of CAD/CAM systems and adicitive manufacturing. Natural anisotropic materials, such as solid wood can also be transformed using CAD/CAM systems, but only with subtractive processes such as machining with CNC routers. Whilst some 3 axis CNC routing machines are affordable to buy and widely available, more flexible 5 axis routing machines still present themselves as a too big investment for small companies. Small refurbished articulated robots can be a cheaper alternative but they require a light end-effector. This paper presents a new lightweight tool changer that converts a small 3kg payload 6 DOF robot into a robot apprentice able to machine wood and similar soft materials.

1. Introduction

In manufacturing, the length of time between the ideation stage and the making of physical objects is decreasing due to the use of CAD/CAM systems. These systems allow small creative industries and workshops to use flexible manufacturing and just-in-time (JIT) production and as a result manufacture products with a price point able to compete with similar products manufactured by larger companies that benefit from economy of scale on pricing. In East London we can find a successful example of a furniture shop that produces solid wood furniture on demand using a CAD/CAM large router and a computer integrated manufacturing system (CIM) [1]. The company, Unto This Last, does not need to stock any products, offers a degree of product customisation to its clients and is able to compete in price with much larger furniture retailers.

Materials suited for fused deposition modelling, such as polymers, a new generation of affordable additive manufacturing machines is providing desktop manufacturing for small creative industries.

The last year also saw the emergence of composite wood plastic compounds being created for additive manufacturing.
Natural anisotropic materials, such as solid wood can also be transformed using CAD/CAM systems, but only with subtractive processes such as machining with CNC routers. Whilst some 3 axis CNC routing machines are affordable to buy and widely available, more flexible 5 axis routing machines are still as a too big investment for small companies. Small refurbished articulated robots can be a cheaper alternative but they require a light end-effector.

Examples of articulated robots used in routing of wood materials are increasingly found in industry [2] (Fig.1).

Figure 1 - Robot at work in Louis Mackall's Breakfast Woodworks cabinet making company [3]

Although articulated robots present less rigidity than CNC machines that can result in inaccurate machining [2], the accuracy is measured in hundreds of a millimetre and therefore suited for wood routing [4]. Articulated robots present increased machining flexibility when compared CNC machines because of the extended work surface by use of tracks [5].

1.1 Tool changers for robot material removal operations

Industry examples used in routing found during this research use large payload robots that borrow motors, spindle and automatic tool changers for the CNC routing industry (Fig.2). Automatic tool holders for CNC routing industry use a drawbar mechanism that clamps tool holders to a spindle. For this type of end effectors the lightest tool changer found weighed 3.5 kg and this is excluding any motor.

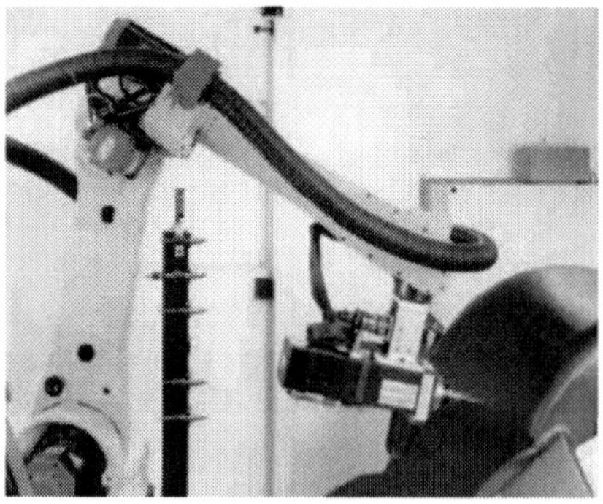

Figure 2 – Robot Trim RT-400 with V-taper automatic tool changer [6]

In mechanical process operations requiring less cutting forces than routing, such as deburring and polishing it is possible to find in the market tool changers that change the spindle motor and tool

(Fig.3). This option applied to this specific project of routing would mean a different motor for each routing tool on the tool changer and that would be a costly option.

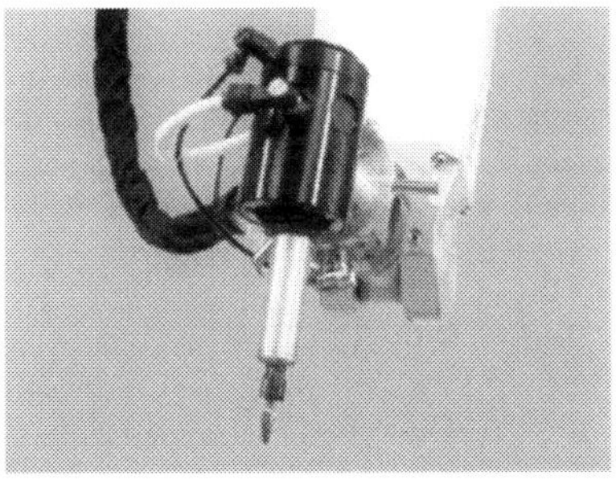

Figure 3 - Application of robotic arm in deburring of metal and plastic moulded pieces [7]

This paper presents a new lightweight tool changer that converts a repurposed 3kg small payload 6 DOF articulated robot into a robot apprentice able to machine wood and similar soft materials.

2. Method

For the design and development of this automatic tool changer presented in this paper the following method was used. A literature review was made in the use of robots in industry to understand the context and review existing patterns in technical specifications of robotic applications similar to the one in this paper. A second literature review examined existing tool holding and tool changing devices, both manual and automatic. This served to provide an image of what is standard in the industry and understand the structure of the mechanism this thesis proposes to develop.

The literature review was followed by the analysis of wood cutting forces (the main material the robot would be asked to work with) and calculations of torque, speed and power parameters to create a clear path for motor research. Following the technical knowledge provided by the above analysis, it was possible to research spindle motor with the best weight, torque speed and price ration to use as motor around which the tool changer was designed.

Following the choice and acquisition of a spindle motor, nine design concepts under the form of sketches were produced, inspired by the research and observation of existing solutions. This phase was followed by a scrutiny of each concept using the weighing chart. A 3D CAD model provided the first detailed outline with the necessary dimensions and visual representation to proceed with the mathematical analysis and finite element analysis of some of the forces that would be affecting the mechanism. A prototype was developed to test the design concept on its ability to engage and clamp a tool holder both when the motor was off and on.

3. The robot routing station

3.1. The robot

The light tool changer was designed for a Thermo F3 [8] desktop articulated robot manipulator with 6 degrees of freedom (DOF) and maximum payload of 3kg or 29.4N mounted on a 3 meter linear track. This implies that the spindle motor, tool changer, tool holder and end-effector adaptor can have, all together a maximum 3kg in weight. This robot is aimed at the pharmaceutical industry and due to being on the lightest payload of articulated robot arms, it is also on the most affordable price range, making it a great example of robot apprentice for small creative companies manufacturing that don't have a large budget. (Fig. 4)

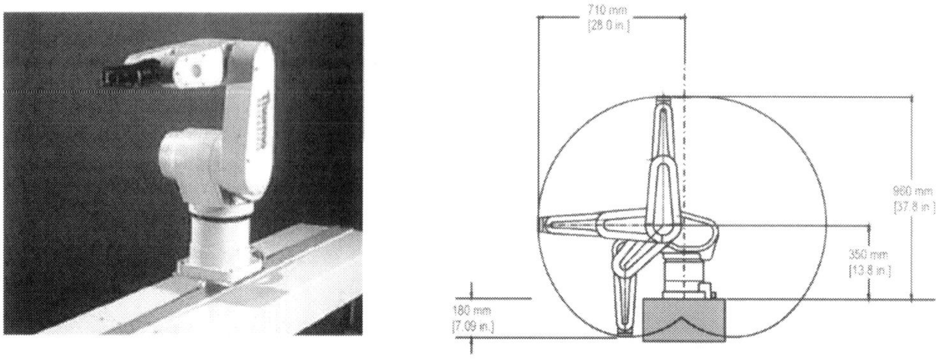

Figure 4 – Thermo F3 robot mounted on track and its work envelope [8]

3.2. Motor selection

Motor systems for machining are composed of two parts: a motor and a spindle with the tool. These two parts connect through belts, pulleys and gears or by direct flange. Belt and pulley systems would interfere in the robot's movement. A routing motor was chosen from a selection or electric and pneumatic motors suitable for routing. Hydraulic motor systems were excluded as they are, per definition, heavy and more expensive than pneumatic and electric motors [9]. The chosen motor was the Unimatic direct drive electric motor suitable for wood routing. The Unimatic motor has an output power of 0.35Kw and on-load speed of up 24,100 RPM and an on-load torque of 0.14Nm. Since the maximum weight allowed for the end effector was 3kg, and motor weighs 1.4kg, the automatic tool changer and the robot end of arm adaptor could have a maximum weight of 1.6kg.

4. Design of the automatic tool changer

The automatic tool changer design presented is a spring loaded clamping mechanism that needs supplied air pressure only when the end effector collects and returns the tool holder to the tool magazine. Each routing tool would be placed in a tool holder and several tool holders can be stored in a tool magazine. The robot arm would select the adequate tool from the tool magazine, grip the tool, perform the routing task, return tool to tool holder and collect the next tool [1].

The spring loaded mechanism had to hold the weight of the tool holder. Thrust bearings located on the tool holder absorb friction, centrifugal forces and cutting forces that were calculated with mathematical models. The multijaw coupling has to cope with torque generated by cutting forces involved in wood routing.

4.1. Self-centring mechanism selection

4.1.1. Forces involved in wood cutting Cutting forces in wood were calculated using Eyma [10] simplified formula for calculating wood cutting forces: $Fc = F1 \times b \times Kh \times Km$, where F1 is related to direction of cut and wood chip = 44.74 N, Kh is the moisture coefficient = 0.12, b is the tool width (diameter) in mm and Km is the material coefficient given by the formula: $Km = (5.73 \times 10{-}5 \times Ec/SG) + (2.57 \times 10{-}5 \times Pf,I)$. For the Km formula Pf,I is coefficient of determination of 0.86 between the Ec or modulus of elasticity in Mpa and and SG is the specific gravity.

As a reference the cutting forces were calculated (Table1) for high density wood Boco (Bocoa prouacensis Aubl.) [10] as this a hardwood that would produce higher cutting forces then softwood.

Boco (Bocoa prouacensis Aubl.) properties	**Boco wood material coefficient**	**Boco wood cutting forces**
F1 = 44.74N Ec = 9940 MPa SG = 1.109	Kmboco = (5.73 × 10–5 ×9940/1.109) + (2.57 × 10–5 × 0.86) → Kmboco = 1.0963 N	Fcboco = 44.74 x b x 0.12 x 1.0963 → Fcboco = b x 5.89 where b is tool width in mm.

Table 1. Boco wood cutting forces [10]

The Fcboco formula was used to calculate the necessary cutting forces for different tool diameters and determine the range of tools the Unimatic motor could cope with to cut Boco hardwood (Table 2)

Tool diameter		**Tool radius**	**Cutting force needed**	**Torque**	**Preal using** a)	**Spindle speed**
in	mm	m	N	Nm	kW	RPM
1/4"	6.3500	0.0032	37	0.12	0.230	12000
3/16"	4.7625	0.0024	28	0.07	0.176	16040
1/8"	3.1750	0.0016	19	0.03	0.120	24060

Table 2. Range of tools the Unimatic motor can cope with for cutting Boco hardwood

4.1.2. Multijaw coupling The self-centring element was achieved using R3208 stainless steel DIN10718 multi-jaw coupling from Automotion [11]. The multijaw coupling can hold a maximum torque of 7.8Nm which is sufficient to cope with the largest cutting tool the Unimatic motor can cope with. (Table2).

5. Design Validation

To test the design a prototype was made. The automatic tool changer prototype weighs 0.345kg and the tool holder weighs 0.176kg.

The prototype was attached to the motor and placed onto a pneumatic rig that has tested the clamping mechanism ability to clamp and release. The clamping mechanism was then loaded with weights up to 120kg

After this first test the mechanism was mounted onto the robot arm (Fig.5) and the robot programed to perform autonomously the task of collecting a tool from the tool magazine, move away from the tool magazine, return a tool the magazine and collect another tool from the tool magazine.

Finally the design was tested with the motor on without cutting loads at its lowest RPM – 10,000.

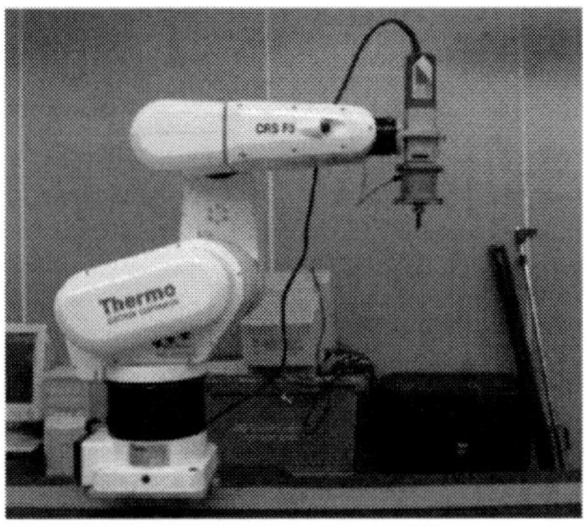

Figure 5 – Automatic tool changer with motor and tool holder as the new end-effector for the F3 robot

6. Results and discussion of the tool changer design

The automatic tool changer clamping mechanism successfully coped with the 120kg without mechanical deterioration of its components. In addition, the robot was able to easily perform the task of collecting, returning and changing tools meaning the self-centring and the clamping mechanisms are both successful.

Some issues with overheating arisen when turning the routing motor. The overheating is considered to be due to an incorrect choice of thrust bearings or imperfections on the mechanical attachment between bearings and tool holder. The next action to be taken is to develop this design iteration to a point where it has solved a number of performance issues and is ready for prototyping.

In addition, total weight of the tool changer and tool holder at this prototyping stage is 0.521kg. With the motor being 1.4kg it leaves 1.9kg of payload. This means that there is the possibility of attaching the automatic tool changer to a larger and more powerful motor of up to 2.4kg weight. This would mean the possibility of increasing the maximum tool diameter for hardwood cutting and therefore increasing the scope of the design.

7. Conclusions and future work

The use of affordable second hand robots in just in time manufacturing with CAD/CAM will allow small companies to compete with larger manufacturers of wood products that benefit from economy of scale. It will also allow these small manufacturers to customise designs and therefore offer added value to the products they sell, creating new business opportunities.

The goal of this work was to design and develop an end-effector with an automatic tool changing mechanism for an articulated robot arm and expand the manufacturing capabilities of small creative businesses.

Through the first part of the validation of this new automatic tool changer mechanism we proved the clamping and self-centring design proposal works well.

Further work is being carried to select and possibly co-design appropriate bearings that will cope with wood routing spindle speeds and the forces generated by the this wood machining process so that this automatic tool changer can be used in industry in the near future.

References

[1] Rehg J.A 2003*Introduction to robotics in CIM systems*(Prentice Hall)
[2] Hardin W 2006 *Robot CNCs: The Path from Tending to Machining* (Embedded)
[3] Aitken R 2007 *The Ecstasy of Possibility: An Interview with Louis Mackall* (Fine Homebuilding)
[4] Papp T and Kovacs Z 2012 *Development of Process Capability Based Fits andTolerances for the Enhancement of Productivity in the Secondary Wood Processing* International Scientific Conference on Sustainable Development & Ecological Footprint
[5] Kim C and Chung J.H 2006 *Robust coordination control of a pneumatic deburring tool* J.Robot.Syst. **22**: S1-S13
[6] 2006 *KMT Cutting Systems launches new tool changer for robotic routers*, Industrial Robot-an International Journal **33** pp 236-237.
[7] ATI http://www.ati-ia.com
[8] 2004 Thermo Electron Corporation. 2004, CRS F3 Robot System User Guide, UMI-F3-400
[9] Asfahl C.R. *Robots and manufacturing automation (*Wiley)
[10] Eyma F, Méausoone P, Martin P. 2004*Strains and cutting forces involved in the solid wood rotating cutting process* JMater.Process.Technol. **148** issue pp220 – 225
[11] Automotion http://www.automotioncomponents.co.uk/en/catalog/rotary/rigid-couplings/rigid-shaft-couplings/r3208

Interval Analysis Approach to Prototype the Robust Control of the Laboratory Overhead Crane

J Smoczek[1] J Szpytko[1] and P Hyla[1]

[1]AGH University of Science and Technology,
Faculty of Mechanical Engineering and Robotics
Mickiewicza Av. 30, PL 30-059 Krakow, Poland

E-mail: szpytko@agh.edu.pl

Abstract. The paper describes the software-hardware equipment and control-measurement solutions elaborated to prototype the laboratory scaled overhead crane control system. The novelty approach to crane dynamic system modelling and fuzzy robust control scheme design is presented. The iterative procedure for designing a fuzzy scheduling control scheme is developed based on the interval analysis of discrete-time closed-loop system characteristic polynomial coefficients in the presence of rope length and mass of a payload variation to select the minimum set of operating points corresponding to the midpoints of membership functions at which the linear controllers are determined through desired poles assignment. The experimental results obtained on the laboratory stand are presented.

Keywords: overhead crane, control system prototyping, fuzzy scheduling, interval arithmetic

1. Introduction

The automation of cranes operations is very important owing to necessity of ensuring the safety and efficiency of the transportation process, that is involved by requirements of enhancing the productivity of manufacturing processes. Those requirements motivate scientists and engineers to develop and implement solutions for crane control system which face up to the following problem: transfer a payload as fast as possible from point to point with avoidance of collision with obstacles, and precise positioning at a final point with reduction of sway of a payload suspended at the end of a rope. The different control schemes, developed for example based on input and command shaping [1], optimal control theory [2, 3], sliding mode control [4], and robust feedback controllers [5] have been addressed to the crane control problem for decades. However both, open- and closed-loop methods cause some well known problems in implementation. The open loop anti-sway crane control are sensitive to disturbances owing to lack of sway angle of a payload feedback. There is not always the possibility to build a reliable sway angle measurement system using encoders, thus, some works report the Hall-effect magnetic sensors [6-7], machine visions [8-10] and stereovision-based systems [11].

Furthermore, the soft computing techniques are widely employed to the considered problem. The fuzzy logic is used to create linguistic-rule-based fuzzy controllers [12], PID gains tuning [13-14] and sliding mode control [13]. Membership functions tuning techniques have been developed based on an inverse dynamic [15], gradient algorithm [16], genetic algorithm [17], fuzzy clustering methods [18], and through applying artificial neural network [19].

The proposed in the literature methods are mostly applied for only membership functions tuning for a fixed number of fuzzy rules, and the robustness to mass of a payload is frequently neglected. Thus, the paper presents the novelty approach to crane dynamic system modelling and prototyping the robust control scheme with fuzzy interpolation of linear controller parameters in the presence of varying rope length and mass of a payload. The control system design is based on the a discrete-time crane dynamic model and interval analysis of closed-loop system characteristic polynomial coefficients. The iterative procedure for membership functions tuning, and creating the complete and coherent rule base is described. The presented method was used to prototype the control system for the laboratory scaled overhead crane. The fuzzy control scheme was implemented using structured text on the PAC system with RX3i controller. The control-measurement solutions, realized on the laboratory stand are described: contact and contactless techniques of sway angle of a payload sensing, including a dynamic vision system employed to detect the sway angle of a rope, and a single camera based stereovision system used for identifying a map of the crane workspace and planning the non-collision trajectory of a payload. The rest of this paper is organized as follows. Section two describes the method of designing an anti-sway fuzzy control scheme. The software-hardware equipment and control-measurement solutions of laboratory stand, and experimental results are presented in section three. Section four delivers the final conclusions.

2. Fuzzy scheduling control scheme design

2.1. *Poles assignment-based fuzzy scheduling scheme design*

The model of a planar crane dynamic system (Figure 1) is assumed in the form first- and second-order discrete-time transfer functions representing the relations between crane speed and input function (1), and sway angle of a payload and crane speed (2), where a_0, a_1, b_0, b_1, c_0, and d_0 are the parameters which vary in relation to the rope length l and mass of a payload m.

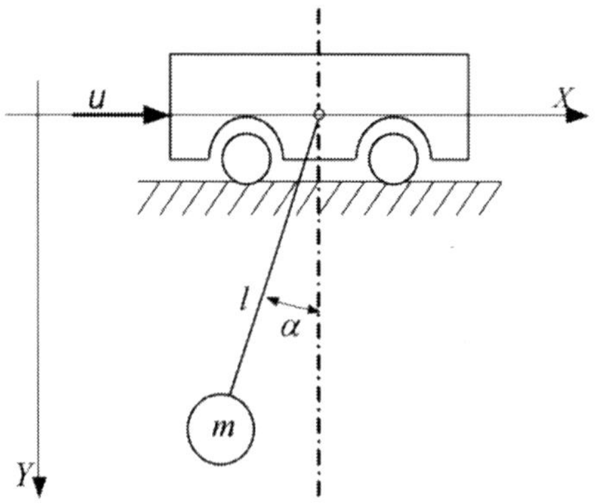

Figure 1. Planar crane system.

$$\frac{\dot{X}(z)}{U(z)} = \frac{d_0}{z + c_0} \tag{1}$$

$$\frac{\alpha(z)}{\dot{X}(z)} = \frac{b_1 z + b_0}{z^2 + a_1 z + a_0} \tag{2}$$

The control system consists of two proportional controllers k_1 and k_2 for crane position and speed, and first-order discrete-time controller of payload sway angle with coefficients denoted q_0, q_1 and s_0

(Figure 2). The closed-loop control system transfer function can be presented as (3), where $\mathbf{z} = [z^4, z^3, z^2, z^1, z^0]$, **R** is the vector of controllers parameters and their multiplications (4), **S** is the matrix consisting of crane model's parameters (5), and T_s is a sample time.

$$\frac{\alpha(z)}{X_r(z)} = \frac{k_1 k_2 d_0 \left(b_1 z^2 + (b_1 s_0 + b_0)z + b_0 s_0\right)}{z^5 + \mathrm{zSR}} \tag{3}$$

$$\mathrm{R} = \left[1, k_2, q_0, q_1, s_0, k_1 k_2, k_2 s_0, k_1 k_2 s_0\right]^T \tag{4}$$

$$\mathrm{S} = \begin{bmatrix} a_1 + c_0 - 1 & \begin{pmatrix} a_0 - a_1 + \\ c_0(a_1 - 1) \end{pmatrix} & \begin{pmatrix} c_0(a_0 - a_1) \\ - a_0 \end{pmatrix} & -a_0 c_0 & 0 \\ d_0 & d_0(a_1 - 1) & d_0(a_0 - a_1) & -d_0 a_0 & 0 \\ 0 & 0 & -d_0 b_1 & d_0(b_1 - b_0) & -d_0 b_0 \\ 0 & -d_0 b_1 & d_0(b_1 - b_0) & d_0 b_0 & 0 \\ 1 & a_1 + c_0 - 1 & \begin{pmatrix} a_0 - a_1 + \\ c_0(a_1 - 1) \end{pmatrix} & \begin{pmatrix} c_0(a_0 - a_1) \\ - a_0 \end{pmatrix} & -a_0 c_0 \\ T_s d_0 & T_s d_0 a_1 & T_s d_0 a_0 & 0 & 0 \\ 0 & d_0 & d_0(a_1 - 1) & d_0(a_0 - a_1) & d_0 a_0 \\ 0 & T_s d_0 & T_s d_0 a_1 & T_s d_0 a_0 & 0 \end{bmatrix}^T \tag{5}$$

By assuming the stable poles z_f for closed-loop control system we obtain the Diophantine equation (6), where $\mathbf{p} = [p_4, p_3, p_2, p_1, p_0]^T$ is the vector of coefficients of desired characteristic equation of closed-loop transfer function.

$$z^5 + \mathrm{zSR} = \prod_{f=1}^{5}\left(z - z_f\right) = z^5 + \mathrm{zp} \tag{6}$$

Therefore, the controllers parameters can be derived from the equations system expressed as follows:

$$\mathrm{SR} = \mathrm{p} \tag{7}$$

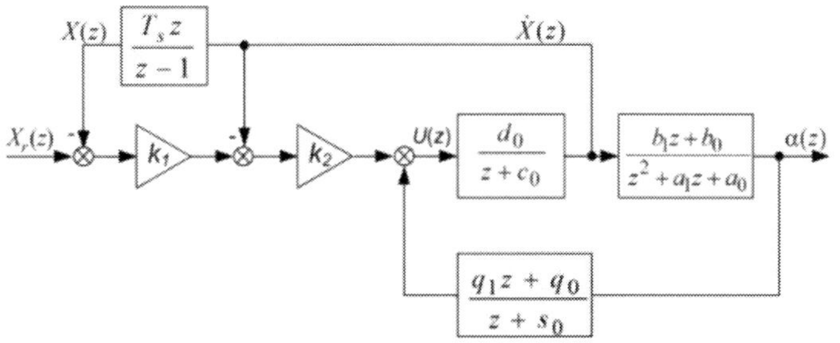

Figure 2. Discrete-time crane control system.

Assuming, that parameters of a model and coefficients of a desired closed-loop characteristic equation are the real numbers, the equation (7) has at least one vector of real roots among three possible solutions. The fuzzy interpolation control scheme can be expressed as a set of controllers determined at N operating points selected within the ranges of scheduling variables $[l_{min}, l_{max}]$ and $[m_{min}, m_{max}]$. Hence, the rule base is a set of N rules expressed as follows:

$$R_k\text{: IF } l \text{ is } \mathbf{A}_i \text{ and } m \text{ is } \mathbf{B}_j \text{ THEN } \mathbf{K}_k = \mathbf{K}(l_i, m_j) \tag{8}$$

where $k = 1, 2, ..., N$, $\mathbf{K}_k = [k_1, k_2, q_0, q_1, s_0]^T$ is the rule output, and $\mathbf{K}(l_i, m_j)$ is the vector of controllers parameters determined at operating point (l_i, m_j) corresponding to the midpoints of membership functions of fuzzy sets $\mathbf{A}_i$ and $\mathbf{B}_j$.

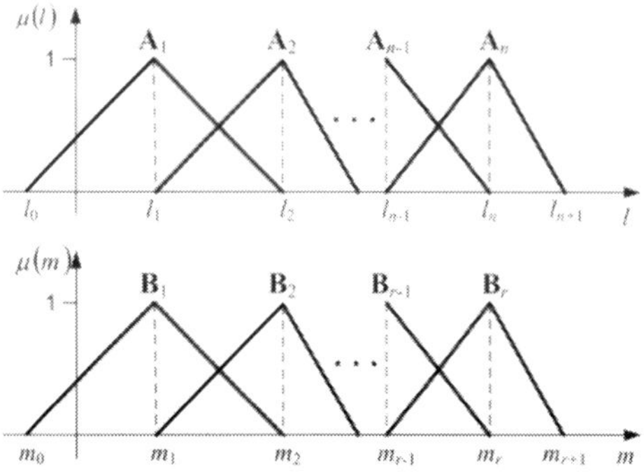

Figure 3. Triangular membership functions specified for rope length and mass of a payload.

The fuzzy scheduler design is based on the following assumptions. The n and r fuzzy sets **A** and **B** correspond to the triangular-shaped (Figure 3) membership functions (9-10) with centre points $[l_1, l_2, ..., l_n]$ and $[m_1, m_2, ..., m_r]$ distributed within the intervals $[l_{min}, l_{max}]$ and $[m_{min}, m_{max}]$ of scheduling variables.

$$\mu_{A_i}(l) = \max\left(\min\left(\frac{l - l_{i-1}}{l_i - l_{i-1}}, \frac{l_{i+1} - l}{l_{i+1} - l_i}\right), 0\right) \tag{9}$$

where: $l_{i-1} \le l_i \le l_{i+1}$, $i = 1, 2, n$.

$$\mu_{B_j}(l) = \max\left(\min\left(\frac{m - m_{j-1}}{m_j - m_{j-1}}, \frac{m_{j+1} - m}{m_{j+1} - m_j}\right), 0\right) \tag{10}$$

where: $m_{j-1} \le m_j \le m_{j+1}$, $j = 1, 2, ..., r$.

The number of rules $N = n \cdot r$ depends on all combinations of fuzzy sets $\{\mathbf{A}_i, \mathbf{B}_j\}$ specified for scheduling variables l and m. The output of fuzzy scheduler is a vector $\mathbf{K}$ (l,m) of controllers parameters interpolated for $l \in [l_{\min}, l_{\max}]$ and $m \in [m_{\min}, m_{\max}]$:

$$\mathrm{K}(l,m) = \sum_{k=1}^{N} w_k \cdot \mathrm{K}_k \tag{11}$$

where w_k is a weight of k-rule calculated as follows:

$$w_k = \mu_{A_i}(l) \cdot \mu_{B_j}(m) \tag{12}$$

For the above assumptions, the fuzzy controller design involves to select the number of membership functions specified for scheduling variables and determine distribution of their center points within the assumed intervals.

2.2 Interval arithmetic-based design of a fuzzy scheduler

The requirements of the control performances can be represented by the interval vector of characteristic equation coefficients **P** corresponding to the desired region of closed-loop poles defined as the intervals:

$$\left[z_f^-, z_f^+\right] = \left\{z_f \in \Re \mid z_f^- \le z_f \le z_f^+\right\} \tag{13}$$

where: $f = 1, 2, ..., 5$.

Thus, the desired closed-loop system characteristic polynomial can be presented as follows:

$$P(z)=\prod_{f=1}^{5}\left(z-\left[z_f^-,z_f^+\right]\right)=z^5+\mathrm{zP} \tag{14}$$

where **P** is the vector of closed-loop characteristic polynomial coefficients intervals:

$$\mathrm{P}=\left[\left[p_4^-,p_4^+\right]\left[p_3^-,p_3^+\right]\ldots,\left[p_0^-,p_0^+\right]\right]^T \tag{15}$$

The control quality requirements for parameters varying within intervals $[l_{min}, l_{max}]$ and $[m_{min}, m_{max}]$ are satisfied if the closed-loop system characteristic polynomial coefficients lie within desired interval vector **P**:

$$\mathrm{S}(l,m)\cdot\mathrm{R}(l,m)\in\mathrm{P} \tag{16}$$

where the vector $\mathbf{R}(l, m)$ consists of controller gains $\mathbf{K}(l,m)$ interpolated by fuzzy inference system according to the formula (7). In practice, it is enough to check the condition (16) for the most hazardous operating points which correspond to the crossover points of membership functions:

$$l_c=\left(l_i+l_{i+1}\right)/2\ ,\ m_c=\left(m_j+m_{j+1}\right)/2 \tag{17}$$

The procedure of designing the fuzzy gain scheduling system involves to identify parameters of crane dynamic model at operating points $\{l_{min}, m_{min}\}$, $\{l_{min}, m_{max}\}$, $\{l_{max}, m_{min}\}$ and $\{l_{max}, m_{max}\}$. The bounds of scheduling variables intervals are assumed as the centre points of triangular membership functions of fuzzy sets denoted $\mathbf{A}_1$, $\mathbf{A}_n$, $\mathbf{B}_1$ and $\mathbf{B}_r$. Assuming desired regions of closed-loop poles for those operating points, the vector of controller gains are determined according to (7) for the midpoints of those regions. Hence, the initial rule base (18) is created for $n = r = 2$ fuzzy sets defined in the universes of l and m.

$$\begin{aligned}
&\mathrm{R_1: IF\ l\ is\ A_1\ and\ m\ is\ B_1\ THEN\ K_1 = K(l_1, m_1)}\\
&\mathrm{R_2: IF\ l\ is\ A_1\ and\ m\ is\ B_r\ THEN\ K_2 = K(l_1, m_r)}\\
&\mathrm{R_3: IF\ l\ is\ A_n\ and\ m\ is\ B_1\ THEN\ K_3 = K(l_n, m_1)}\\
&\mathrm{R_4: IF\ l\ is\ A_n\ and\ m\ is\ B_r\ THEN\ K_4 = K(l_n, m_r)}
\end{aligned} \tag{18}$$

Selecting the number of sample points N_l and N_m within the scheduling variables intervals $[l_{min}, l_{max}]$ and $[m_{min}, m_{max}]$, the fuzzy scheduler design is realized in a two-stage process through firstly incrementing the sample points l_i ($i = 1, 2, ..., N_l$) within the interval $[l_{min}, l_{max}]$, and in the next stage incrementing sample points m_j ($j = 1, 2, ..., N_m$) within the range $[m_{min}, m_{max}]$. In the first stage, assuming $n = 2$, the sample points of rope length are incremented from l_{min}. At each iteration the n is incremented and the sample point l_i is temporarily assumed as the centre point of membership function of fuzzy set $\mathbf{A}_{n-1}$. The parameters of a crane's model are linearly interpolated based on the parameters identified at operating points $\{l_{min}, m_{min}\}$, $\{l_{min}, m_{max}\}$, $\{l_{max}, m_{min}\}$ and $\{l_{max}, m_{max}\}$ for the operating points corresponding to l_i and l_c, where l_c is the crossover point between neighboring membership functions $\mathbf{A}_{n-2}$ and $\mathbf{A}_{n-1}$. The condition (16) is tested for each crossover point l_c to check that coefficients of closed-loop characteristic polynomial lie in a restricted region for gains interpolated by the fuzzy system. If any violation of the condition (16) occurs, then the new membership function is specified with the center point l_{i-1} tested successfully in the previous iteration. The iterative procedure for selecting center points of membership functions distributed within the interval $[l_{min}, l_{max}]$ is realized in the following steps:

1. Increment i and n, and assume that the current value l_i is a center point l_{n-1} of temporary membership function of fuzzy set $\mathbf{A}_{n-1}$.
2. Approximate the parameters of crane's model at operating points $\{l_{n-1}, m_1\}$ and $\{l_{n-1}, m_r\}$ through linear interpolation between operating points $\{l_{min}, m_{max}\}$ and $\{l_{max}, m_{max}\}$. Specify for those two

points desired regions of closed-loop poles and calculate the controllers gains $\mathbf{K}(l_{n-1}, m_1)$ and $\mathbf{K}(l_{n-1}, m_r)$ for midpoints of those regions according to (7).

3. Add to the rule base (19) the two new rules:
   R_{k+1}: IF l is $\mathbf{A}_{n-1}$ and m is $\mathbf{B}_1$ THEN $\mathbf{K}_{k+1}$=$\mathbf{K}(l_{n-1}, m_1)$
   R_{k+2}: IF l is $\mathbf{A}_{n-1}$ and m is $\mathbf{B}_r$ THEN $\mathbf{K}_{k+2}$=$\mathbf{K}(l_{n-1}, m_r)$
   where $k = N$, and N is equal to the number of rules in the previous iteration.
4. Linearly interpolate the system parameters for operating points $\{l_c, m_1\}$ and $\{l_c, m_r\}$, and test the condition (16) at those operating points, for vectors of gains $\mathbf{R}(l_c, m_1)$ and $\mathbf{R}(l_c, m_r)$ interpolated by a fuzzy system.
5. If the condition (16) is satisfied, remove the rules created in the step 3, decrement n, and repeat the steps from 1 to 4 until $i = N_l$.
6. If any violation of the condition (16) occurs, then decrement i, and specify the new triangular membership function for a fuzzy set $\mathbf{A}_{n-1}$ with the center point l_n defined at l_{i-1}, which has satisfied the condition (16) at previous iteration. Remove rules created in the step 3, and repeat the steps 2-3, to create the rules for a new set $\mathbf{A}_{n-1}$. Repeat the steps 1-5 until $i = N_l$.

The similar procedure is used to find the number of membership functions for scheduling variable m. Assuming initial number of fuzzy sets $r = 2$, the sample points m_j are incremented starting from m_{min}:

1. Increment j and r, and assume that current value m_j is a center point m_{r-1} of temporary membership function of a fuzzy set $\mathbf{B}_{r-1}$.
2. Linearly interpolate the parameters of crane model at the operating points $\{l_1, m_{r-1}\}$, $\{l_2, m_{r-1}\}$, ..., $\{l_n, m_{r-1}\}$. Specify desired regions of closed-loop poles for those operating points, and calculate the controllers gains $\mathbf{K}(l_1, m_{r-1})$, $\mathbf{K}(l_2, m_{r-1})$, ..., $\mathbf{K}(l_n, m_{r-1})$ according to (7).
3. Add to the rule base the following n rules:

R_{k+1}: IF l is $\mathbf{A}_1$ and m is $\mathbf{B}_{r-1}$ THEN $\mathbf{K}_{k+1}$=$\mathbf{K}(l_1, m_{r-1})$
R_{k+2}: IF l is $\mathbf{A}_2$ and m is $\mathbf{B}_{r-1}$ THEN $\mathbf{K}_{k+2}$=$\mathbf{K}(l_2, m_{r-1})$
...
R_{k+n}: IF l is $\mathbf{A}_n$ and m is $\mathbf{B}_{r-1}$ THEN $\mathbf{K}_{k+n}$=$\mathbf{K}(l_n, m_{r-1})$

4. Test the condition (16) for operating points corresponding to the crossover points (17) of membership functions of fuzzy sets $\mathbf{A}_1$, $\mathbf{A}_2$, ..., $\mathbf{A}_n$, and $\mathbf{B}_{r-2}$ and $\mathbf{B}_{r-1}$.
5. If the condition (16) is satisfied, remove the rules created in the step 3 and remove fuzzy set $\mathbf{B}_{r-1}$, decrement r, and repeat the steps from 1 to 5 until $j = N_m$.
6. If any violation of the condition (16) occurs, then decrement j, and specify the new fuzzy set $\mathbf{B}_{r-1}$ with the center point $m_{r-1} = m_{j-1}$ of triangular-shaped membership function. Remove rules created in the step 3 and repeat the steps 2-3, to create the rules for new set $\mathbf{B}_{r-1}$. Repeat the steps from 1-5 until $j = N_m$.

The iterative procedure results in selecting the suitable number of membership functions and their distribution within scheduling variables intervals. Nevertheless, the crane's model parameters should be identified at selected operating points and the condition (16) should be tested again for all crossover points of membership functions $\{l_c, m_c\}$ to confirm the robustness of a control scheme.

3. Control system prototyping for laboratory scaled overhead travelling crane

3.1 The crane control equipment and measurement solutions

In this section the laboratory stand and control-measurement solutions addressing the problems of anti-sway crane control, non-collision path planning and trajectory tracking are presented. The overhead crane consists of motions mechanisms driven by DC motors. The control system can be configured using different platforms: PAC system with RX3i controller or PC with I/O board and Matlab software (Figure 4). The measurement systems implemented on the laboratory scaled overhead crane, are based on contact and contactless techniques. The position and speed of crane motion mechanisms are measured using incremental encoders. The sway angle of a payload can be measured using three methods: i) encoder used to measure the rotary angle of the fork-bottomed arm embracing a rope, ii) Hall-effect magnetic field sensors UV-3HF mounted on the hook assembly to measure its deflection from the equilibrium, iii) the intelligent camera type of Sony XCI V3 installed under the trolley and directed to the rope. In the last technique, the vision software integrated with the embedded system, allows to extract the specified rectangular region of snapshot with the isolated rope straight line, and determine the angle between the detected rope edge and parallel line of image height [10]. The strain-gauge based measurement system is employed to identify the mass of payload.

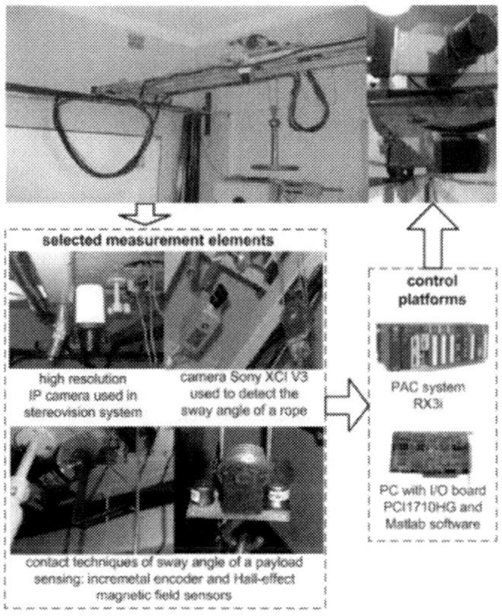

Figure 4. The control-measurement equipment of laboratory scaled overhead crane

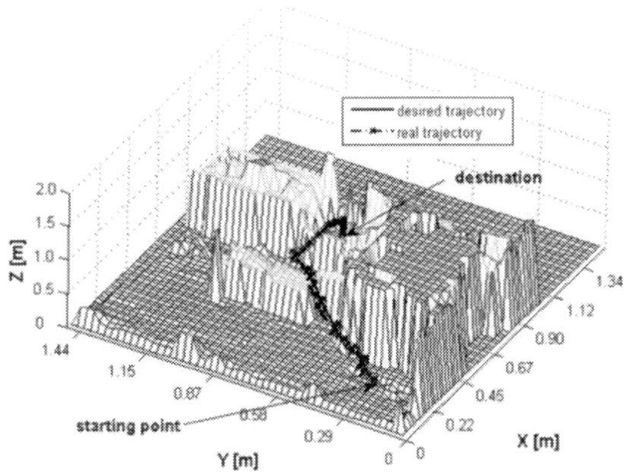

Figure 5. The map of 3D crane workspace, desired trajectory determined using A-star algorithm, and real trajectory of a payload transferred by a crane.

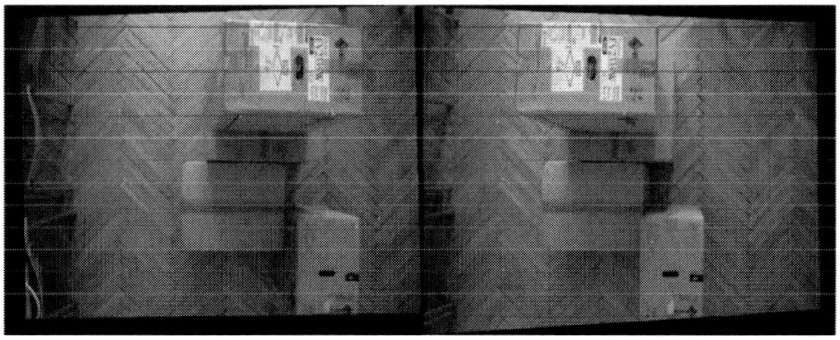

Figure 6. The epipolar transformation of stereo pair of images.

Moreover, the stereovision based method with a single camera was implemented on the laboratory stand to identify a map of transportation workspace. The camera is installed under the trolley, and used to capture the image pair at different position of the crane. The three-dimensional map of crane workspace is obtained through rectification of images leading to obtaining epipolar lines, dense disparity map calculation, and determination of coordinates for each of the image pixel. The example of a crane workspace map for a rectified image pair (Figure 6) is presented in figure 5, where the non-collision path of a payload between two points was designed using well known A-star graph search

IOP Conf. Series: Materials Science and Engineering **65** (2014) 012024 doi:10.1088/1757-899X/65/1/012024

algorithm with the objective function defined based on the average velocities of crane motion mechanisms [20], and compared with trajectory obtained in experiment.

3.2 Examples of experiments

The method of designing the fuzzy gain scheduling anti-sway crane control system, which is described in section 2, was tested on a laboratory scaled overhead crane. The control scheme was designed for scheduling variables intervals $[l_{min}, l_{max}] = [0.8, 2.2]$ m and $[m_{min}, m_{max}] = [10, 90]$ kg. The parameters of a crane model (1-2) were identified using OE method for the operating points: $\{l_{min}, m_{max}\}$, $\{l_{min}, m_{min}\}$, $\{l_{max}, m_{max}\}$, $\{l_{max}, m_{min}\}$. The regions of desired poles were assumed as:

$$z_f = \left[\exp\left(\left(-\omega_n \mp \omega_n (1-\xi)\right)T_s\right)\right] \tag{19}$$

where sample time and a parameter determining the width of desired poles interval were assumed as $T_s = 0.1$ s and $\xi = 0.9$, respectively. Taking into account constraints of a control system imposed by limitation of crane speed and controlling signal, and assuming that reference signal $x_r = 1$ m should be obtained with tolerance ±0.02 m of crane and payload position, the closed-loop poles interval specified for $\omega_n \in [2.06, 2.95]$ rad/s corresponds to the expected settle time interval [5.5, 6.5] s. The iterative procedure resulted in designing the fuzzy scheduler with six rules, and membership functions with centre points at $l = \{0.8, 1.5, 2.2\}$ m and $m = \{10, 90\}$ kg. The control algorithm was written in the form of structured text on the PAC system. Figures 7-9 show the examples of experiments carried out on the laboratory object for operating points corresponding to the centre points (Figure 7-8) and crossover points (Figure 9) of membership functions.

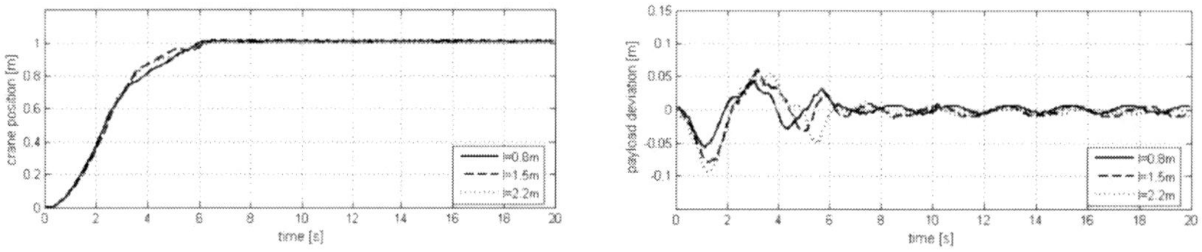

Figure 7. Crane position and payload deviation - experiments conducted for $m = 10$ kg.

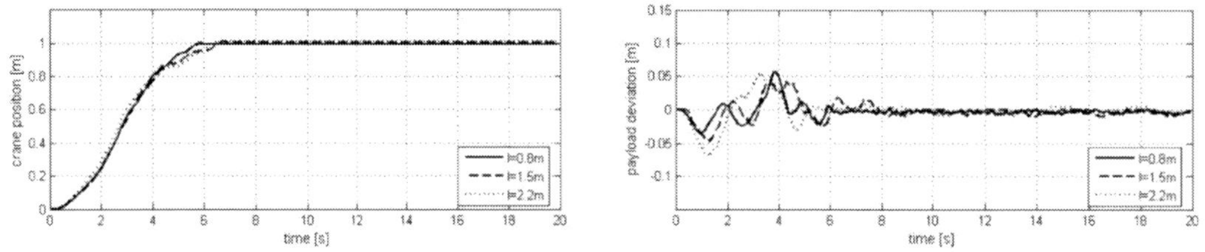

Figure 8. Crane position and payload deviation - experiments conducted for $m = 90$ kg.

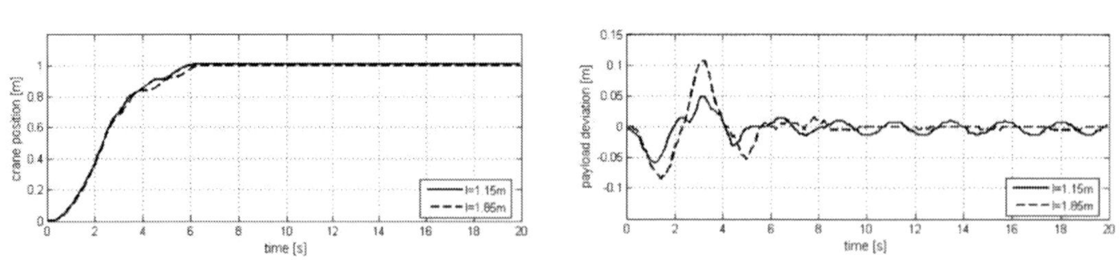

Figure 9. Crane position and payload deviation - experiments conducted for $m = 50$ kg.

The experiments conducted on the laboratory scaled overhead crane shown satisfactory results. The settle time for experiments conducted for operating points lying within intervals [0.8, 2.2] m and [10, 90] kg was between 5.5 and 6.5 seconds, and reduction of payload deflection was in the expected range ±0.02 m. Thus, the results of the experiments prove the effectiveness of the proposed method for designing a fuzzy logic-based gain scheduling control scheme.

Figure 10 presents the comparison of system responses in three experiments carried out at operating point corresponding to the midpoints of intervals [0.8, 2.2] m and [10, 90] kg. The first experiment (solid line) was conducted with fuzzy control scheme satisfying the desired condition for closed-loop system characteristic polynomial coefficients that resulted in determining the center points of triangular shaped membership functions at l = {0.8, 1.5, 2.2} m and m = {10, 90} kg. The crane position and reduction of payload deflection with tolerance ±0.02 m is achieved at about 6 s. In the next experiments, carried out for fuzzy system interpolating the controllers parameters within interval [0.8, 2.2] m (dashed line), and for linear closed-loop control system designed at operating point {2.2 m, 10 kg} (dotted line), the condition (16) has been violated for desired region of closed-loop system poles (19) resulting in control system performances deterioration: settle time in both experiments is 6.8 s and 7.1 s, respectively.

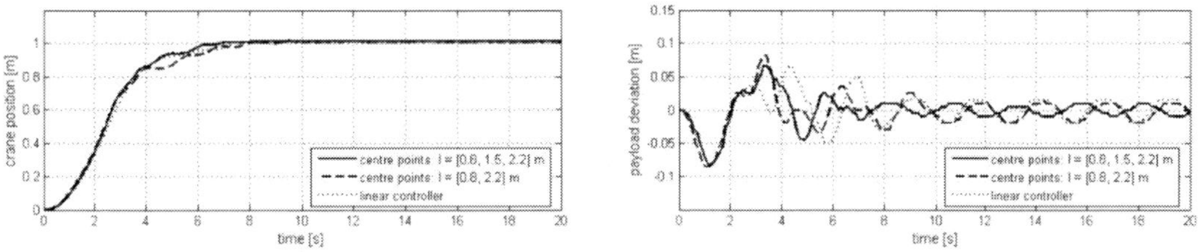

Figure 10. Crane position and payload deviation - experiments conducted for l = 1.5 m and m = 50 kg.

4. Conclusions

The fast, precise and safe transfer of goods in crane operations requires a control application to solve the problems, including shortest and non-collision trajectory planning and limitation of payload oscilations. This paper delivers solutions which can be employed to create such complex control systems. The control-measuremnt equipment elaborated for laboratory overhead crane are presented. The contact and contactless techniques were implemented for measuring the sway angle of a payload, including a dynamic vision system based on a camera detector of sway angle of a rope. The single camera based stereovision system was elaborated to identify a crane workspace map, which has been succesfully tested to design the non-collision trajectory of a payload.

The novelty approach to crane dynamic system modeling and the fuzzy robust control scheme design utilizing interval analysis of closed-loop system characteristic polynomial coefficients is proposed in the form of the iterative procedure for selecting a suitable number of fuzzy sets and membership functions parameters distribution within the expected ranges of scheduling variables. Good control performances obtained during the experiments carried out on the laboratory scaled device proved that the proposed solution can be applicable using software-hardware equipment's used frequently in the industrial practice

Acknowledgment
The research project is financed from the Polish Science budget in the year 2014.

References

[1] Sorensen K L, Singhose W and Dickerson S 2007 A controller enabling precise positioning and sway reduction in bridge and gantry cranes *Control Engineering Practice.* **15** 825-837

[2] Sakawa Y and Shindo Y 1982 Optimal control of container cranes *Automatica.* **18(3)** 257-266
[3] Volckaert M Diehl M and Swevers J 2013 Generalization of norm optimal ILC for nonlinear systems with constraints *Mechanical Systems and Signal Processing.* **39** 280-296
[4] Park H Chwa D and Hong K S 2008 Antisway tracking control of overhead cranes with system uncertainty and actuator nonlinearity using an adaptive fuzzy sliding-mode control *IEEE Transactions on Industrial Electronics.* **55(11)** 3972-3984
[5] Yoshida Y and Tabata H 2008 Visual feedback control of an overhead crane and its combination with time-optimal control, IEEE/ASME Int. Conf. on Advanced Intelligent Mechatronics, Xi'an China 1114-1119
[6] Kim Y S, Hong K S and Sul S K 2004 Anti-sway control of container cranes: inclinometer, observer, and state feedback *International Journal of Control, Automation, and Systems.* **2(4)** 435-449
[7] Smoczek J and Szpytko J 2008 A mechatronics approach in intelligent control systems of the overhead traveling cranes prototyping *Information Technology and Control* **37(2)** 154-156
[8] Miyamoto D Nara S and Takahashi S 2006 Visual feedback control with laser for the position detection of crane hook. *SICE-ICASE Int. Joint Conf*erence. Busan Korea 2079-2083
[9] Kawai H Choi Y Kim Y B and Kubota Y 2009 Measurement system design for sway motion based on image sensor *IEEE Int. Conf. on Networking, Sensing and Control.* Okayama Japan 185-188
[10] Smoczek J Szpytko J and Hyla P 2013 The anti-sway crane control system with using dynamic vision system *Solid State Phenomena.* **198** 589-593
[11] Yoshida Y and Tabata H 2008 Visual feedback control of an overhead crane and its combination with time-optimal control *IEEE/ASME Int. Conf. on Advanced Intelligent Mechatronics*. Xi'an China 1114-1119
[12] Benhidjeb A and Gissinger G L 1995 Fuzzy control of an overhead crane performance comparison with classic control *Control Engineering Practice.* **3(12)** 1687-1696
[13] Liu D Yi J Zhao D and Wang W 2005 Adaptive sliding mode fuzzy control for a two-dimensional overhead crane *Mechatronics.* **15(5)** 505-522
[14] Solihin M I Wahyudi and Legowo A 2010 Fuzzy-tuned antiswing control of automatic gantry crane *Journal of Vibration and Control.* **16(1)** 127-145
[15] Trabia MB Renno JM and Moustafa K A F 2008 Generalized design of an anti-swing fuzzy logic controller for an overhead crane with hoist *Journal of Vibration and Control.* **14(3)** 319-346
[16] Chang C Y 2007 Adaptive fuzzy controller of the overhead crane with nonlinear disturbances *IEEE Transactions on Industrial Informatics.* **3(2)** 164-172
[17] Smoczek J and Szpytko J 2014 Evolutionary algorithm-based design of a fuzzy TBF predictive model and fuzzy anti-sway crane contrl system *Engineering Applications of Artificial Intelligence.* **28** 190-200
[18] Sadati N and Hooshmand A 2006. Design of a gain-scheduling anti-sway controller for tower cranes using fuzzy clustering techniques Int. Conf. on Computational Intelligence for Modeling, Control and Automation, Sydney Australia
[19] Yu W Moreno-Wrmendariz A and Rodriguez F O 2011 Stable adaptive compensation with fuzzy CMAC for an overhead crane *Information Sciences.* **181(21)** 4895-4907
[20] Smoczek, J Szpytko J and Hyla P 2013Non-collision path planning of a payload in crane operating space *Solid State Phenomena.* **198** 559-564

EXPLORING MANUFACTURING COMPETENCIES OF A TWO WHEELER MANUFACTURING UNIT

Chandan Deep Singh[1*], Jaimal Singh Khamba[2], Rajdeep Singh[3] and Navdeep Singh[4]

Department of Mechanical Engineering, Punjabi University Patiala – 147002, India.

[1]email id: er.chandandeep@gmail.com
[2]email id: jskhamba@gmail.com
[3]email id: rajdeepsingh.87@gmail.com
[4]email id: navdeep079@gmail.com

[*]corresponding author

Keywords- Competitiveness, Manufacturing Competency, Two-wheeler Manufacturing Unit

Abstract-The two wheeler industry of India is one of the most dependable industries as every person has at least a two wheeler with him, if not any four wheeler. Earlier there were scooters like Bajaj Chetak, Priya but with evolution of motorcycles like splendor, splendor+, etc. the scooter market started declining but with arrival of gearless scooters like Honda Activa, Scooty Pep, etc. the market place has become increasingly competitive in recent time and industries are facing tough test of improving products and thus market share. The competitiveness among industries is an important issue. Competency development is a vital tool to enhance the competitiveness of industries. Based, on aggregate performance of a firm, it comprehensively explores the varying importance of manufacturing competencies and drives of industrial competitiveness. Hence by, exploring the manufacturing competencies of a two wheeler industry, one can reflect the competitiveness of two wheeler manufacturing industry as a whole. This study presents various factors of manufacturing competencies affecting industrial competitiveness as the significance of these competencies is increasing day by day in two wheeler manufacturing industry.

Introduction

This research is about the competencies existing in manufacturing industries. As a result, R&D and technology executives are increasingly being asked to define the core technical competencies of their companies; unfortunately, they often fail to come up with convincing answers. This is not surprising, as the best methods of employing the thinking in organizations have not been elucidated. The method described here is applicable to almost all core competencies, whether they are technical or non-technical in nature and whether they are currently available to the company or will need to be developed in future. It will allow many organizations to put core competency thinking into practice – to great competitive advantage- without necessary struggle and disruption to business activities. [14]

The term "competency" refers to a combination of skills, attributes and behavior that are directly related to successful performance on the job, which are considered important for all staff of the Organization, regardless of their function or level. Competencies are important both for the Organization and for staff. Competencies are forward-looking. They describe the skills and attributes staff and managers will need in order to build a new organizational culture and meet future challenges. They help organizations clarify expectations, define future development needs, and do more focused recruitment and development planning. Competencies provide a sound basis for consistent and objective performance standards by creating shared language about what is needed and expected in an Organization. In most circumstances, competent persons are appointed to specific areas of the workplace, and their competence can be demonstrated by means such as:

- Work place training,
- Formal and informal qualification's
- Experience gained over periods of time under direct supervision of a competent person performing the task.

Demonstration of competence can also be met by the use of teams to develop and demonstrate competence. It is also important that all persons appointed should know, understand and operate within the limits of their own individual competence. 'Competence development' involves a firm's measures to develop their present competence status, and includes both traditional teaching and learning methods (such as courses or literature) as well as more action-learning and on-the-job oriented methods (such as work rotation or networking). Organizational market performance', is here understood as a firm's performance on the market as compared to its own specific line of industry. [8]

"A firm's competitiveness is a set of differentiated technological skills, complementary assets, and organizational routines and capacities that provide the basis for a firm's competitive capacities in one or more businesses"

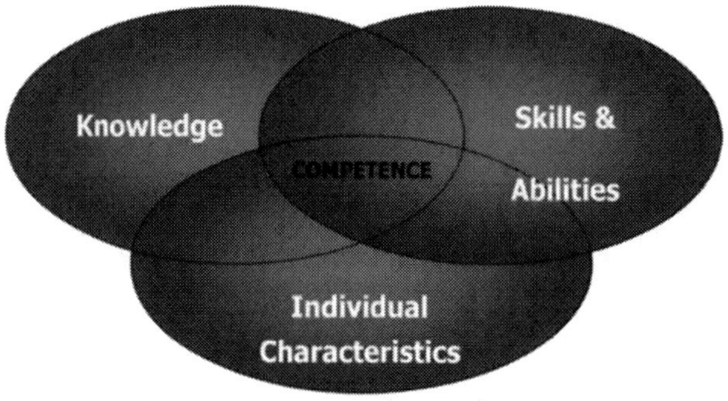

Figure 1 Competency characteristics [27]

In common dictionaries the meaning is often explained by (sufficient) skills, capability, authority and qualification. Scientific literature emphasizes knowledge as an important component of competence, but also suggests components such as aptitude, attitude, commitment and motivation. Definitions are normally also related to specific tasks or a specific context. [27]

Competency Model

A competency model is a collection of multiple competencies that together define successful performance in a defined work setting. A model provides a clear description of what a person needs to know and be able to do – the knowledge, skills, and abilities – to perform well in a specific job, occupation, or industry.

Competency and Skill Standard

A competency is the capacity to draw upon and apply a set of related knowledge, skills, and abilities to successfully perform a work role, function, or task. Competencies often serve as the basis for skill standards that specify the level of knowledge, skills, and abilities required for success in the workplace. Skill standards in turn form the basis for measurement criteria to assess competency attainment. A competency model differs from a set of skill standards in that skill standards define levels of skills (or competencies) required in a given job or role, while competency models typically do not. [27]

Competitive Priorities

The competitive priorities are listed in below. Basically, competitive priorities are the operating advantages that a firm's processes must possess to outperform its competitors.

- **Cost:** Lowering prices can increase the demand for products or services, but it also reduces profit margins if the product or service cannot be produced at a lower cost.
- **Quality**: Quality is a dimension of a product or service that is defined by customer. Today, more than ever, quality has important market implications. As for operations, two competitive priorities deal with quality: *high performance design* (It is the determination of the level of operations performance required in making a product or performing a service) and *consistent quality* (It is the measurement of the frequency with which the product or service meets the design specifications).
- **Time:** as the saying goes "Time is Money". Some companies do business at internet speed while others thrive on consistently meeting delivery promises. These competitive priorities deal with time: *fast delivery* (The time elapsed between receiving a customer's order and filling it), *on-time delivery* (measurement of the frequency with which delivery time promises are met), and *delivery speed* (Measurement of how quickly a new product or service is introduced, covering the elapsed time from idea generation through final design and production).
- **Flexibility**: Flexibility is a characteristic of a firm's operations that enables it to react to customer needs quickly and efficiently. Some firms give top priority to types of flexibility: customization (The ability to satisfy the unique needs of each customer by changing product or service designs) and volume flexibility (The ability to accelerate or decelerate the rate of production quickly to handle large fluctuations in demand). [15]

Need for Present Work

To study and propose manufacturing competencies that contributes to the competitiveness and success of two wheeler manufacturing unit, especially in north India.

Objectives

1. To synthesis the concept of manufacturing competencies two wheeler manufacturing unit.
2. To explore the manufacturing competencies in two wheeler manufacturing unit.

Manufacturing Competencies

Competencies refer to skills or knowledge that lead to superior performance. These are formed through an individual/organization's knowledge, skills and abilities and provide a framework for distinguishing between poor performances through to exceptional performance. Competencies can apply at organizational, individual, team, and occupational and functional levels. Competencies are individual abilities or characteristics that are key to effectiveness in work. Competencies are the characteristics of a manager that lead to the demonstration of skills and abilities, which result in effective performance within an organizational area. [1]

Once the job requirements have been clarified (and competencies provide a framework for doing this), then competency interviewing helps interviewers look for evidence of those requirements in each candidate. For people already in jobs, competencies provide a way to help identify opportunities for growth within their jobs. [9]. Competencies are not "fixed"–they can usually be developed with effort and support (though some are harder to develop than others). Employees and their managers together can identify which competencies would be most helpful to work on to improve the employee's effectiveness. They can then integrate that into a learning plan that may include on-the-job experience, classroom training, or other developmental activities. [12]

Factors of Manufacturing Competencies

Competencies in the area of manufacturing involving steps like conceptualization, designing, development, process planning ,raw material, production and control and quality control.

Creativity

Creativity can be defined as the tendency to produce or recognize ideas, possibilities or alternatives that may be useful in solving difficulties and communicating. Creativity is the ability to think up and design new inventions, produce works of art, solve problems in new ways, or develop an idea based on an original, novel, or unconventional approach. Ability to produce something new through imaginative skill, whether a new solution to a problem, or a new artistic object or form. The term generally refers to a richness of ideas and originality of thinking. Psychological studies of highly creative people have shown that many have a strong interest in apparent disorder, contradiction, and imbalance, which seem to be perceived as challenges.

Innovation

Innovation is the basic driving force behind entrepreneurship and the creation of small businesses. When an individual comes up with an idea that has not previously been explored, he or she may be able to turn that idea into a successful business venture. "Ideas are the fuel that keep entrepreneurial fires blazing," Of course, not every new idea has the potential to become a successful business. And in many cases, individuals with good, marketable ideas fail to come up with the capital needed to turn their ideas into reality. But innovation is still a necessary first step for small business success in many instances. Moreover, entrepreneurs cannot afford to stop innovating once they have established a successful business. Innovation applies not only to new business and product ideas, but also to the internal workings of a company.

Evolution

Evolution is a process of gradual, progressive change and development, as in a social or economic structure with the passage of time. a process in which something passes by degrees to a different stage (especially a more advanced or mature stage); "the development of his ideas took many years"; "the evolution of Greek civilization"; "the slow development of her skill as a writer.

Computer Aided Design

Computer-aided design (CAD) is the use of computer systems to assist in the creation, modification, analysis, or optimization of a design. CAD software is used to increase the productivity of the designer, improve the quality of design, improve communications through documentation, and to create a database for manufacturing. CAD may be used to design curves and figures in two-dimensional (2D) space; or curves, surfaces, and solids in three-dimensional (3D) space.CAD is an important industrial art extensively used in many applications, including automotive, shipbuilding, and aerospace industries, industrial and architectural design.

Aesthetics

Aesthetics is the branch of philosophy which deals with the nature of art and of artistic judgment. There are two traditional views concerning what constitutes aesthetic values. The first finds beauty to be objective, that is, inherent in the entity itself. The second position holds that beauty is subjective, in that it depends on the attitude of the observer. Immanuel Kant argued that judgments of taste, as he called aesthetic judgments, rest on feelings, which, though subjective, have universal validity. The instrumental theory of value, an extension of subjectivism, holds that the value of art consists in its capacity to produce an aesthetic experience.

Ergonomics

Ergonomics is the process of changing the work environment (equipment, furniture, pace of work, etc.) to fit the physical requirements and limitations of employees, rather than forcing workers to adapt to jobs that can, over time, have a debilitating effect on their physical well-being. Companies of all shapes and sizes have increasingly recognized that establishing an ergonomically sensitive work environment for employees can produce bottom-line benefits in cutting absenteeism, reducing health care costs, and increasing productivity. The most progressive of these firms have after careful analysis of the workplace environment and the tasks that their employees have to perform taken steps to modify that environment (whether in a shop floor or an office) to better fit the physical needs and abilities of workers.

Process Planning

Process planning involves determining the sequence of operations to perform to manufacture a part given its description and the specification of the resources in the workshop. It should take into account both technological and economic considerations, some of which are hard constraints and some preferences. This knowledge often represents both the experience and know-how of engineers/specialists, which differ from one company to another. Many approaches have been proposed for automating process planning which focus on generating a single plan that is optimal with respect to some predefined criteria.

Production And Control

Once the entrepreneur has taken the decisions regarding the product design and production processes and system, his next task is to take steps for production planning and control, as this function is essentially required for efficient and economical production. One of the major problems of small scale enterprises is that of low productivity small scale industries can utilize natural resources, which are otherwise lying. Small scale sector can play an important role, similar to the one played by small scale industries in other developed countries. Planned production is an important feature of the small industry. The small entrepreneur possessing the ability to look ahead, organize and coordinate and having plenty of driving force and capacity to lead and ability to supervise and coordinate work and simulates his associates by means of a program of human relation and organization of employees, he would be able to get the best out of his small industrial unit.

Precision Knowledge

The degree of refinement in the performance of an operation, or the degree of perfection in the instruments and methods used to obtain a result. Precision relates to the quality of an operation by which a result is obtained, and is distinguished from accuracy, which relates to the quality of the result. the marksman has achieved a uniformity, although it is inaccurate. This uniformity may have been achieved by using a sighting scope, or some sort of, the analysis of precision can be misleading if a certain degree of precision is implied but not actually attained.

Quality

If a product fulfils the customer's expectations, the customer will be pleased and consider that the product is of acceptable or even high quality. If his or her expectations are not fulfilled, the customer will consider that the product is of low quality. This means that the quality of a product may be defined as "its ability to fulfill the customer's needs and expectations. Quality needs to be defined firstly in terms of parameters or characteristics, which vary from product to product. For example, for a mechanical or electronic product these are performance, reliability, safety and appearance. For pharmaceutical products, parameters such as physical and chemical characteristics, medicinal effect, toxicity, taste and shelf life may be important. For a food product they will include taste, nutritional properties, texture, shelf life and so on.

Case Study at Honda Motorcycle & Scooter India, Private Limited

Being the leader in product and process technologies in the manufacturing sector, two wheeler industry has been recognized as one of the drivers of economic growth. During the last decade, well directed efforts have been made to provide a new look to the automobile policy for realizing the sector's full potential for the economy. Aggressive marketing by the auto finance companies have also played a significant role in boosting automobile demand, especially from the population in the middle income group. Honda is the world's largest manufacturer of two wheelers. Its symbol, the Wings, represents the company's unwavering dedication in achieving goals that are unique and above all, conforming to international norms. These wings are now in India as Honda Motorcycle & Scooter India Pvt. Ltd. (HMSI), a wholly owned subsidiary of Honda Motor Company Ltd., Japan. These wings are here to initiate a change and make a difference in the Indian two wheeler industry. Honda's dream for India is to not only manufacture two wheelers of global quality, but also meet and exceed the expectations of Indian customers with outstanding after sales support. Following its separation from erstwhile partner Hero Group, Japanese auto major Honda, through its subsidiary Honda Motorcycle & Scooter India (HMSI), is pursuing its India strategy with renewed vigour. The company has been in the market for over a decade now, and its independent foray in the world's second largest two-wheeler market is knitted around a strong product and production strategy, with an eye on claiming the top spot in years to come

Company Profile

The HMSI factory is spread over 52 acres, with a covered area of about 85, 815 square meters at Manesar, Gurgaon district of Haryana. The foundation stone for the factory was laid on 14th December 1999 and the factory was completed in January 2001. The initial installed capacity was 100,000 scooters per year, which has reached 6,00,000 scooters by the year by 2007 and motorcycle capacity shall be 4,00,000 per annum. The total investment outlay for the initial capacity was Rs. 215 crores and now the accumulated investment is 800 cores.

Company Strategy

HMSI(Honda Motorcycle & Scooter India, Private Limited) operates on a principle, which is followed worldwide by all Honda companies. Maintaining a global viewpoint, we are dedicated to supplying products of the highest quality, yet at a reasonable price for worldwide customer satisfaction. Honda's philosophy is based on the company's guiding principle.

Respect for the Individual

Honda recognizes and respects individual differences. The respect for individual stems from the following three points: Initiative, Equality, Trust. It is the contribution from each individual in the company that has made our company what it is today and that, which will take us into the future.

The Three Joys

In line with Honda's Philosophy, HMSI conducts all its daily activities in pursuit of the following joys:

- The joy of manufacturing high quality products.
- The joy of selling high quality products.
- The joy of buying high quality products.

Techonologies used at HMSI

1. Cutting-edge technology and the spirit of Honda
2. Combi Break System
3. Honda Matic Transmission
4. Automation

Quality

Honda provides excellent quality for their products .Numerous checks during the manufacturing process ensure that every part meets the required quality. An interesting process at the facility is where the engines merge with the frame assembly line. The built-up engine is transferred to a lift, which takes it underground and is then transferred through a belt to another lift on the frame assembly line, from where it comes up to an operator. The operators then begin the integration process of the body and power train. Having automated this process saves significant amount of time and adds efficiency to the process. The engineers ensure implementation of the most efficient manufacturing practices, and in the best way possible for the Indian market.

Conclusion

The case study was prepared at Honda manufacturing unit at Manesar, Gurgaon district of Haryana. During this work, manufacturing competencies have been investigated. It has been analyzed that the sales have improved with an improvement in competitiveness of manufacturing unit.
Following conclusions have been made after preparation of the case study:
1. The manufacturing unit is technically sound and is competent with the latest machinery.
2. Profit of the unit has been improved.
3. Market share has enhanced.

Based on the above study, the following aspects for future work have been presented:
1. The work can be carried out in other areas of India.
2. The work can be extended in other automobile sectors.
3. Strategy issues can be studied with this work.

References

[1] Anders Drejer (2001) How can we define and understand competencies and their development? *The journal Technovation 21*

[2] Athey, T. R., & Orth, M. S. (1999). Emerging competency methods for the future. *Human Resource Management*.

[3] Cardy, R. L., & Selvarajan, T. T. (2006). Competencies: Alternative frameworks for competitive advantage. Business Horizons.

[4] Caroline mothe, Bertrand Quelin (2000) Creating Competencies Through Collaboration.The Case of Eureka R&D Consortia. *European Management Journal Vol. 18.*

[5] Eric Bonjour, Jean-Pierre Micaelli . (2010) 'Design Core Competence Diagnosis: A Case From the Automotive Industry',IEEE Transactions on Engineering Management, Vol 57, pp 323-337.

[6] Erik Schlie, George Yip (2000) Regional Follows Global: Strategy Mixes in the World Automotive Industry. *European Management Journal Vol. 18.*

[7] Felicia Fai a, Nicholas von Tunzelmann (2002) Industry-specific competencies and converging technological systems: evidence from patents. Structural Change and Economic Dynamics

[8] G.S. Dangayach, S.C. Pathak, A.D. Sharma.(2006) 'Advanced Manufacturing Technology: A Way of Improving Technological Competitiveness',. International Journal of Global Business and Competitiveness, Vol. 2, No 1, pp 1-8.

[9] John P. Millikin, Peter W. Hom, Charles C. Manz .(2010) 'Self-management competencies in self-managing teams: Their impact on multi-team system productivity',.Elsevier Publisher,vol 21, pp 687-702.

[10] K. Pavitt(1990) What We Know About the Strategic Management of Technology, *California Management Review Reprint Series 32 17-26. [40] R.D. Pearce, The Internationalization.*

[11] Krisztina Demeter .(2002) 'Manufacturing strategy and competitiveness (International Journal of Production Economics', Elsevier Science B.V,pp 205–213.

[12] Kwasi Amoako-Gyampah, Moses Acquaah. (2008) 'Manufacturing strategy, competitive strategy and firm performance: An empirical study in a developing economy environment', .international Journal of Production Economics, Elsevier B.V 111, pp 575-592.

[13] Lei, D., Hitt, M. A., & Bettis, R. (1996). Dynamic core competencies through metal earning and strategic context. *Journal of Management.*

[14] Mark R. Gallon, Harold M. Stillman, David Coates. (1999) 'Putting Core Competency thinking into practice', .RTM Journal, pp 1-12.

[15] Monica Sharma, Rambabu Kodali. (2008) 'Development of a framework for manufacturing excellence', Measuring Business Excellence, Emerald Group Publishing Limited, vol 12 no. 4 , pp 50-66.

[16] P. Pari and K. Pavitt,(1990) Large Firms in the Production ofthe World's Technology: An Important Case of Non-Globalization, *Journal of International Business Studies.*

[17] Qingyu Zhang , Mark A. Vonderembse , Jeen-Su Lim (2003) Manufacturing flexibility: defining and analyzing relationships among competence, capability, and customer satisfaction. *Journal of Operations Management 21.*

[18] Sanjib K. Dutta .(2007) 'Enhancing competitiveness of India Inc. International Journal of Social Economics', Emarald Group Publishing Limited, Vol. 34 No. 9, pp 679-711.

[19] Sajee B. Sirikrai , John C.S. (2006) 'Tang Industrial competitiveness analysis: Using the analytic hierarchy process', Journal of High Technology Management Research, Vol 17, pp 71-83.

[20] T. Laosirihongthong, G.S. Dangayach (2005) A Comparative Study of Implementation of Manufacturing Strategies in Thai and Indian Automotive Manufacturing Companies. *Journal of Manufacturing Systems Vol. 24.*

[21] T.S. Nagabhushana, Janat Shah. (1999) ' Manufacturing priorities and action programmes in the changing environment', International Journal of Operations and Production management, MCB University Press, Vol. 19 No. 4, 1999, pp 389-398.

[22] WANG Zhi-yu, QIU Yan-lin, GUI Shi-he.(2006).Quality competence:a source of sustained competitive advantage',. Journal of china universities of posts and telecommunications vol 13 no 1.

[23] Yu ting lee .(2010) 'Exploring high-performers' required competencies',An International

Journal, Expert Systems with Applications, pp 434-439.

[24] Yonggui Wang, Hing-Po Lo, Yongheng Yang.(2004) 'The constituents of core competencies and firm performance: evidence from high-technology firms in china.Journal of Engineering and Technology Management', Elsevier B.V, vol 21, pp 249-280.

[25] Yonggui Wanga, Hing-Po Lob, Yongheng Yangc (2004) The constituents of core competencies and firm performance: evidence from high-technology firms in china. J. Eng. Technol. Manage.

[26] Yonggui Wanga, Hing-Po Lob, Yongheng Yang (2004) The constituents of core competencies and firm performance: evidence from high-technology firms in china. Y. Wang et al. / J. Eng. Technol. Manage. 21..

[27] *http://:*www.slideshare.net (accessed on January 12, 2013

Enhancing technological innovation in small firms: Role of collaboration

D Singh[1], J S Khamba[2] and T Nanda[3]

[1]Assistant Professor, Department of Mechanical Engineering, Punjabi University, Patiala. Email: davinder5206@yahoo.co.in

[2]Professor and Head, Department of Mechanical Engineering, Punjabi University, Patiala. Email: jskhamba@gmail.com

[3]Assistant Professor, Department of Mechanical Engineering, Thapar University, Patiala.

Email: tarunnanda@thapar.edu

Abstract. Contribution of Micro-Small and Medium Enterprises (MSMEs) is highly remarkable in the overall industrial economy of the country. In recent years, the MSME sector has consistently registered higher growth rate compared to the overall industrial sector. With its agility and dynamism, the sector has shown admirable innovativeness and adaptability to survive the recent economic downturn and recession. However, MSMEs growth rate is still at low level. Therefore, it becomes essential for organizations to adopt new technologies or upgrade existing setup to meet continuously changing global market and fulfill customer needs. This paper explores the relationships between different collaboration networks and technological innovation of small firms through an extensive review of literature. The study finds that collaboration with larger enterprises, R&D institutions, universities and government agencies play a significant role in enhancing technological innovation in small firms.

Keywords: MSMEs, Technological Innovation, Collaboration, R&D.

1. Introduction

However, MSMEs contribution is phenomenal in the growth of Indian economy, simultaneously; they are facing intense pressure and constraints to sustain their competitiveness in globalized world. Some other issues such as recession, low demand, finance, heavy competition from MNCs etc. are becoming exposed problem to MSMEs in India [1]. Therefore, it becomes necessary for organization to adopt new technologies or upgrade existing setup to meet continuously changing global market and fulfills customer needs. The complexity of innovation processes led to a tremendous growth in the use of external networks by MSMEs [2].

Firms invest in the creation of knowledge not only for their own direct use, but also to more easily engage with others to create network ties. Indeed, collaborative ties between firms can lead to mutual or shared learning and the development of new or enhanced internal capabilities leading to technological growth, diversification and hence further innovative opportunities [3].

1.1 Basic Concept of Industrial Collaboration

Collaboration is working with each other to do a task [4]. It is a recursive process where two or more people or organizations work together to realize shared goals. This is more than the intersection of common goals seen in co-operative ventures, but a deep, collective, determination to reach an identical objective for example an endeavour that is creative in nature-by sharing knowledge, learning and building consensus [5]. In particular, teams that work collaboratively can obtain greater resources, recognition and reward when facing competition for finite resources [6].

In its traditional form, collaboration refers to the process in which geographically proximate producers, suppliers, buyers, and other actors develop and intensify cooperation with mutually beneficial effects.

However, in its most advanced form, according to a widely accepted definition proposed by [7], a cluster is a geographically proximate group of interconnected enterprises and associated institutions in a particular field, linked by commonality and complementarities. Under this definition, a cluster may include suppliers of inputs, or extend downstream to regular buyers or exporters. It also includes government institutions, business associations, providers of business services, and agencies that support clustered enterprises in such fields as product development, production process improvement technology, marketing information (for example, on new market and designs), vocational training, and so on.

2. Literature review

Innovation studies have underlined the crucial role played by the interaction of different organizations in fostering the innovation process [8]. Firm's innovation activity is enhanced and sometimes depends upon, collaboration with other organizations [9, 10] such as universities and research centers [11], suppliers and users and even competitors [12]. In order to carry out innovative activities, firms must accumulate and process internal and external knowledge, establishing learning processes based on different sources [13].

2.1. Technological Collaboration with Large Firms

Over the last few years, trade liberalization and globalization processes have increased customer expectations and competition between companies [14]. The development of new and improved products rather requires an active search-process involving several firms and institutions to tap new sources of knowledge and technology [15]. Exchange of information and resources with different partners are important factors in the innovation process. Inter-firm collaborations occur especially within technology-based industries. Inter-firm collaborations occur especially within technology-based industries. Arora and Gambardella [16] demonstrate the high importance of R&D collaborations for large US chemical and pharmaceutical companies in the biotechnology sector. Colombo [17] provides empirical evidence of complementary relationship between inter-firm cooperative arrangements and R&D intensity for a representative sample of international firms in the information technology industries.

2.2. Technological Collaboration with R&D Institutes

Forrest [18] has investigated the role of technology collaboration with R&D institutes in the German automobile industry. Formal R&D arrangements with universities are preferred because automobile suppliers can enhance their in house capacities and use their automobile-specific potentials more efficiently. In the specific case of technological collaboration, the alliances include collaborative R&D agreements, university and/or research institute agreements and technology licensing.

MSMEs, however, have alternatives to internal development that may enable them to bridge the resource gap that exists with large firms [19]. In this respect, the literature on innovation stresses the role of cooperative R&D in overcoming the lack of internal resources and in improving innovativeness and competitiveness [20].

2.3. Collaboration with Academic Institutions, Universities and Govt. Aided Institutes

Rogers [21] defended the strategic importance of the innovation process for the companies themselves. The relationship between the competitive capacity of companies and countries with the knowledge and its management have made the importance of structures like scientific and technological research through partnerships between universities or institutes and companies increasingly evident. These partnerships contribute to the development of science, its applications and the appropriation of innovations resulting thereof, and progress from the paradigm of industrial society to the society of knowledge can be seen as a crucial factor in the economic, social and cultural growth of nations and organizations [22].

According to Etzkowitz [23], the term "innovation has taken on a wider meaning. More than the development of new products in companies, it is also the creation of new arrangements between institutional spheres that provide the conditions necessary for innovation".

In this sense, collaboration between university, institution and company is also an important instrument in the generation of science and technology in a country. Since, by splitting costs and sharing risks between the two institutions, collaborative research allows for greater investment in the development of new technologies for products and processes that guarantee greater competitiveness to organizations and the broadening of the scientific knowledge of the nation.

3. Conclusion

In particular, MSMEs have to undergo some change when they compete with global companies and other large buyers, as they are dependent on supply contracts from the same. This puts considerable pressure on them to control both their costs and quality and meet the different legal requirements. This is a serious challenge for MSMEs, especially for those operating in developing countries. The need today is to enable MSMEs tide-over these problems and enhance their access to new technologies through the process of increased communication, cooperation and linkages, both within the enterprise as well as across enterprises and knowledge-producing organizations.

In this study, an attempt has been made to identify the factors affecting technological innovation of MSMEs. The present work presents four key issues under technological collaboration; which include *TC with larger enterprises, TC with R&D institutes, TC with academic institutions & universities, TC with government aided institutes* for the improvement in technological innovation of MSMEs.

4. References

[1] R. Jeswal, in: Role of SME in Indian Economy, National Conference on Emerging Challenges for Sustainable Business (2012).
[2] S.X. Zeng, X.M. Xie and C.M. Tam, in: Relationship between cooperation networks and innovation performance of SMEs, Technovation Vol. 30 (2010), p. 181–194.
[3] S. Brusoni, A. Prencipe and K. Pavitt, in: Knowledge specialization, organizational coupling and the boundaries of the firm: why do firms know more than they make?' Administrative Science Quarterly Vol. 46 (2001), p. 597–662.
[4] M. I. J. Moyano, in: Exploring the Dynamics of Collaboration in Inter organizational Settings, Creating a Culture of Collaboration, Jossey-Bass, Schuman (Editor) ISBN 0-7879-8116-8 (2006).
[5] Spence and U. Muneera, in: Graphic Design: Collaborative Processes-Understanding Self and Others (lecture) Art 325: Collaborative Processes. Fairbanks Hall, Oregon State University (2006).
[6] C. S. Wagner and L. Leydesdorff, in: The diffusion of international collaboration and the formation of a core group, Globalisation in the network of science (2005).
[7] M. E. Porter, in: Location, Competition and Economic Development: Local Clusters in a Global Economy, Economic Development Quarterly Vol.14 (2000), p. 7-20.
[8] M. Dodgson and R. Rothwell: The Handbook of Industrial Innovation, Cheltenham: Edward Elgar (1994).
[9] C. Freeman, in: Networks of Innovators: A Synthesis of Research Issues, Research Policy Vol. 20

(1991), p. 499–514.
[10] C. Freeman, in: The Economics of Technical Change, Cambridge Journal of Economics Vol. 18 (1994), p. 463–514.
[11] W. Cohen, R. R. Nelson and J. Walsh in: Links and Impacts: the Influence of Public Research on Industrial R&D, Management Science Vol. 48 (2002), p. 1–23.
[12] R. Coombs, A. Richards, P. P. Saviotti and V. Walsh, in: Technological Collaboration: The Dynamics of Cooperation in Industrial Innovation, Cheltenham: Edward Elgar (1996).
[13] S. Kline and N. Rosenberg: An Overview of Innovation. In: Landau, R. and Rosenberg, N. (eds.), The Positive Sum Strategy: Harnessing Technology for Economic Growth, Washington: National Academy Press (1986).
[14] J. R. Gradzol, C. J. Gradzol and S. T. Rippey, in: An emerging framework for global strategy, International Journal of Manufacturing Technology & Management Vol. 7 (2005), p. 11.
[15] B. Nooteboom, in: Innovation and inter-firm linkages: New implications for policy, Research Policy Vol. 28 (1999), p. 793-805.
[16] A. Arora and A. Gambardella, in: *Complementarity and external linkages: The strategies of the large firms in biotechnology,* Journal of Industrial Economics Vol. 38 (1990), p. 361-379.
[17] M. Colombo, in: Firm Size and Cooperation: The determinants of cooperative agreements in information technology industries, International Journal of the Economics of Business Vol. 2 (1995), p. 3-29.
[18] J. E. Forrest, in: *Strategic Alliances and the Small Technology-Based Firm,* Journal of Small Business Management Vol. 28 (1990), p. 37-45.
[19] G. Miles, S. B. Preece and M. C. Baetz, in: *Dangers of Dependence: The Impact of Strategic Alliance Use by Small Technology-Based Firms,* Journal of Small Business Management Vol. 37 (1999), p. 20-29.
[20] M. Rogers, in: *Networks, Firm Size and Innovation*, Small Business Economics Vol. 22 (2004), p. 141-153.
[21] E. Rogers, in: *Diffusion of innovations*, New York: Free Press (2003).
[22] H. Etzkowitz and L. Leydesdorff, in: *The dynamics of innovation: from national systems and 'mode 2' to a triple helix of university-industry-government relations,* Research Policy Vol. 29 (2000), p. 411-424.
[23] H. Etzkowitz, in: Research groups as 'quasi-firms': the invention of the entrepreneurial university, Research Policy Vol. 32 (2003), p. 109-121.

IOP Conf. Series: Materials Science and Engineering **65** (2014) 012027 doi:10.1088/1757-899X/65/1/012027

Developing an experimental case in aluminium foils 1100 to determine the maximum angle of formability in a piece by Dieless-SPIF process.

Paramo Gabriel[1], **Benitez Adrian**[2]

[1]Production Technologies group, Faculty of Engineering EAFIT University, Medellin, Colombia. E-mail: gparamo@eafit.edu.co

[2]Production Technologies group, Faculty of Engineering EAFIT University, Medellin, Colombia. E-mail: abenite2@eafit.edu.co

Abstract. Incremental sheet forming by the method of single point incremental forming Dieless-SPIF, is a widely studied process, experimented and developed in countries with high manufacturing technologies, with friendly costs when the productive configuration in a productivity system is based in small production batches. United states, United kingdom and France lead this type of studies and cases, developing various proof with experimental geometries, different from the national environment such as Colombia, Bolivia, Chile, Ecuador and Peru where this process where discretely studied. Previously mentioned, it pretends develop an experimental case of a particular geometry, identifying the maximum formability angle of material permissible for the forming of a piece in one pass, the analysis of forming limit curve (FLC), with the objective to emphasizes in this innovative method based in CAD-CAM technologies, compare with other analogous process of deformation sheet metal like embossing, take correct decisions about the viability and applicability of this process (Dieless) in a particular industrial piece, which responses to the necessities of productive configurations mentioned and be highly taken like a manufacturing alternative to the other conventional process of forming sheet metal like embossing, for systems with slow batches production.

1. Introduction

The incremental forming process by the method of supporting a single point Dieless (SPIF), involves the transformation of a metal sheet by a mechanical stress which produces a progressive deformation in the sheet. The process is done in several stages: it starts with a CAD (Computer aided design) modeling which represents the experimental geometry in three dimensions of the particular piece, the second stage is fed by the CAD file, assigning parameters such as the advance, RPM (Rev/Min),

[1] Production Technologies group, Faculty of Engineering EAFIT University, Medellin, Colombia. E-mail: gparamo@eafit.edu.co.
[2]Production Technologies group, Faculty of Engineering EAFIT University, Medellin, Colombia. E-mail: abenite2@eafit.edu.co.

IOP Conf. Series: Materials Science and Engineering **65** (2014) 012027 doi:10.1088/1757-899X/65/1/012027

diameter tool, step depth, to a CAM (Computer aided manufacturing) system, you get a programming tool path, expressed in machine code known as G code[3]. This code is entered into a machine with CNC technology which reproduces the toolpath on the surface and deforms the end of the geometry piece, which is the final stage. Figure 1, illustrates a representative form of this process.

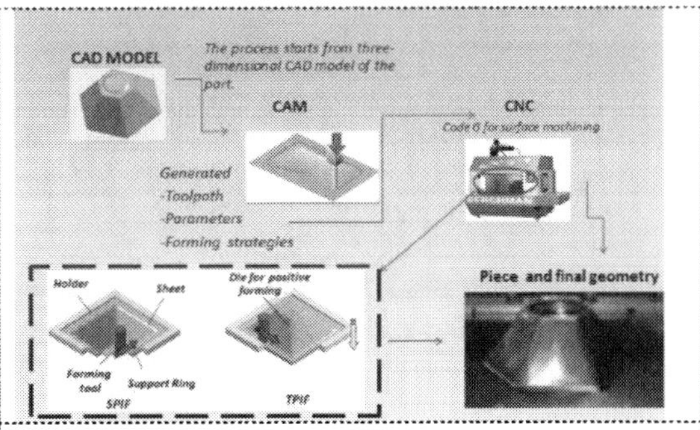

Figure 1. Dieless representative illustration

The incremental deformation process without matrix (Dieless), is a recent process (its inception refer to 1994) [1] with respect to other techniques of conventional foil strain such as embossing, stamping, superforming and hydroforming, which are costly and involve working tooling and high volume production runs for its construction and operation.

The purpose of this paper is to study an experimental case associated with a particular one, which is due to a conical geometry, to analyze the deformation of the same building through the forming limit curve diagram (FLD), identify the maximum angle of formability obtained for this geometry, through the test assembly with various inclinations. The formability of a sheet or foil, such as its ability to be deformed by forming a specific process from its original form to the final piece flat, without the occurrence of failure in the material, either broken or necking, the ease of a material to plastic deformation without defects. [1]- [3].

The above is intended to orient the investigation to work with geometries that approximate and apply to a principle or Industrial application case, making decisions on the viability argued acceptable or not the process and have a specific comparison with the filling process. With the ultimate goal of making a functional prototype tested mechanical and dynamic evaluations of this process, to ensure the functionality and usability of a product manufactured under this innovative manufacturing process such as Dieless.

2. Procedure and method

For the experimental procedure followed the next methodology:

A. As a starting point, we determine the geometry of the part: settled for the experimental case geometry conical differential characteristic angle formability, as shown in Figure 2.

---

[3] **G-Code**

The G code is a language trough which people can tell computer-controlled machine tools do and how. These "what" and "how" are defined mostly by instructions on where to move, how fast moving and toolpath or follow. Typical machines are controlled is code are milling machines, cutting machines, lathes and 3d printers.

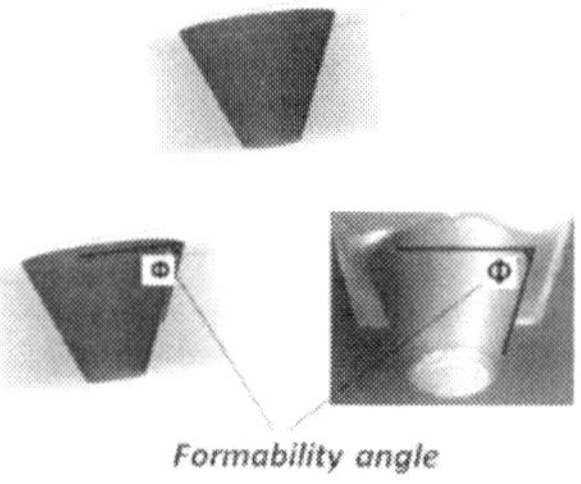

Figure 2. Geometry associated an experimental case

The blank plates or training are square sheets of 100 mm 1100 aluminium side and 1 mm thick.

B. For construction of diagrams-forming limit curves, it was necessary to perform the marking of the pieces, which consists of a grid of continuous circles of 2.5 mm diameter. In Figure 3, this system can be observed on the piece of lined, which was made into a laser marking machine.

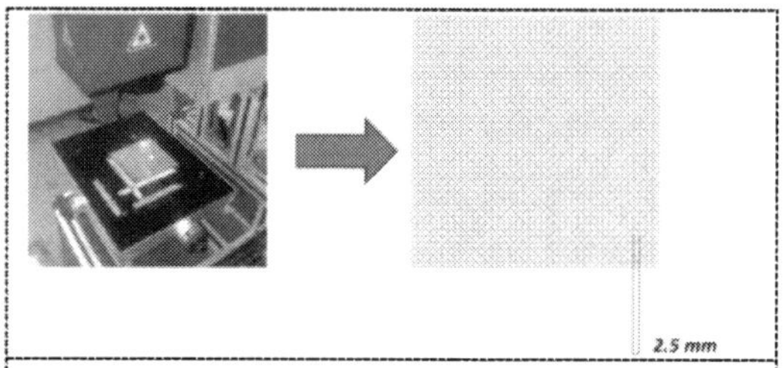

Figure 3. Geometry associated an experimental case

C. Once the marking of parts is carried generation strategy and toolpath: The helical toolpath responds to a helical scan is made along the inner surface of the sheet deformed, it was necessary to work with CAM software, which is mentioned below.

   Was subsequently carried determining process parameters Dieless SPIF. Both for design (CAD) of the experimental geometry, generating cutting strategies CAM, and assigning values to parameters involved in the process, it has high engineering software.

D. Machine Preparation and installation: Dieless-SPIF device, Figure 4 shows the assembly of the part-SPIF Dieless device.

Figure 4. Assembly of a piece in SPIF device work center laboratories EAFIT University

3. Results and discussion

The testing and subsequently different experimental cases with different angles of formability: were specified angles between 64 ° and 76 °, which are detailed below. The formation and fracture criterion answers the formability evaluation part as mechanical forming; Figure 5 shows the visual appearance of a formed part and a fractured part formation.

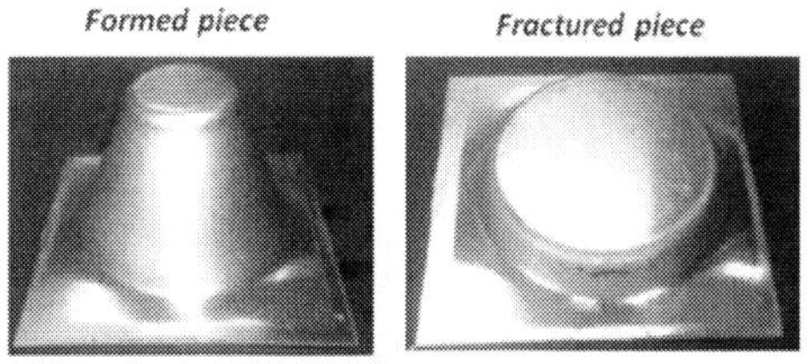

Figure 5. Aspect for a formed and fractured piece

In Table 1 are recorded angles and training criteria, fracture, as the case:

Table 1. Forming criteria respect to the formability angle obtained

FORMABILITY ANGLE	
Formabitity angle	**Piece (Formed/Fractured)**
ϕ1 76	Fractured
ϕ2 74	Fractured
ϕ3 72	Fractured
φ4 68	***Formed***
ϕ5 70	Fractured
ϕ6 69	Fractured
ϕ7 64	Formed
ϕ8 66	Formed
ϕ9 72	Fractured

Looking closely at table 1, the largest angle of formability obtained without breaking the piece present, is 68 °, from this angle values above such as 69 °, 70 °, 72, 74 ° and 76 ° . For angles less than this value, such as 66 ° and 64 °, the piece does not show break and trained properly, as modeled in CAD geometry. This is presented as the first fundamental result of this work, as it is known in principle the maximum angle of taper formability for experimental geometry, which will aim to make decisions about a close and future applicability to an industrial part obedient geometries angles such limits established in this work.

From a distance of 20 mm depth, measured from the surface of the workpiece, such crack had fracture in the same place. As a hypothesis sheet thinning located at this point, which together with the formability angles than 68 ° are two extreme conditions which result in the presentation physically cracks and long shapes as shown in Figure 6, which is then analyzed.

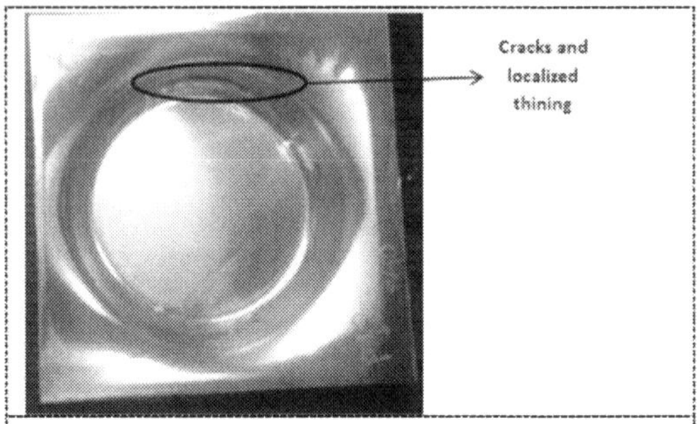

Figure 6. Cracks for fracture piece

The cracks occur as a result of residual stresses produced in the work piece. These residual stresses are a major feature in the sheet forming operations and are commonly caused by non-uniform deformation during forming, causing a partial distortion when cut, and producing said cracking [4]-[9]. The piece, with maximum angle of formability determined not presented wrinkles across the study areas.

The parameters presented in Table 2, are best suited and recommended for this geometry work, several authors in their research and procedures used tool diameters between 8 and 12 mm, high RPM and feed rates relative to the depth of cut [9] - [12], has been shown that the lower step, the surface quality and formability of the material are much better, considering cutting times are greater.

Table 2. Nominal Values – Parameters for experimental cases SPIF process

FORMING PARAMETERS PROCESS	
Tool diameter (mm)	8,7
RPM	3000
Cutting feed (mm/min)	4000
Step depth (mm)	0,7-1-1,2

Observing the process parameters shown in Table 2, it is important to note that the progress, which has a value of 4000 mm / min, corresponds to approximately 50% of the maximum speed of the CNC machining center used (maximum advance machining center used 7200 mm / min). The strategy or tool path tool path was helical steps and depths of 0.7-1 and 1.2 mm.

FLD Curve construction

The diagram limit shaped for a sheet, is a graphical representation of the boundaries of main strains, where it may arise failure in plastic deformation during the forming process. From the above definition, we can identify areas along the deformed surface of the test piece, and account for the formability of the material.

For construction of the limit curves forming or diagrams FLD is necessary to calculate the result of the conventional deformation suffered by the piece, which is the relationship between the change in length of a specimen in the direction that applies strength and the original length of the sample considered [5]. In Equation 1 shows how to calculate it:

$$\%Stretching = \frac{Lf - Lo}{Lo} x100 \quad (1)$$

From the equation above, it is important to take data in different areas of the deformed part, before and after the process in order to obtain readings of initial lengths Lo, and final lengths of the same Lf.

Data for construction of curves FLD

Table 3 shows the data measured once deformed the workpiece, with the aid of the grid circles, ellipses now to determine the major and minor deformities in different areas of the piece, to analyze the formability of it.

Table 3. Datum for FLD construction curve

Datum for FLD construction curve				
Datum	Major Length (mm)	Minor Length (mm)	% Major strain	% Minor strain
0	2,5	2,5	0%	0%
1	2,8	2,45	12%	-2%
2	4,8	2,2	92%	-12%
3	7,95	2,6	218%	4%
4	6,6	2,25	164%	-10%
5	6,5	2,3	160%	-8%
6	6,8	2,5	172%	0%
7	6,8	2,35	172%	-6%
8	6,6	2,65	164%	6%
9	7	2,8	180%	12%
10	4,25	2	70%	-20%

The above data obtained were plotted on the graph of FLD curve for the product obtained with maximum angle of formability.

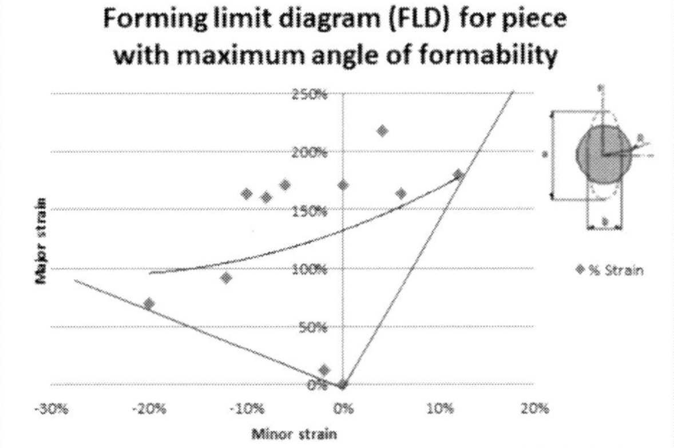

Figure 7. Forming limit diagram for piece with maximum angle of formability

As seen in figure 7, shown at the top right, a circular geometrical figure, shaded continuous contour which represents the circle inscribed in the aluminum foil before being deformed, and dashed ellipse in which due to the two-dimensional geometry of the elongated circle (ellipse) once the piece has been deformed. The dimensions of the coordinates shown with the letter b, the lengths are larger and smaller, respectively, which are the final readings (Lf) to determine the percent elongation and major and minor strains required for the construction of the curve obtained FLC above.

Greater deformation has been shown in axis Y, always positive due to stretching occurs and less deformation can be negative or positive, as can occur when stretched narrowing (Poisson effect) the initial specimen. It can be stated that under bounded or curved line are safe values and above the fault. For the particular case, there is an extreme value for which the strain is 70% higher and lower distortion is -20%, suffered in aluminium alloy 1100, which represents a good plastic deformation of the material obtained by through this process. The formability of a particular material will be better when the FLD curve is higher [4].

Comparing analogous to the process of drawing, on what is commonly used in the manufacture of hollow pieces such as this geometry experienced in this work and responds to the fundamental principle of sheet metal processing, are shown in Table 4, a general comparative Dieless process [12]-[16] and stamping-drawing, for performing the piece addressed in this paper.

Table 4. Process comparative – stamping and dieless

	STAMPING-DEEP DRAWING	DIELESS
Principle	Deforming a metal sheet, which is pressed against a die by a punch which generates deformation and make a hollow.	Deformation a metal sheet with a round tool that **presses** the sheet, outlining its surface until the final geometry.
Types	**HOT STAMPING:** The sheet is heated to a temperature of 800 ° to 850 ° **COLD DRAWING:** Practice at room temperature.	**SPIF:** (Single point incremental forming, Forming negative) fixed frame and draw the shape tool inward, increasing the Z axis is not required matrix **DPIF:** (double point incremental forming, forming positive) the frame moves up and down and draw the shape tool outward, increasing the Z axis, it is necessary to use a matrix
Tools	Requires a large and expensive tooling with a die, a punch, a metal foil and a device for tightening the foil and avoid folds.	It takes a round ball mill, a matrix if necessary (DPIF), a metal foil and a support device to hold and tighten the foil.
Production	There is no flexibility in the design, restricting the complex geometries, is used for high-volume production or for a unique product.	Flexibility to changes in design, allows for complex geometries, is used for low volume production scale prototypes and functional models
Speed	Fast process cycles	Fast process cycles with post-processing
Quality	Good	SPIF: moderate surface quality

With the overview of processes of forming two sheet metal details the analogy to perform the experimental part, for which respective comparisons are described based on the observations in Table 5 which shows a comparison of the performance part case study worked through the two processes in question, according to the calculations made for the part geometry and particular design parameters for the filling process [4], and any of the parameter values specified for Dieless-SPIF.

Table 5. Comparative to obtain geometric part by dieless and stamping- deep drawing process

Comparative to obtain geometric part		
	Dieless	Drawing-stamping
Die for forming part	No	Yes
Blank holder for forming part	No	Yes
Speed (mm/min)	4000	30000
Steps Number	1	3
Process press (Kg/cm^2)	Low (Max. 4)	High(10)
Tooling Costs	Low	High

According to the above tables, IV and V, that for high-speed production cycles or production rate for the manufacture of the part studied in this work, the drawing process would be the most convenient alternative, but requires investment in tooling and tooling to ensure the formability of the sheet, such as matrix, blank holders to ensure the strength and pressure of the process described above. Dieless-SPIF process, contrary to what stated above, does not require blank holders, receivers and since the force and pressure in the process are relatively low [17]-[20], only one pass is required to obtain the shaping of the work piece but the speed is much lower compared to the stamping –drawing process.

4. Conclusions

As a conclusive principle and fundamental objective of this work was obtained aluminium sheets 1100, the value of 68 ° maximum angle of formability an obedient taper experimental geometry. With the above provides a starting point and also an important step in finding a near future implementation of the process and a piece of industrial applicability.

The results shown in this paper and demonstrate, FLC curves constructed for the product obtained with the maximum angle of formability argue the ability of the material, in this case aluminium that it deforms plastically without presenting any kind of defects for the critical angle obtained in this work, which is 68 °. This will make safe and acceptable of the application of Dieless-SPIF process to national and international manufacturing environment.

Dieless-Spif process is a relatively new process compared with the sausage pieces, allows greater flexibility in the design of various surfaces, even complex, requires high labor costs in tooling and dies. Parts formability is acceptable, as demonstrated in the results of this work.

Acknowledgments
This work was supported by the EAFIT Manufacturing department and its laboratories. The author's thanks EAFIT, production technologies group and it's I&D program.

References

[1] MATSUBARA, S. Incremental backward bulge forming of a sheet metal with a hemispherical tool. J.JSTP, (1994), Vol.35. P 1311-1316.
[2] CAMPOS, A.; ESCAMILLA, A.; SILVA, J.; GONZÁLEZ, A. Diagramas límite de conformado. 8° congreso Iberoamericano de Ingeniería Mecánica (2007), P 1-8.
[3] DEJARDIN, S.; GELIN, J.C.; THIBAUD, S. Experimental investigations and numerical analysis for improving knowledge of incremental sheet forming process for sheet metal parts. Journal of Materials Processing Technology 2010, 201 (2), 363–369.
[4] KALPAKJIAN, SEROPE (2008). *Manufactura, ingeniería y tecnología*. México. Quinta edición. Pearson educación.
[5] [5] SMITH, WILLIAM (2004). *Ciencia e ingeniería de materiales*. España. Tercera edición. McGraw-Hill.
[6] GARCÍA, JA., PARAMO G. Análisis y caracterización del método Single Point Incremental Forming en la tecnología de control numérico para un caso de estudio. Universidad EAFIT, (2011).
[7] AMINO, H, Lu, Y., Ozawa, S., Fukuda, K. and Maki, T. Dieless NC forming of automotive service panels. Advd Technol. Of Plasticity, (2002), Vol.2. P. 1015-1020
[8] JESWIET, J., HAGAN, E., SZEKERES, A. Forming parameters for incremental forming of sheet metal, CIRP Annals. Manufacturing technology, Volume 54, Issue 2, (2005). P 1367-1371.
[9] XIAOFAN SHI, L. GAO, HOSEIN KHALATBARI, YAN XU, HUI WANG, LINGLING JIN. Electric hot incremental forming of low carbon steel sheet: accuracy improvement. International Journal of Advanced Manufacturing Technology, Volume 3, Issue 4, (2013). P 1-8
[10] RAJIV MALHOTRA, LIANG XUE, TED BELYTSCHKO, JIAN CAO. Mechanics of fracture in single point incremental forming. Journal of Materials Processing Technology, Volume 212,Issue: 7, (2012). P 1573-1590
[11] FRANCESCO SETTI, RUGGERO BINI, MASSIMO LUNARDELLI, PAOLO BOSETTI, STEFANIA BRUSCHI, MARIOLINO DE CECCO. Shape measurement system for single point incremental forming (SPIF) manufacts by using trinocular vision and random pattern. Measurement science and technology, Volume 23, Issue 11, (2012). P 1-7.

[12] GHULAM HUSSAIN, NASIR HAYAT, GAO LIN. Pyramid as test geometry to evaluate formability in incremental forming: Recent results. Journal of Mechanical Science and Technology, Volume 26, Issue 8, (2012). P 2337-2345

[13] M. HAM, B. M. POWERS. Roughness Parameter Selection for Novel Manufacturing Processes. Scanning, Volume 8, Issue 2, (2013). P 1-9

[14] M. C. RADU, I. CRISTEA. Processing metal sheets by SPIF and analysis of parts quality. Materials and Manufacturing Processes, Volume 28, Issue 3, (2013). P 287-293

[15] G. HUSSAIN , H. R. KHAN , L. GAO , N. HAYAT B. Guidelines for tool-size selection for single-point incremental forming of an aerospace alloy. Materials and Manufacturing Processes, Volume 28, Issue 3, (2013). P 324-329

[16] M. B. SILVA, T. M. MARTINHO, P. A. F. MARTINS. Incremental forming of hole-flanges in polymer sheets. Materials and Manufacturing Processes, Volume 28, Issue 3, (2013). P 330-335

[17] BEHERA, A.K. , VERBERT, J. , LAUWERS, B. , DUFLOU, J.R. Tool path compensation strategies for single point incremental sheet forming using multivariate adaptive regression splines. Computer Aided Design, Volume 45, (2013).P 575-590

[18] M.B. SILVA, P.A.F. MARTINS. Two-Point Incremental Forming with Partial Die: Theory and Experimentation. Journal of Materials Engineering and Performance, Volume 22, Issue 4, (2012). P 1018-1027

[19] ADAMS, D.W., JESWIET, J. Energy consumption in single point incremental forming. Transactions of the North American Manufacturing Research Institution of SME, Volume 40, (2012). P 738-742

[20] GARCIA-ROMEU, CERETTI E, FIORENTINO, GIARDINI. Forming force prediction in two point incremental forming using Backpropagation neural networks in combination with Genetic Algorithms. ASME 2010 International Manufacturing Science and Engineering Conference, (2010) P 99-106

Impact of Company Size on Manufacturing Improvement Practices: *An empirical study*

C S Syan and K Ramoutar

The University of the West Indies, St. Augustine, Trinidad and Tobago

E-mail: chanan.syan@sta.uwi.edu

Abstract: There is a constant search for ways to achieve a competitive advantage through new manufacturing techniques. Best performing manufacturing companies tend to use world-class manufacturing (WCM) practices. Although the last few years have witnessed phenomenal growth in the use of WCM techniques, their effectiveness is not well understood specifically in the context of less developed countries. This paper presents an empirical study to investigate the impact of company size on improving manufacturing performance in manufacturing organizations based in Trinidad and Tobago (T&T). Empirical data were collected via a questionnaire survey which was send to 218 manufacturing firms in T&T. Five different company sizes and seven different industry sectors were studied. The analysis of survey data was performed with the aid of Statistical Package for Social Sciences (SPSS) software. The study signified facilitating and impeding factors towards improving manufacturing performance. Their relative impact/importance is dependent on varying company size and industry sectors. Findings indicate that T&T manufacturers are still practicing traditional approaches, when compared with world class manufacturers. In the majority of organizations, these practices were not 100% implemented even though they started the implementation process more than 5 years ago. The findings provided some insights in formulating more optimal operational strategies, and later develop action plans towards more effective implementation of WCM in T&T manufacturers.

Keywords: Manufacturing Improvement, Manufacturing Performance, Trinidad and Tobago

1. Introduction

World class manufacturing (WCM) is under much discussion today. Regardless of this, there is no definition that is universally accepted [1]. Hayes and Wheelwright (1984) introduced the term to describe organizations that used their manufacturing competences to achieve a global competitive advantage [2]. World-class status is a never-ending journey since the concepts are constantly being improved upon. Despite the many shortcomings, the positive impact of WCM has been emphasized in numerous studies [3][4]. Other factors that have an impact on company's performance such as type of industry and company size are generally not considered in many of these studies.

The government of T&T needs to diversify the economy to ensure its survival in this rapidly changing, international environment. T&T's economy comprises of the energy and the non-energy sectors with the major contributor to the country's economy being the energy sector. Globally, T&T competitiveness was ranked 84 out of 139 countries [5]. The contribution of the energy sector declined to 40.2 percent of total GDP in 2012. Manufacturing is the second largest non-energy sub-sector. The estimated contributions to a country's gross domestic product for manufacturing industry was computed and it was found that the manufacturing industry only contributed for 5 % of GDP.

The manufacturing sector in T&T today, still faces many challenges. Hence, T&T manufacturers should be forging ahead and implementing practices to improve its performance. Research of this nature is not prevalent in developing or less developed countries. This paper presents an empirical study to investigate the impact of company size on improving manufacturing performance in manufacturing organizations based in Trinidad and Tobago (T&T).

2. Literature Review

Many organizations implement practices to improve performance effectively and achieve benefits such as a reduction in errors/waste, improvements of profit margins competitive advantage and increase in productivity [6]. On the other hand, many encounter problems and are less successful [7][8]. Therefore there is a need to investigate why some organizations are less successful in their improvement initiatives and determine what can be done to facilitate these improvement plans.

2.1. Existing Frameworks/Models – Overview

Many techniques have been designed and justified by researchers to improve manufacturing performance. However, the dynamics between practitioners, consultants and researchers may deter organizations from understanding which best suits their organization [9]. Different techniques probably are continuously tried (partially implemented) and abandoned without any analysis as to why the implementation was unsuccessful [10]. The actual process of implementation is not sufficiently addressed in the literature. Ironically, the implementation process has been reiterated as crucial and needs more focus.

Before detailed guidelines are recommended, it is necessary to consider what are the facilitating factors which persuade an organization to obtain a sustainable advantage over the competition and the impeding factors which inhibit an organization from gaining advantage over the competition.

2.2. 2.2 Facilitating and Impeding Factors

Researchers have classified facilitating and impeding factors in various ways. For example, some authors discuss external facilitating factors such as global competition and international customers' needs [11][12]. Chan and Swatman (2000) include internal facilitating factors such as organizational strategies and saving cost [13]. Many researchers have argued that the implementation process has many limitations such as lack of training and education, overly optimistic expectations and cultural resistance to change [14][15]. These facilitating and impeding factors would differ depending on cultural, economical, social and political situations [16][17]. Many researchers conclude that impeding factors are more prevalent in developing countries than developed countries.

Based on a comprehensive literature review, four (4) facilitating and seven (7) impeding factors were identified. Each of these facilitating and impeding factors was then broken down into forty (40) sub-factors [18]. These are listed in the Appendix A.

2.3. 2.3 The Influence of Organization Size

The literature has presented different views on the effect of organizational size on WCM implementation. Haar and Spell (2008) and Temtime (2003) have concurred that the adoption of

WCM practices is affected by the size of an organization [19][20]. An increase in size means an increase in resources, investment and expertise, which positively affects the plans for improving performance. Conversely, Gulbro et al. (2000) argued that small firms are lagging behind large firms with regards to continuous improvement activities [21]. This is in contrast to Ahire et al. (1996) who found that the size of an organization does not affect their ability to effectively implement WCM practices [22].

3. An Empirical Study

Based on previous studies conducted in this research area, quantitative research was more prominent. The quantitative method requires the use of predetermined response categories. A standardized research instrument is employed to ensure that varying perspectives can be easily compared. Numbers/values are assigned instead of open-ended comments. The main advantages of this method are that it can collect data for a large sample size, data collection is fast and low cost [23]. However, Poolton (1994) emphasized that the design of the survey is critical because answers are final. Surveys result in a broad, generalized set of findings which is presented concisely [23].

3.1. Design of Survey Instrument

Two consecutive rounds of pre-testing were conducted. First, the questionnaire was reviewed by academic researchers from various developing countries (England and Egypt) experienced in questionnaire design. They returned the questionnaires with their comments and revisions were made based on their recommendations.

Next, the questionnaire was formally pre-tested (piloted) with two manufacturing firms in T&T. The pilot took the form of an interview. The participants were first asked to complete the questionnaire. Then the representatives were asked to provide feedback on ease of comprehension, clarity of the questions, possible changes and suggestions for additional items. The representatives' suggestions were then carefully evaluated and revisions were made to the questionnaire modified. The formal survey was started in 2010.

3.2. Actual Survey Process

A total of 218 light-manufacturing firms in T&T were selected to participate in this survey. Respondents were selected from the Trinidad and Tobago Manufacturers' Association membership as well as Central Statistical Office database. To ensure that the responses obtained were representative of the larger population, the manufacturing companies were stratified.

The target respondent was the production/operation manager or chief executive officer/managing director. Data collecting involved a face-to-face administered survey questionnaire where respondents are allowed to complete the questionnaire with the researcher there. If the respondent had any questions pertaining to the questionnaire, the matter could be resolved there. A face-to-face administered questionnaire was the method used in order to get a response rate that was acceptable for data analysis.

4. Findings

A total of 218 light-manufacturing firms in Trinidad and Tobago were selected as the sample and a total of 47 completed questionnaires were received yielding a response rate of 21.6%. According to Flynn et al (1994) a sample size (n) of 30 or more is statistically sufficient for the analysis.

4.1. Valid Responses – Descriptive Analysis

Table 1 shows more details of the general characteristics of the surveyed companies, based on the firm's size, the manufacturing sector, type of ownership and the certification attained. The respondent's role in the firm is also included. Most of the responses were obtained from Group 1 (1-49 employees) and Group 3 (100 – 249 employees). Pertaining to industry sectors, majority of responses were received from the Food, Drink and Tobacco sub-sector (21 %). A great majority of surveyed organizations (77%) were of local private ownership. Majority (52.8 %) of the respondents were involved in implementation plans to improve their performance for more than 5 years. Only 8.3% were younger than 20 years old and the majority (91.7%) were more than 20 years old. Unfortunately, the majority of organizations had no certification (47%) such as ISO 9000 certified or other forms of certifications.

Table 1: Descriptive Analysis of Valid Responses

Size	%	Sector	%	Ownership	%	Certification	%	Respondents role in firms	%
1-49	32	Food, Drink & Tobacco	21	Local Private	77	ISO 9000 / 1/2 /3	28	Engineer	26
50 - 99	23	Textiles, garments, footwear & headwear	5	Local & Overseas	2	No Certification	47	CEO/MD	32
100 - 249	30	Printing, publishing & paper converters	17	Overseas	13	Other Certification	25	PM/OM	42
250-500	11	Wood & related products	4	Government	4				
> 500	4	Chemicals & non-metallic minerals	15	Joint Private & Govt	4				
		Assembly type & related industries	19						
		Miscellaneous manufacturing	19						

4.2. Analysis based on Company Size – Facilitating/Impeding Factors

Respondents rated the facilitating and the impeding factors towards initiatives to improve manufacturing performance. Table 2 (a) shows the mean and standard deviations (SD) of the facilitating factors and Table 2 (b) shows the impeding factors towards improving manufacturing performance. The data in both tables are arranged according to company size.

Overall, the top five facilitating factors were 'Improving quality of Product/s', 'Local Competitiveness', 'Change in Customer Needs', 'Reducing Operating Cost' and 'Avoid Losing Market Shares to Competitors'. Findings also showed that Group 1 (1-49 Employees) and Group 2 (50 – 99 Employees) were encouraged to improve their performance because of 'improving quality of Product/s'. The small manufacturers seem to be focused on quality improvement in order to compete. However, Group 5 (>500 Employees) were encouraged to improve manufacturing performance based on 'Change in Customer Needs' and 'Reducing Operating Cost'.

Table 2 (a): Ratings of Facilitating Factors by Company Size

…tatin g Facto	Company Size (Employees)											
	1 – 49		50 – 99		100 – 249		250 – 500		> 500		Total	
	Mean	SD	Mean	SD	Mean	SD	Mean	SD	Mean	SD	Mean	SD
IQP	4.17	0.994	4.27	0.704	4.00	0.734	4.44	0.727	4.16	0.898	4.208	1.570
LCm	3.96	0.991	4.00	0.926	4.25	0.930	3.71	1.490	4.00	1.118	3.984	0.983
EJ	3.06	1.130	3.59	1.121	4.06	0.772	3.64	1.151	3.40	0.876	3.550	0.899
ROC	3.80	1.110	4.00	1.054	3.47	1.060	3.79	0.579	4.41	0.998	3.894	1.673
CICN	3.77	1.091	4.00	0.907	3.85	0.899	3.36	0.842	4.56	1.890	3.908	1.117
COS	2.21	0.933	2.75	1.251	4.36	0.745	2.86	0.663	3.79	1.142	3.194	1.098
ALMS	3.92	0.973	4.06	0.998	3.43	1.016	3.93	0.884	3.84	1.167	3.836	0.908
GC	3.13	1.242	3.85	0.988	3.56	1.263	3.54	1.267	2.76	0.799	3.368	1.450
GP	3.53	0.998	3.75	0.910	3.75	1.000	3.92	0.954	2.99	3.017	3.588	1.021
BEA	3.58	1.256	3.31	0.946	4.00	0.966	3.08	1.188	3.96	1.309	3.586	0.914
DIT	3.60	1.397	3.76	0.903	2.93	1.438	3.92	0.881	3.50	1.552	3.542	1.402

Food, Drink and Tobacco are larger more prominent manufacturers in T&T. This explains their main facilitating factor as 'Global Competition'. These manufacturers are at the level to compete globally. Printing, publishing & paper converters manufacturers are mainly small manufacturers struggling to survive. Hence, 'Change in Organizational Strategy', their main facilitating factor was necessary to attempt to improve manufacturing performance for survival.

The top five impeding factors were 'Lack of Suitable Organizational Culture for Change', 'Improper Identification of Training Needs', 'Relevant Documents / Systems Unavailable', 'Insufficient Knowledge Base' and 'Need for Cost Justification'.

The findings also showed that Groups 1 (1-49 Employees) and 2 (50 – 99 Employees) highlighted 'Lack of Knowledge', 'Unavailability of Resources' and 'Lack of Communication from those in Authority' as impeding factors. Group 5 (>500 Employees) were deterred by 'Lack of Implementation Knowledge', 'Managers/Employees Resistance to Change' and 'Insufficient level of education and skills of workers'. Both categories of company size on opposite sides of the spectrum ranked lack of knowledge whether it is implementation knowledge or the education level of their workers as impeding factors. Education seems to be major impeding factor for all sizes of companies. This fact can suggest why there is resistance to change within T&T manufacturers: lack of knowledge on these improvement practices / techniques.

Table 2 (b): Ratings of Impeding Factors by Company Size

Impeding Factors	Company Size (Employees)											
	1 – 49		50 – 99		100 – 249		250 – 500		> 500		Total	
	Mean	SD	Mean	SD	Mean	SD	Mean	SD	Mean	SD	Mean	SD
LEIIP	2.62	1.031	2.52	0.151	2.50	1.414	2.04	0.887	2.76	0.928	2.48	1.908
ILESW	2.60	2.850	2.59	0.418	2.80	1.424	1.86	0.963	3.04	1.930	2.57	0.934
LSOCC	3.51	0.983	2.91	0.155	3.00	1.195	3.02	1.059	2.98	1.089	3.09	2.019
MERC	2.49	1.109	2.21	1.311	3.33	1.397	2.43	1.176	3.10	0.941	2.71	0.998
ITIP	2.47	1.085	2.82	1.468	2.75	1.511	2.30	1.082	2.17	0.689	2.50	1.309
UEE	2.00	1.108	2.85	1.214	2.93	0.884	2.62	1.031	2.64	1.307	2.60	1.893
INP	1.57	0.824	2.64	1.216	2.69	0.947	2.60	2.850	2.93	1.223	2.48	1.016
LIK	1.99	0.944	2.64	1.307	2.64	1.084	2.51	0.983	3.22	1.166	2.60	0.875
UR	2.98	0.892	2.93	1.223	3.69	1.182	2.49	1.109	2.00	1.414	2.61	1.411
LK	3.00	1.103	3.22	1.166	2.64	1.284	2.47	1.085	2.40	0.986	2.74	1.605
IRS	2.13	1.038	2.00	1.414	3.69	1.251	3.42	1.240	2.50	1.414	2.74	0.908
IITN	2.30	0.932	3.40	0.986	3.57	1.284	3.08	1.165	2.80	1.424	3.03	0.795
IKB	2.47	1.089	2.93	1.335	3.31	1.251	2.93	0.829	3.00	1.195	2.95	1.905
NCJ	2.28	1.089	2.73	1.163	3.08	1.311	3.07	1.033	3.33	1.397	2.89	1.449
LCA	2.51	0.980	3.00	1.134	3.33	1.557	3.13	1.258	2.05	1.059	2.80	1.296

ED	2.11	0.946	3.02	1.240	2.44	1.365	3.17	0.937	1.99	0.943	2.62	0.893
LMM	2.69	0.905	3.08	1.165	2.75	1.571	3.56	1.365	1.57	0.826	2.73	0.947
LSS	2.04	0.887	2.93	0.829	2.60	1.298	2.86	0.939	1.98	0.890	2.48	0.947
LIIDC	1.86	0.963	3.00	1.033	3.23	1.015	3.63	1.147	1.86	0.963	2.71	1.013
LKSTM	2.02	1.059	3.13	1.258	2.83	1.264	3.22	1.166	2.02	1.059	2.64	1.519
LEE	2.43	1.176	3.17	0.937	2.09	0.980	2.09	0.980	2.43	1.176	2.44	0.989
LAMI	2.30	1.082	2.86	0.949	2.21	1.109	2.21	1.109	2.30	1.082	2.37	0.932
FACR	2.28	1.103	3.00	0.655	2.76	0.928	2.76	0.928	2.28	1.103	2.61	0.867
ISI	2.15	1.067	2.86	0.949	3.04	1.930	3.04	1.930	2.60	1.298	2.73	0.879
RDU	2.02	1.051	3.56	1.365	2.98	1.089	2.98	1.089	3.16	1.015	2.94	1.221
OOEM	2.05	1.059	2.86	0.939	3.10	0.941	2.00	1.103	2.83	1.264	2.56	0.889
LMI	1.99	0.943	3.63	1.147	2.17	0.689	2.13	1.038	2.09	0.980	2.40	0.889
BGIIG	1.57	0.826	3.22	1.166	1.84	0.779	2.30	0.932	2.21	1.109	2.22	1.097
LTMCS	1.98	0.890	2.67	1.155	1.99	1.326	2.47	1.089	2.76	0.928	2.37	0.990

4.3. Impact of Implementing Practices towards Improving Manufacturing Performance

The most modern practices/techniques are less implemented by the T&T manufacturing firms. This will lead T&T manufacturers to apply older techniques that lie under umbrella of mass production approach. From the data, 52.8% of T&T manufacturers had more than 5 years of experience with implementing practices and 30.6 % had 3 – 5 years' experience. The most benefit the organization had from implementing these practices is increasing their 'Competitiveness' and being more 'Quality oriented'.

5. Conclusions / Future Work

Adopting world class approaches to improve manufacturing performance has been studied but not common in the Caribbean. This research focused on the factors facilitating and impeding the implementation of such approaches. An assessment of the approach currently implemented within manufacturing organizations in T&T was also conducted to determine what is required to improve performance.

Research work of this nature has not been conducted in T&T so this information would be used to obtain a better understanding of the current status of the manufacturing sector and the direction these companies can take to improve their performance

The findings provided some insights in diagnosing and developing improvement plans/strategies for manufacturing organizations. This would be of value for both manufacturers and researchers in the areas of WCM.

This study focused mainly on organization size. Other factors on improving manufacturing performance such as industry type, type of ownership, age of firms and geographical location could be investigated. More analysis can also be carried out for the factors (organization size and industry type) studied in this research work. A hypothetical approach to improve manufacturing performance was utilized in this research. Longitudinal studies / case studies for assessing manufacturing performance can be researched. This type of detailed approach would provide the analysis required for developing model/framework.

6. References

[1] Kodali R B, K S Sangwan and V K Sunnapwar 2004 Performance value analysis for the justification of world-class manufacturing systems *Journal of Advanced Manufacturing Systems* 3(1): 85–102.

[2] Hayes, R H and S C Wheelwright 1984 *Restoring our Competitive Edge: Competing through Manufacturing* (John Wiley: New York)

[3] Salaheldin I S and R Eid 2007 The implementation of world class manufacturing techniques in Egyptian manufacturing firms: an empirical study *Industrial Management & Data Systems* 107(4):551–566.

[4] Sharma, M and R Kodali 2008 TQM implementation elements for manufacturing excellence *The*

TQM Journal 20(6):599–621.
[5] World Economic Forum 2010 *The Global Competitiveness Report 2010-2011* (World Economic Forum: SRO-Kundig, Switzerland)
[6] Tobon, F 2007 New survey reveals positive impact of ISO 9001:2000 in Colombia *ISO Management Systems Magazine* May/April:41-44.
[7] Aarts, F M and E Vos 2001 The impact of ISO registration on New Zealand firms' performance: a financial perspective *The TQM Magazine*, 13(3):180-191.
[8] Tannock, J, L Krasachol and S Ruangpermpool 2002 The development of total quality management in Thai manufacturing SMEs *International Journal of Quality and Reliability Management* 19(4):380-395.
[9] Hermel, P and J Ramis-Pujol 2003 An evoluation of excellence: some main trends *The TQM Magazine* 15(4):230–243.
[10] Miller, T R and B J Vaughan 2001 Messages from the management past: Classic writers and contemporary problems *SAM Advanced Management Journal* 66:4–11.
[11] Assiri A, M Zairi and R Eid 2006 How to profit from the balanced scorecard: an implementation raodmap *Industrial Management & Data Systems* 106(7):937–952.
[12] Eid, R and M Trueman 2004 Factors affecting the success of business-to-business international internet marketing (B-to-B IIM): an empirical study of UK companies *Industrial Management and Data Systems* 104(1):16–30.
[13] Chan, C and P M C Swatman 2000 From EDI to internet commerce: the BHL steel experiences *Internet Research: Electronic Networking Applications and Policy* 10(1):72–82.
[14] Doyle, K 1992 Who's killing total quality? *Incentive* 16(8):12–19.
[15] Westphal, J D, R Gulati and S M Shortell 1997 Customization or conformity? An institutional and network perspective on the content and consequences of TQM adoption *Adminstrative Science Quarterly* 42(2):366–394.
[16] Mersha, T 1997 TQM implementation in LDCs: driving and restraining forces *International Journal of Operations and Production Management* 17(2):164-183.
[17] Szirmai, A 1997 *Economic and Social Development* (Prentice Hall: London)
[18] Ramoutar, K and C S Syan, 2010 Drivers and Barriers for WCM In Manufacturing Firms in Trinidad and Tobago *Proceedings of the 25th ISPE International Conference CAD/CAM, Robotics and Factories of the Future 2010, South Africa*
[19] Haar, J M and Spell, C S 2008 Predicting total quality adoption in New Zealand *Journal of Enterprise Information Management*, 21(2):162-178.
[20] Temtime, Z T 2003 The moderating impacts of business planning and firm size on total quality management practices *The TQM Magazine* 15(1):52-60.
[21] Gulbro, R D, L Shonesy and P Dreyfus 2000 Are small manufacturers failing the quality tests? *Industrial Management and Data Systems* 100(6)76-80.
[22] Ahire S L 1996 TQM age versus quality: An empirical investigation *Production and Inventory Management Journal* 1: 18–23.
[23] Poolton, J 1994 *Concurrent Engineering Establishment: A Framework Proposal* (PhD Thesis, University of Liverpool)

Appendix A – List of Facilitating and Impeding Sub-factors

This list the facilitating and impeding sub-factors (abbreviations) used in paper.

#	FACILITATING SUB-FACTORS	
1	CICN	Change in Customer Needs
2	IQP	Improving Quality of Product
3	LC	Local Competitiveness

IOP Conf. Series: Materials Science and Engineering **65** (2014) 012028 doi:10.1088/1757-899X/65/1/012028

4	ALMS	Avoid Losing Market Shares to Competitors who already implemented practices
5	ROC	Reducing Operating Costs
6	DIT	Development in Technology
7	GC	Global Competition
8	EJ	Economical Justification
9	COS	Change in Organizational Strategies
10	BEA	Being an Early Adopter
11	GP	Government Pressures

#		**IMPEDING SUB-FACTORS**
12	LSOCC	Lack of Suitable Organizational Culture for Change
13	IITN	Improper Identification of Training Needs
14	ITIP	Insufficient Training in Implementation Process
15	IKB	Insufficient Knowledge Base
16	INP	Indiscipline - Non-conformance with Procedures
17	LIIDC	Lack of Inter and Intra Departmental Communication
18	LPK	Lack of Practices Knowledge
19	IESLW	Insufficient Education and Skills Level of Workers
20	LEE	Lack of Employee Empowerment
21	IRS	Insufficient Rewards Systems
22	LIK	Lack of Implementation Knowledge
23	LEI	Lack of Employee Involvement
24	LCA	Lack of Communication from those in Authority
25	RDU	Relevant Documents Unavailable
26	MERC	Managers/Employees Resistance to Change
27	LMM	Lack of Motivation from Management
28	UEE	Un-sustained Enthusiasm by Employees
29	LTMCS	Lack of Top Management Commitment and Support
30	UR	Unavailability of Resources
31	LKSTM	Lack of Knowledge and Skills of Top Management
32	LSS	Lack of Structural Support
33	LAMC	Lack of Appropriate Monitoring and Control
34	FACR	Failure to Analyze the Company's Readiness
35	OOEM	Overly Optimistic Expectations by Management
36	NCJ	Need for Cost Justification - Planning
37	ISI	Improper System Initiation
38	LMI	Lack of Management Involvement
39	ED	Empowerment Difficulties
40	BPGIIG	Business Performance Goals Inconsistent with Implementation Goals

A framework for development of an intelligent system for design and manufacturing of stamping dies.

H.M.A. Hussein[1] and S. Kumar[2]
[1]Advanced Manufacturing Institute, King Saud University, Riyadh 11421, Saudi Arabia.
[2]Dept. of Mechanical Engineering, S.V. National Institute of Technology, Surat-395007, Gujarat, India.

E-mail: hhussein@ksu.edu.sa, skbudhwar@med.svnit.ac.in

Abstract. An integration of computer aided design (CAD), computer aided process planning (CAPP) and computer aided manufacturing (CAM) is required for development of an intelligent system to design and manufacture stamping dies in sheet metal industries. In this paper, a framework for development of an intelligent system for design and manufacturing of stamping dies is proposed. In the proposed framework, the intelligent system is structured in form of various expert system modules for different activities of design and manufacturing of dies. All system modules are integrated with each other. The proposed system takes its input in form of a CAD file of sheet metal part, and then system modules automate all tasks related to design and manufacturing of stamping dies. Modules are coded using Visual Basic (VB) and developed on the platform of AutoCAD software.

1. Introduction

In sheet metal industries, design of stamping dies is generally considered an art rather than science. Since the early seventies, there has been a remarkable advancement in computer aided design (CAD) of stamping dies. But till today, not a single CAD system is available in the market which is capable to fully automate the task of design of stamping dies. From early seventies, many researchers applied efforts to develop CAD systems for design of specific type of stamping die such as blanking die, progressive die, and deep drawing die. The developed systems were not integrated in nature and required lot of human intervention. Within the mid of nineteen's, the research work in this area inclined towards development of expert systems to simulate the experts' logical thinking in die design. With the start of 21st century, there has been a remarkable research interests towards the integration of CAD, computer aided process planning (CAPP) and computer aided manufacturing (CAM) in the area of design and manufacturing of stamping dies. In the absence of an automated system for design and manufacturing of stamping dies, experienced die designers and process planners are compelled to use many general purpose and costly CAD and CAM software's/packages. Besides this, the quality of stamping die designed using these systems is much dependent on the experience and skill of die designers.

Development of an intelligent system for automated design and manufacturing of stamping dies is an urgent requirement of sheet metal industries. This type of system can be developed using

[1] H.M.A. Hussein, "hhussein@ksu.edu.sa".

applications of expert systems in CAD, CAPP and CAM of stamping dies. In this paper, a framework for development of an intelligent system for design and manufacturing of stamping dies is proposed. According to the proposed framework, the system initially receives the CAD file of sheet metal part. Then it automatically extracts design features from the CAD file of part and identifies the required operations and their sequence for manufacturing the part. Next, the system selects the required type of stamping dies, and its arrangement for carrying out identified sheet metal operations. For each type of stamping die, expert system is to be developed for optimal die design. Once the task of die design is completed, the system displays its final output in form of three-dimensional (3-D) drawings of die components and assembled die. From 3-D drawing of die components, required data is extracted automatically for execution of processes planning module. The proposed system is to be coded using Visual Basic on the platform of AutoCAD software. Microsoft products such as excel and accesses are required to be used as a package database.

2. Literature review

Some researchers have applied efforts to develop computer aided system for design of stamping dies. For example, Cheok and Nee [1] from National University of Singapore developed a computer aided intelligent system for design of progressive dies. The system covers sheet metal unfolding, strip-layout, punch recognition, and die design. Lee et al. [2] proposed a CAPP system for progressive die. Wang [3] from Gintic Institute of Manufacturing Technology developed a CAD system for progressive die using feature based design approach. This system has five modules which covers all applications of progressive dies. In his further work [4], he constructed a CAD system which covers three types of dies namely progressive, compound and deep drawing dies. Hwang et al. [5] from Busan National University developed an automated progressive design system with multiple processes (piercing, bending, and deep drawing).

Park [6, 7] from Busan National University discussed how the cost estimation of the die has important role in decision making in the tool and die factories. He developed an integrated CAD/CAPP/CAM for both sheet metal dies and plastic moulds. He used Taylor made software's such as "In-MouldTM for initial estimating calculations, Pro-MessTM for detailed estimate, and Pro-PESSTM-for press Die. The integrated system covers all activities of design and manufacturing of dies and moulds. It generates die cost estimation and reports to help the workshop administration to take the right decision.

The new trends in Europe are focused more on advanced technical machines instead of dies. Cser et al. [8] suggested a work cell consisting of cutting machine (laser or punching nippling machine) and bending machine. The sheet metal part automatically transferred from one machine to other using a belt conveyor. The system controls the part position during moving through computer vision technique. Recently, researchers have applied efforts to develop integration in many industrial applications. For example, integration in prismatic machined parts [9], integration in cylindrical machined parts [10], Integration in manufacturing using CNC machines based on STEP [11] and Integration in sheet metal machining applications [12]. In fact, all these kind of integrations has their impact on the intelligent integrated system in sheet metal die design and manufacturing.

From the critical review of available literature, it is concluded that no intelligent system is available which is capable to fully automate the design and manufacturing tasks of all types of stamping dies. Authors have applied efforts to develop a framework for development of such type of system.

3. The proposed framework of Intelligent system

The proposed framework of intelligent system is depicted in Figure 1. This framework consists of four main sections. The first section is concerned with the feature extraction and process planning. The system takes its input in the form of 3-D CAD model of sheet metal part. Then it analyzes part data base -STEP file to extract and classify the part data. Next, it recognizes part features and prepares process plan for the part. The system also identifies the required operations and their sequence for manufacturing the part. Next, the system selects the required type of stamping dies, and its arrangement for carrying out identified sheet metal operations. It must be noted that once the sheet metal part is inserted in the database with all its related data, many sheet metal applications could be achieved such as cost and time estimation, manufacturability assessment, nesting and strip layout, die type selection, part comparison, die block design, preparation of wire cutting program, writing of CNC laser cutting G-code, features recognition, flame cutting, folding and unfolding, punch arrangements, punch recognition, process planning, indexing and retrieving, CNC drawing preparing, computer vision, CNC nippling machine program preparation, die design etc.

The second section of proposed framework focuses on development of CAD systems for automatic design of stamping dies. The system deals with eight types of stamping dies namely, piercing, blanking, progressive, compound, bending, deep drawing, fine-blanking, and forming. Each system for a specific type of die is structured in form of various modules for design and modeling of die components. Production rule based expert system approach is used to develop system modules. The system finally gives its outputs in form of 3-D drawings of major die components and die assembly.

The third section of proposed framework receives 3-D models of die components automatically generated by the previous modules. The system automatically reads the 3-D model component data base in STEP format. It extracts the component data, recognizes the machined features, and then prepares the process plan for manufacturing die components.

The fourth section of proposed framework is concerned with CAM. It takes inputs in form of 3-D models of die components and process plan generated by the previous modules. It also invites the user to enter the required data of machine which will be used for manufacturing of die components. A database file of that machine can be prepared and recalled by the proposed system. The system automatically prepares G-code for manufacturing die components, and simulates the machining operation.

The proposed system is to be developed as a package on AutoCAD platform using visual Basic. The databases are constructed using MS-Excel and MS-Access.

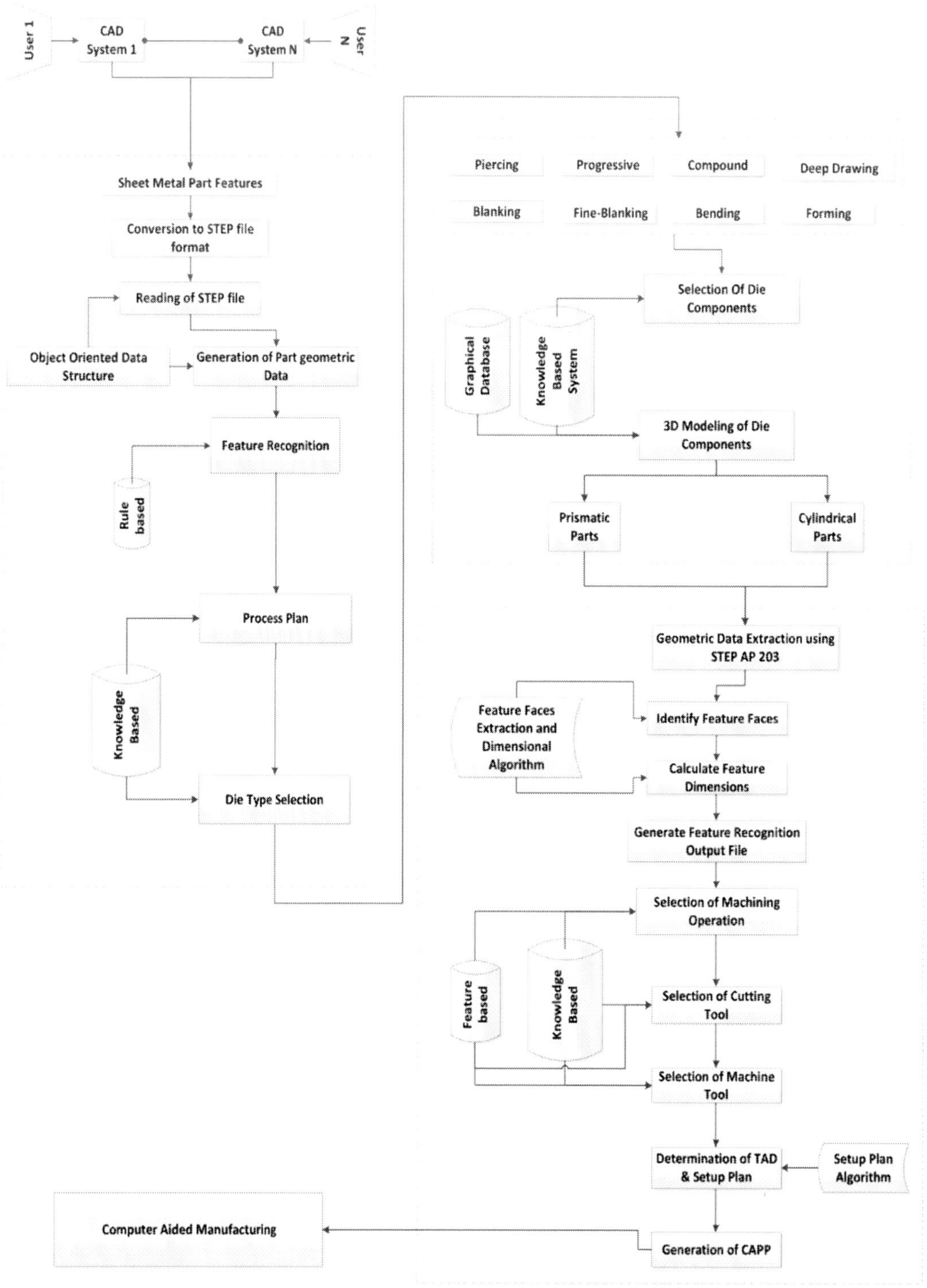

Fig. 1 Proposed framework of intelligent system

Conclusions

The present paper discussed the necessity of an intelligent system for automated design and manufacturing of stamping dies. A framework for development of such system has been proposed. This framework recommends developing the whole system in four sections. The developed system should be capable to automate the design and manufacturing tasks of almost all types of stamping dies without the involvement of experienced and skilled die designers. Authors are engaged in the development of system modules in line with the proposed framework.

Acknowledgment

This work is supported by NSTIP strategic technologies programs, Grant number (12-INF2816-02) in the Kingdom of Saudi Arabia

References

[1] Cheok, B.T, Nee, A.Y.C, 1999" Computer applications in intelligent progressive dies design (IPD)", *Computer Applications in Near-Net-shape Operations*, Advanced Manufacturing Series, Springer Verlag, Chapter 8, pp285-313.
[2] Lee, I.B.H, Lim, B.S., and Nee, A.Y.C., 1993, *international Journal of Production Research*, Vol 31, no.2, pp.251-278, 1993.
[3] Wang, F., 1998, "SM-Design: A CAD Tool for designing die structure based on a feature-based approach", GINTIC Institute of Manufacturing Technology, SMCAD User's Guide for Progressive Die Design.
[4] Wang, F., 2005, "Composer-CAD User's Guide for Sheet Metal Die Design.
[5] Hwang, B.C., Han, S.M., Bae, W.B., and Chul Kim 2009, *International Journal of advanced Technology*, Vol.43, pp.644-653.
[6] Park, S.B., Ramin, N, 2003, *Proceeding of SMEIR conference*, Iran.
[7] Park, S.B., January 4-6, 2004, *Proceeding MDP-8 Cairo University Conference on Mechanical Design and Production,* Cairo, Egypt, pp.999-1006.
[8] Cser, L., Cseleny, J., Geiger, M., Mantyla, M., Korhonen, A.S., 2000, *Journal of Materials Processing Technology*, Vol. 103, pp6-13.
[9] Amaitik, S.M., 3-6,July,2012,*Proceedings of the 2012 International Conference on Industrial Engineering and Operations Management*, Istanbul, Turkey, pp.665:673.
[10] Hussein, H.M.A., Abouel Nasr, E.S. and Khan, A.A. 26-29 June 2013, *Sustainable Intelligent Manufacturing International Conference, SIM2013, Leiria, Lisbon, Portugal*, Green design, Materials and Manufacturing Processes published by Taylor & Francis, pp.233-238.
[11] Valilai, O.F. and Houshmand, M., 20-22 October 2010, *Proceedings of the World Congress on Engineering and Computer Science* , San Francisco, USA., Vol II, WCECS 2010,
[12] Xie, S., 2011 "Compound Machining Method for OKP Product Development", *One-of-a-kind Product Development*, Springer Verlag, London, chapter 5pp.105-124.

Micromachining Lithium Niobate for Rapid Prototyping of Resonant Biosensors

Zeyad Yousif Abdoon Al-Shibaany[1], John Hedley[1], Dehong Huo[1], Zhongxu Hu[1]

[1] School of Mechanical and Systems Engineering, Newcastle University, UK
E-mail: z.y.a.al-shibaany@newcastle.ac.uk; zeyad.al-shibaany@ieee.org

Abstract. Lithium niobate material is widely used in MEMS application due to its piezoelectric properties. This paper presents the micromachining process of lithium niobate to rapid prototype a resonant biosensor design. A high precision CNC machine was used to machine a sample of lithium niobate material at 5 different spindle speeds to find out the best conditions to machine this brittle material. A qualitative visual check of the surface was performed by using scanning electron microscopy, surface roughness was quantitatively investigated using an optical surface profiler and Raman spectroscopy to check the strain of the surface. Results show that the surface quality of the lithium niobate was significantly affected by the spindle speed with optimum conditions at 70k rpm giving a strained surface with 500 nm rms roughness.

1. Introduction

The design and fabrication of point–of–care biosensors attracts the attention of many researchers due to their applicability as they have high sensitivity, low price and the ability for being used outside the specialised laboratories. Mass biosensors are one of the most attractive designs for many applications especially the resonant based MEMS (microelectromechanical sensors) [1] which offers superlative sensitivity. As the main part of the diagnostic device, the resonant based MEMS biosensor needs to be designed and fabricated precisely ideally from crystalline material to give a high quality factor resonator.

The first attempt at developing a degenerate mode mass biosensor [2] at Newcastle University utilised microfabrication of crystalline silicon wafers to produce high quality factor resonators [3] although signal recovery from this capacitively driven and sensed device proved complex for the intended point-of-care applications [4]. To overcome this issue, a composite resonator was designed with thin film PZT for actuation leading to simplified signal recovery at the expense of resonator quality [1], thereby ultimately limiting device sensitivity. The group's aim is to further develop this sensor system, moving away from a silicon based technology to materials that give high mechanical and electrical performance.

The ability to produce MEMS based sensors using microfabrication techniques offers a low cost technology provided production volumes are high. The high overhead costs associated with microfabrication results in prototyping of designs typically costing in the range of £10000 to £100000 and with process flows not yet optimised for such designs, yields can be low. Currently a number of

[1] Zeyad Yousif Abdoon Al-Shibaany.

alternative micro manufacturing processes, such as micro milling, micro laser machining, micro water jet machining, can be used to produce micro components. Micro milling is the most flexible and versatile micro machining process and capable of generating a wide variety of complex micro components and microstructures due to its strengths in achievable precision and surface finish in addition to its simple set-up and versatility [5]. Micro milling as a newly emerged micro machining process has been successfully applied in processing various engineering materials [6].

The fabrication of piezoelectric based micro components is attracting more attention nowadays because of their properties such as compact size, high sensitivity and rapid response [7]. The crystalline form of lithium niobate coupled with its piezoelectric coefficients and chemical stability give this material excellent performance both mechanically and electrically. These characteristics make this material widely used in many applications especially in the field of sensor fabrication [8], applications include ultrahigh speed modulators and acoustic sensors [9]. The surface quality of the lithium niobate is very important aspect during any micromachining process especially when this material is used in precise devices such as biosensors [10]. However due to its low fracture toughness, lithium niobate has a lower material removal rate when machining compared to other materials and it is more likely to be scratched during the machining process [11]. Despite its mechanical and electrical properties, there are a limited number of publications available about micromachining lithium niobate for the purpose of manufacturing sensors.

Fabrication of micro scale devices made from other crystalline materials, such as silicon and graphite, by using conventional machining is reported by several groups. The reason for attempting to machine such brittle materials is to attempt to overcome the high costs and initial low productivity that can result from microfabrication approaches [12]. Mohammed Arif et al used a micro ball end machining process to machine silicon using a cubic boron nitride tool. Their results showed that there is a threshold of the tool feed rate, which had a significant effect on the surface quality, and they proved that it is possible to have a crack free surface by using this micromachining process on silicon [12]. Rusnaldy et al reported that even though there is a high possibility of surface damage when machining a brittle material such as silicon, it is possible to cut the silicon by using a micromilling process under various conditions [13]. They had implemented the micromachining process using diamond–coated end mills in order to machine grooves into silicon [13]. Kovalchenko presented a comprehensive report on machining brittle materials and showed that it is applicable in industry to machine silicon traditionally in order to increase the level of productivity and maintain the surface quality at acceptable levels [14]. In this paper, the micromachining of lithium niobate for the purpose of rapid prototyping a circular diaphragm resonant biosensor, akin to the design of Ismail et al [2], is assessed. The application of micro machining is expected to open new industrial avenues for prototyping biosensors.

2. Methodology

2.1 Micromachining lithium niobate using CNC machine

Experiments for this work were carried out on a NANOWAVE MTS5R – ML CNC machine as shown in figure 1. Standard CNC machine tools were used to ensure that the experimental results are applicable in industry. SAW grade, Z-cut lithium niobate samples of dimensions 10 *mm* × 10 *mm* × 0.5 *mm* thick were used in the experiment. As the samples were to be drilled from the back side down to thicknesses of around 50 μm, the front surface required rigid supported to prevent the crystal from fracturing under the applied drilling pressure. Samples were glued to a holder by using a nail – lacquer glue and then left for three days to ensure they were rigidly bonded.

The tool used in this experiment was a 0.5 mm diameter cubic boron nitride (CBN) square end mill, manufactured by Union Tools. A G – code program was written and performed in order to machine holes in the lithium niobate samples with a depth of 450 μm. Holes were machined at constant spindle speed starting from 40k rpm to 80k rpm in steps of 10k rpm. The reason for choosing five different spindle speeds was to determine the most suitable condition for machining this brittle material and to check the surface quality at each speed. Five holes were drilled at each speed. After finishing the machining process, the samples were removed from the support by submerging in acetone and then cleaned electronically by using a JPL Ultra-8060 Pro Ultrasonic Cleaner filled with water to ensure that the surface of each hole was cleaned before checking the surface quality.

Figure 1. Nanowave MTS5R – ML CNC machine.

2.2 Characterisation

2.2.1 Scanning electron microscopy (SEM)
A Hitachi TM3030 scanning electron microscopy was used to provide a qualitative visual check of the surface quality of each hole. A 15kV accelerating voltage was used in this experiment to provide high resolution topographical imaging at up to 30000x magnifications.

2.2.2 Optical surface profiler
To provide a quantitative evaluation of the quality of each machined surface, a Zygo NewView 5020 optical surface profiler with 10x objective was used. As lithium niobate is transparent, a 30 nm gold coating was evaporated onto the surface to improve measurement quality. A depth measurement of a hole confirmed a drill depth of 444 μm. Profiles were then taken of each drilled surface, figure 3 shows an example for a 60k rpm machined hole, and a plane mathematically subtract from the data. The rms value of the surface was then calculated and the number of points used in this analysis recorded, this number being indicative of the amount of visible (debris free) surface.

2.2.3 Raman system setup
It has been shown that Raman spectroscopy may be used to give a quantitative measure of crystalline quality and stoichiometry of lithium niobate after crystals are processed [11]. To quantify the micromachining process reported here, Raman measurements were taken of the samples using a Horiba Jobin Yvon FHR1000 system. The system utilised a beam from a 632.8 nm helium neon laser, this being focussed with a 50× long working distance objective giving a spot size of 4 μm. The Stokes Raman signal was collected, directed through the spectrometer and recorded on a liquid nitrogen cooled CCD camera set at -140 °C. The system resolution was 0.25 cm^{-1}.
Spurious spikes (defined as greater than 5% increase in intensity from the neighbouring 6 points) were automatically removed from the spectrum and a Voigt profile (a convolution of a Gaussian and Lorentzian profile) was fit to the data using a Matlab nonlinear least-squares fitting routine. The Raman spectra of lithium niobate contain multiple peaks [15]. Work by Gallinetto et al [11] showed that for the 872 cm^{-1} line, the linewidth is indicative of composition whilst peak energy position reflects strains induced in the machined surface. However this peak is asymmetric showing a distinct

shoulder at around 900 cm^{-1}. For this work, only the lower energy side of the peak was fit, as shown in figure 2, between 820 cm^{-1} to 880 cm^{-1} to give an indicative assessment of peak position and width of this line resulting from the machining process.

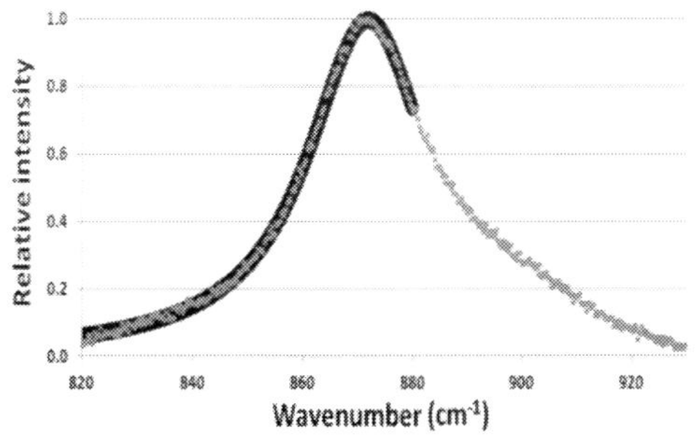

Figure 2. The 872 cm^{-1} spectral line shows a pronounced shoulder at higher energies. The peak was fit using the data points recorded on the lower energy side.

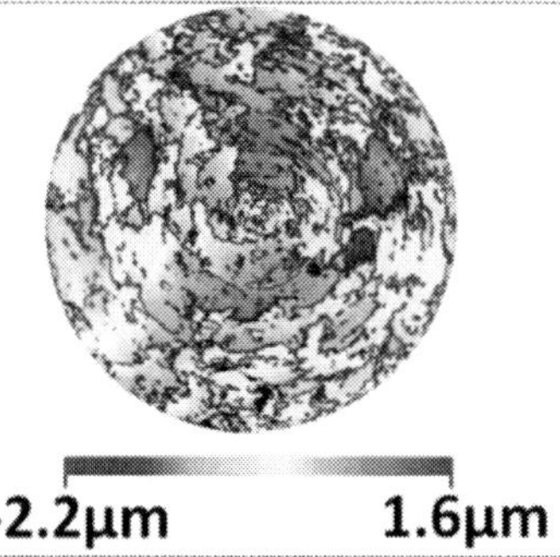

Figure 3. Surface profile measurement of the lithium surface machined at a speed of 60k rpm.

3. Results and Discussion

3.1 SEM results

After finishing the micromachining process to drill 25 holes (5 holes × 5 speeds) in the lithium niobate, the scanning electron microscopy was used to check the surface quality of each hole. Table 1 contains representative images of the surfaces obtained for each speed. The most significant damage occurred at a speed of 40K rpm, the results demonstrating that the slower machining speeds are unsuitable for this material. Above this, surface quality showed more consistency with no cracks being evident on the surface. Despite the ultrasonic clean, residual debris remained on the surface, this appearing to reduce as the machine speed increased.

Table 1. A qualitative comparison of the results of various machine speeds was performed with scanning electron microscope images of the surfaces.

Machining speed	40k rpm	50k rpm	60k rpm	70k rpm	80k rpm
	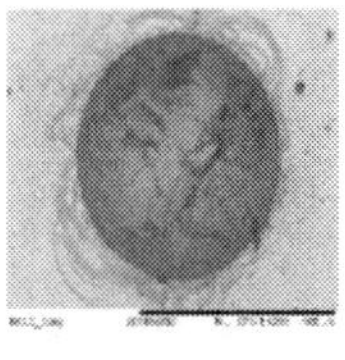	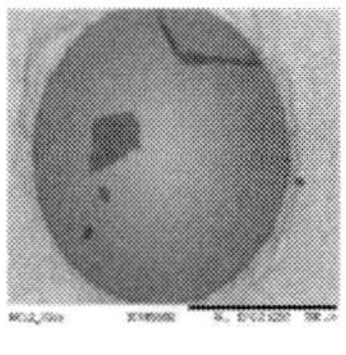	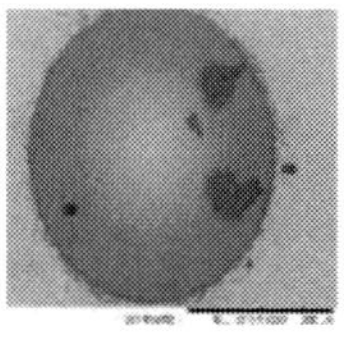	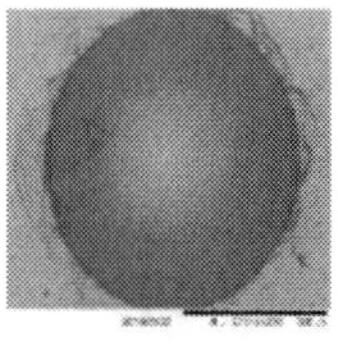	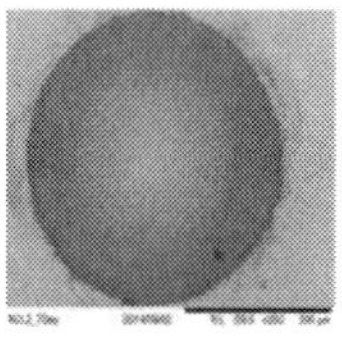
	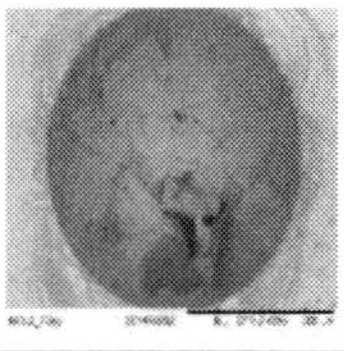	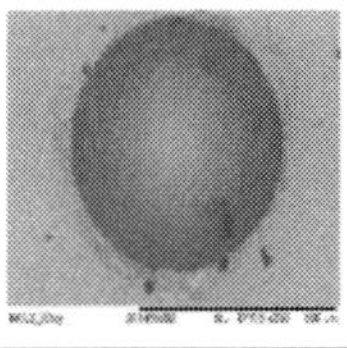	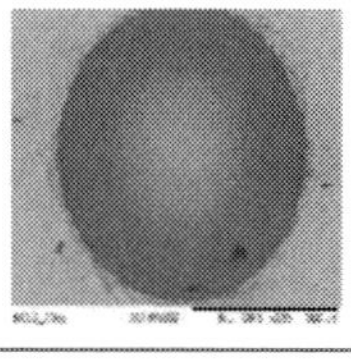	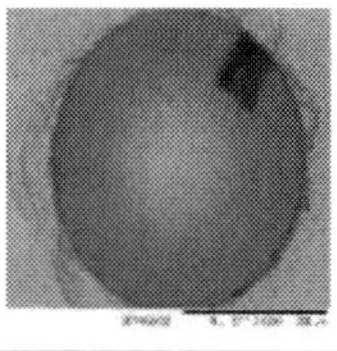	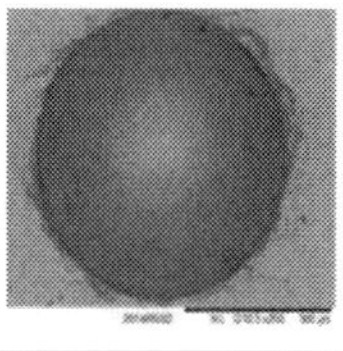

3.2 Surface profiler results

The results of the surface profiler analysis are shown in figure 4. At 40k rpm surface roughness is excessive at over 1500 nm and the large amount of debris after processing is evident from the limited amount of surface that can be measured. At 50k rpm and above, results remain quite consistent with rms values around 500 nm with residual debris minimising at the higher speeds. Whilst most surfaces demonstrated a random variation of height across their surfaces, at 80k rpm, some of the machined surfaces exhibited a cos 2θ undulation resulting in the larger rms.

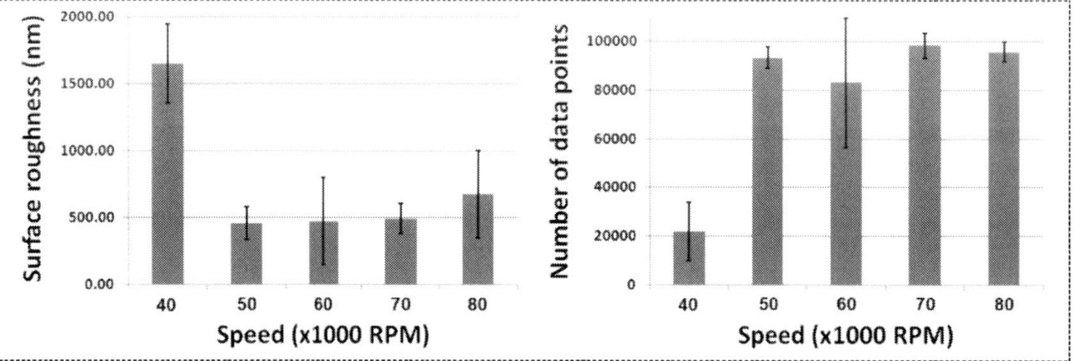

Figure 4. The rms values of the machined surface at various drill speeds. The low number of data points used at 40k rpm highlights the extensive surface debris remaining at this speed. Error bars are 1 standard deviation.

3.3 Raman results

The results of the Raman fitting is given in figure 5 with 10 measurements at each of the different drill speeds compared to the polished front side of the lithium niobate crystal and the unpolished back face. Unstrained lithium niobate shows a Raman peak at 872 cm^{-1}, the peak position measured indicates that the fabricated surface becomes increasing strained at the slower machining rates. The line widths of the Raman peak resulting from the drilled surfaces indicate that stoichiometry remains consistent to within measurement accuracy and is comparable to the unpolished side of the lithium niobate crystal.

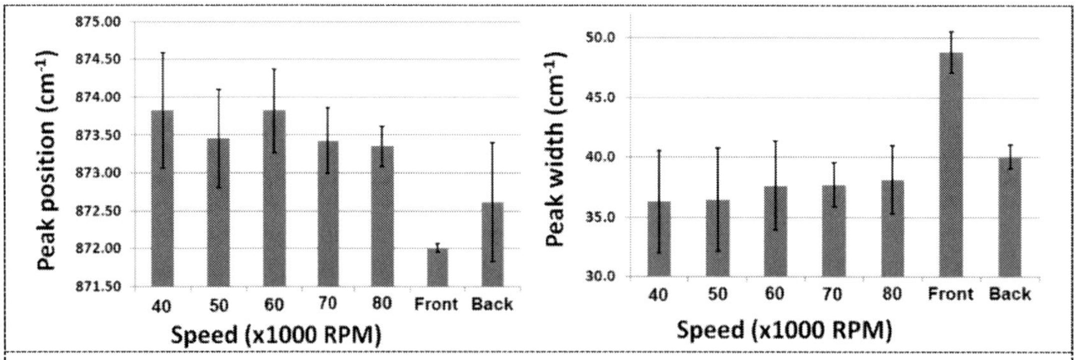

Figure 5. Peak position and line width fitted parameters for the 5 various drill speeds assessed. Data is based on 10 measurements per speed. Data compared with the polished front and unpolished rear surface of crystal. Error bars are 1 standard deviation.

4. Conclusions

Comparing the results of varying the spindle speed from 40k rpm to 80k rpm, it can be clearly seen that the spindle speed has a significant impact on the surface quality of the lithium niobate. Low spindle speeds are redundant for machining this material due to the excessive damage that results. Above 50k rpm, results are generally consistent. Quantification by surface profilometry implies a machining speed of 70k rpm is optimum resulting in an rms value of surface roughness of 500 nm with minimal debris. Raman spectroscopy of the 872 cm^{-1} indicated that the surface is strained due to this process. Work is continuing in optimising the drilling process and further work on analysing the Raman spectra in conjunction with additional surface analysis techniques will aim to better quantify the process. To reduce debris and strain in the surface, post machining treatments will be investigated.

Acknowledgements
The authors wish to thank the Engineering and Physical Sciences Research Council (EPSRC) for financial support for this work (EP/K031953/1 and EP/J02161X/1). Mr Zeyad Al-Shibaany acknowledges the Higher Committee for Education Development (HCED) in Iraq for the scholarship award.

References

[1] Hu Z, Hedley J, Keegan N, Spoors J, Waugh W, Gallacher B, et al. Design, fabrication and characterization of a piezoelectric MEMS diaphragm resonator mass sensor. Journal of Micromechanics and Microengineering. 2013;23(12).
[2] Ismail AK, Burdess JS, Harris AJ, Suarez G, Keegan N, Spoors JA, et al. The fabrication, characterization and testing of a MEMS circular diaphragm mass sensor. Journal of Micromechanics and Microengineering. 2008;18(2).
[3] Ortiz P, Burnett R, Keegan N, Spoors J, Hedley J, Harris A, et al. Issues associated with scaling up production of a lab demonstrated MEMS mass sensor. Journal of Micromechanics and Microengineering. 2012;22(11).
[4] Burnett R, Harris A, Ortiz P, Hedley J, Burdess J, Keegan N, et al. Electronic detection strategies for a MEMS-based biosensor. Journal of Microelectromechanical Systems. 2013;22(2):276-84.
[5] Chae J, Park SS, Freiheit T. Investigation of micro-cutting operations. International Journal of Machine Tools and Manufacture. 2006;46(3-4):313-32.
[6] Cheng K, Huo D. Micro Cutting: Fundamentals and Applications. Wiley: Wiley; 2013.
[7] Lalinský T, Hudek P, Vanko G, Dzuba J, Kuti V, Srnánek R, et al. Micromachined membrane structures for pressure sensors based on AlGaN/GaN circular HEMT sensing device. Microelectronic Engineering. 2012;98:578-81.
[8] Zhang Z, Yang S, Xu C, Wang B, Duan N. Deformation and stress at pop-in of lithium niobate induced by nanoindentation. Scripta Materialia. 2014;77:56-9.
[9] Kurihara R, Nakamura M, Yoshida M, Sumomogi T, editors. Study on micro machining of lithium niobate using single abrasive grinding2006; Monterey, CA.
[10] Jeong S, Lee H, Cho H, Lee S, Kim H, Kim S, et al. Effect of additives for higher removal rate in lithium niobate chemical mechanical planarization. Applied Surface Science. 2010;256(6):1683-8.
[11] Galinetto P, Marinone M, Grando D, Samoggia G, Caccavale F, Morbiato A, et al. Micro-Raman analysis on LiNbO3 substrates and surfaces: Compositional homogeneity and effects of etching and polishing processes on structural properties. Optics and Lasers in Engineering. 2007;45(3):380-4.
[12] Arif M, Rahman M, San WY. An experimental investigation into micro ball end-milling of silicon. Journal of Manufacturing Processes. 2012;14(1):52-61.
[13] Rusnaldy, Ko TJ, Kim HS. An experimental study on microcutting of silicon using a micromilling machine. International Journal of Advanced Manufacturing Technology. 2008;39(1-2):85-91.
[14] Kovalchenko AM. Studies of the ductile mode of cutting brittle materials (A review). Journal of Superhard Materials. 2013;35(5):259-76.
[15] Schaufele RF, Weber MJ. Raman Scattering by Lithium Niobate. Physical Review. 1966;152(2):705-8.

IOP Conf. Series: Materials Science and Engineering **65** (2014) 012031 doi:10.1088/1757-899X/65/1/012031

HUMAN ERROR IN MAINTENANCE: AN INVESTIGATIVE STUDY FOR THE FACTORIES OF THE FUTURE

B.S Dhillon

Department of Mechanical Engineering

University of Ottawa, Ottawa, Ontario, K1N 6N5

CANADA

Dhillon@genie.uottawa.ca

Abstract- This paper presents a study of human error in maintenance. Many different aspects of human error in maintenance considered useful for the factories of the future are studied, including facts, figures, and examples; occurrence of maintenance error in equipment life cycle, elements of a maintenance person's time, maintenance environment and the causes for the occurrence of maintenance error, types and typical maintenance errors, common maintainability design errors and useful design guidelines to reduce equipment maintenance errors, maintenance work instructions, and maintenance error analysis methods.

1. Introduction

Humans play a pivotal role during system/equipment design, production, operation, ad maintenance phases. Although, the degree of their role may vary from one phase to another, but their interactions are subject to deterioration because of human error. Human error may simply be described as the failure to carry out a given task (or the performance of a forbidden action) that could result in disruption of scheduled operations or damage to equipment and property [1 – 3].

The occurrence of human error in the maintenance activity can impact equipment performance and safety in various ways. For example, poor repairs can play an instrumental role in increasing the number of equipment breakdowns which in turn can significantly increase the risk associated with equipment failures and the occurrence of personal accidents [4]. Maintenance error, occurs basically due to wrong preventive actions or repairs and usually the occurrence of maintenance error increases as the equipment/system ages because of the increase in maintenance frequency.

2. Facts, Figures, and Examples

Some of the facts, figures, and examples, directly or indirectly, concerned with human error in maintenance are as follows:

- Over 50% of all equipment fails prematurely after the performance of maintenance work [5].
- A study of electronic equipment reported that around 30% failures were the result of operation and maintenance error [6].
- In 1988, 30 persons died and 69 were injured seriously at the Clapham Junction Railway accident due to a maintenance error in wiring, in the Untied Kingdom [7].
- In 1989, the explosion at the Phillips 66 Houston Chemical Complex in Pasadena, Texas was the result of a maintenance error [8].
- In 1993, a study of 122 maintenance-related occurrences classified maintenance error under four distinct categories: wrong installations (30%), omissions (56%), wrong parts (8%), and miscellaneous (6%) [9, 10].
- A study of an incident that involved the blow-out preventer (assembly of valves) at the Ekofisk oil field in the North Sea, reported that the incident was caused by the upside-down installation of the device. The total cost of the incident was estimated to be approximately $50 million [11].
- A study of maintenance tasks such as remove, adjust, and align reported a human reliability mean of 0.9871 [12]. It simply means that management should expect human errors by people involved with the maintenance activity on the order of 13 times in 1000 attempts [11].
- A study of maintenance-related errors in missile operations reported a number of causes: wrong installation (28%), dials and controls (misread, miss-set) (38%), loose nuts/fittings (14%), inaccessibility (3%), and miscellaneous (17%) [11, 13].

3. Occurrence of Maintenance Error in Equipment Life Cycle and Elements of a Maintenance Person's Time

The occurrence of maintenance error during the system/equipment life cycle (i.e., from the time of system/equipment acceptance to the beginning of its phase-out period) is an important factor. Approximate breakdowns of the occurrence of human error in a system/equipment life cycle are shown in Fig. 1 [11, 14].

A good understanding of time spent by maintenance personnel in performing various maintenance tasks, can be quite useful to analyze the occurrence of maintenance errors. Various studies performed over the years indicate that the most of their time is spent in the area of fault-diagnosis. However, according to one study [11], the maintenance person's time in the area of electronic equipment can be classified under three categories: diagnosis, remedial actions, and verification. The percentage breakdowns of the time for these three categories are as follows [11]:

- Diagnosis: 65 – 75%
- Remedial actions: 15 – 25%
- Verification: 5 – 15%

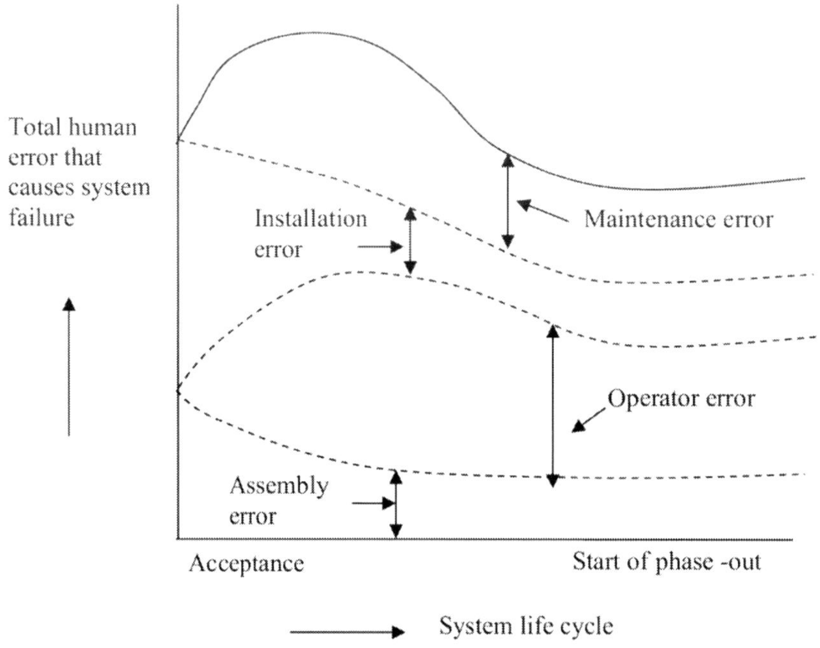

Fig. 1 System life cycle versus four types of human error that causes system failure

4. Maintenance Environment and Causes for the Occurrence of Maintenance Error

As maintenance personnel work directly on equipment, the location of equipment and its design features directly dictate many of the parameters of their work environment. Maintenance environments are susceptible to factors such as noise, poor illumination, and temperature variations. Each of these three factors is described below, separately [15].

Noise

Maintenance environments can be quite noisy as many are not properly sound-controlled. Ambient noise from ongoing activities can interfere with maintenance personnels' tasks. More specifically, sounds can distract maintenance personnel and interfere with their job performance and sufficiently loud sounds can limit the ability of maintenance personnel to converse or to hear verbal instructions.

Finally, it is added that although maintenance personnel can wear protective devices to limit to a certain degree adverse noise effects, these devices can interfere with the performance of their assigned tasks if they are uncomfortable, restrict movement, or hinder conversation.

Poor Illumination

Lighting deficiencies occur because the external light that maintenance personnel rely on is frequently designed to illuminate the general work area, not the specific areas on which they actually focus. More specifically, illumination-related deficiencies can exist in enclosed or confined spaces, or in places where the primary source of illumination is the overhead lighting.

Finally, it is added that maintenance personnel could use portable lighting fixtures to overcome deficiencies such as these, however, if hand-free operations are not possible, their ability to work effectively will be impeded.

Temperature Variations

Maintenance personnel may be exposed to wide variations in temperature because they often perform their tasks in outdoor environments or in environments that are not fully climate controlled. Past experiences indicate that maintenance workers and people in general perform effectively at a fairly narrow temperature range. Furthermore, as per some studies [15 – 18] as the temperature extends beyond a fairly narrow range (i.e., from around 15° C/60° F to about 35° C/90° F), it becomes a stressor that affects the performance of involved individuals.

Over the years, various studies have identified many different causes for the occurrence of maintenance errors. Some of the important ones are shown in Fig. 2 [11, 13, 19]. In particular, with regard to training and experience, a study of maintenance personnel reported that those who ranked highest possessed characteristics such as higher aptitude, greater satisfaction with the work group, higher morale, and greater emotional stability [11, 12].

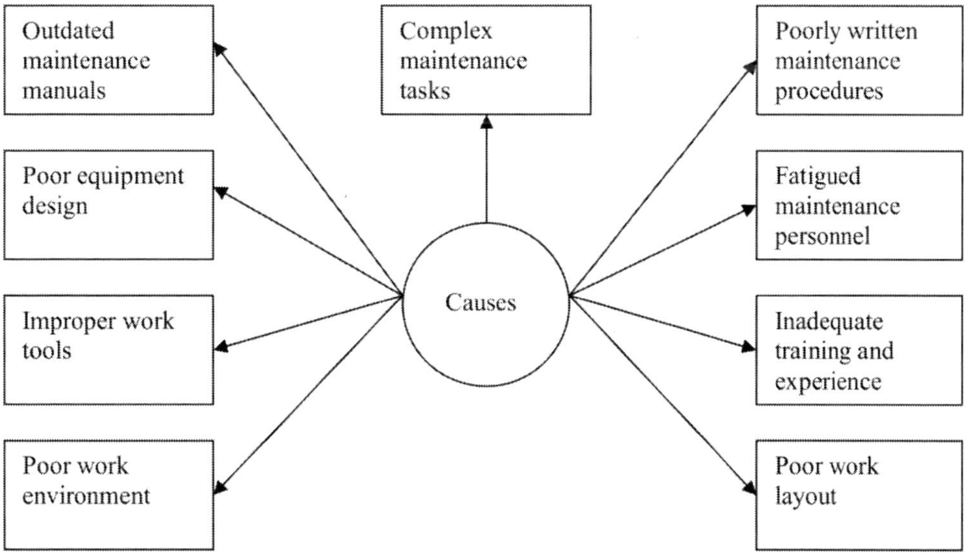

Fig. 2 Causes for the occurrence of maintenance errors

5. Types of Maintenance Errors and Typical Maintenance Errors

There are basically six types of maintenance errors [5]: recognition failures, memory failures, skill-based slips, knowledge-based errors, rule-based slips, and violation errors.

Recognition failures include items such as non-detection of problem states and misidentification of objects, signals, and messages. Memory failures include items such as input failure (i.e., poor attention is paid to the to-be-remembered item), storage failure (i.e., remembered material decays or suffers interference), premature exit (i.e., terminating a job prior to completing all the necessary actions), and omission following interruptions (i.e., rejoining a sequence of actions and omitting certain necessary steps).

Skill-based slips are usually associated with "automatic" routines and they can include branching errors and overshoot errors. Knowledge-based errors occur when maintenance personnel perform unusual tasks for the first time. Rule-based slips are concerned with misapplying a good rule (i.e., applying a rule in a situation where it is not appropriate) and applying a bad rule (i.e., the rule may get job/task done under certain conditions, but it can have various consequences).

Finally, violation errors are the deliberate acts which violate procedures. These include thrill-seeking violations (they are frequently committed simply to avoid boredom or win peer praise), routine violations (they are committed to avoid unnecessary effort, get the job/task accomplished quickly, to demonstrate skill acquired, or avoid what is considered as an unnecessarily lengthy procedure /process), and situational violations (they are committed when it is impossible to get the job done if specified procedures are strictly adhered to). Additional information on all the above six types of maintenance errors is available in Ref. [5].

Some of the typical maintenance errors experienced in the industrial sector are as follows [20]:

- Parts installed backward.
- Use of incorrect greases, lubricants, or fluids.
- Installing incorrect part.
- Failure to follow specified procedures and instructions.
- Failure to align, check, or calibrate.
- Omitting a component or part.
- Failure to close or seal properly.
- Failure to act on indicators of problems due to factors such as time constraints, priorities, or workload.
- Failure to lubricate.
- Error resulting from failure to complete task properly because of shift change.

6. Common Maintainability Design Errors and Useful Design Improvement Guidelines to Reduce Equipment Maintenance Errors

Past experiences indicate that during the equipment design phase often errors are made that adversely affect equipment maintainability and, directly or indirectly, the occurrence of maintenance errors. Some of the common maintainability design errors are as follows [21, 22]:

- Providing poor reliability built-in test equipment.
- Placing poor reliability parts beneath other parts.
- Placing adjustable screws close to a hot part or an exposed power supply terminal.
- Providing inadequate space for maintenance personnel to get their gloved hands into the unit to perform necessary adjustments.
- Omitting necessary handles and placing an adjustment out of arm's reach.
- Placing adjustable screws in locations difficult for maintenance personnel to find.
- Using access doors with numerous small screws and placing screwdriver-related adjustments underneath modules.

There are many useful design improvement guidelines for reducing equipment maintenance errors. Some of the important ones are as follows [20]:

- Use operational interlocks in such a way that subsystems cannot be turned on if they are incorrectly assembled-installed.
- Design to facilitate detection of errors and improve warning devices, readouts, and indicators to reduce human decision making.
- Improve fault isolation design by providing appropriate built-in test capability, clearly indicating the direction of fault, and designating test points and procedures.
- Use decision guides to reduce human guesswork by providing appropriate arrows for indicating direction of flow, correct type of fluids/lubricants, and correct hydraulic pressures.
- Improve part-equipment interface by designing interfaces in such a way that the part can only be installed correctly and provide correct mounting pins and other devices for supporting a part/component while it is being bolted or unbolted.

7. Maintenance Work Instructions

Over the years various studies have indicated that omissions account for over 50% of all human factors-related problems in the area of maintenance. Thus, the development and use of effective maintenance work instructions is very essential in managing these types of errors. Some of the characteristics of good maintenance work instructions are as follows [5]:

- They focus on the risks that may prevent the task/job being carried out safely and to specified quality standards.
- They incorporate sufficient independent inspections at important appropriate points in the instruction.
- They incorporate appropriate and conspicuous reminders for ensuring that important steps are not omitted.
- They group together complex work-related instructions into phases, with each and every phase consisting of many, related tasks/jobs.
- They make use of appropriate pictures and graphics at appropriate places.
- They are written with maintenance personnel who are going to read the instruction in mind.
- They are written clearly and make use of simple and consistent language.

Additional information, on the above characteristics is available in Ref. [5].

8. Maintenance Error Analysis Methods

Over the years many methods and techniques have been developed to perform various types of analysis in the areas of reliability, quality, and safety. Some of these methods can also be used to perform maintenance error analysis. Four of these methods are presented below.

8.1. Probability Tree Method

This is one of the commonly used methods to perform human reliability analysis. It is considered a quite useful approach to perform task analysis in maintenance work. In performing task analysis, the approach diagrammatically represents human actions. Thus, diagrammatic task analysis is denoted by the probability three branches.

More specifically, the branching limbs denote outcomes (i.e., success or failure) of each event or action associated with a problem under consideration. Also, each branch of the probability tree is assigned an occurrence probability.

The method is described in detail in Refs. [13, 21]. Its application to perform maintenance error analysis is demonstrated through the example presented below.

Example 1

Assume that a maintenance person performs two independent tasks, say, m and n. Task m is performed before task n and each of these two tasks can be either performed correctly or incorrectly. Draw the probability tree for the example and obtain probability expressions for the followings:

i) Successfully accomplishing the overall mission by the maintenance person.
ii) Not successfully accomplishing the overall mission by the maintenance person.

In this case, the maintenance person first performs task m correctly or incorrectly and then proceeds to performing task n. This complete scenario is represented by the Fig. 3 probability tree diagram.

The four symbols used in Fig. 3 diagram are defined below.

m denotes the event that task m is performed correctly by the maintenance person.
$\bar{m}$ denotes the event that task m is performed incorrectly by the maintenance person.
n denotes the event that task n is performed correctly by the maintenance person.
$\bar{n}$ denotes the event that task n is performed incorrectly by the maintenance person.

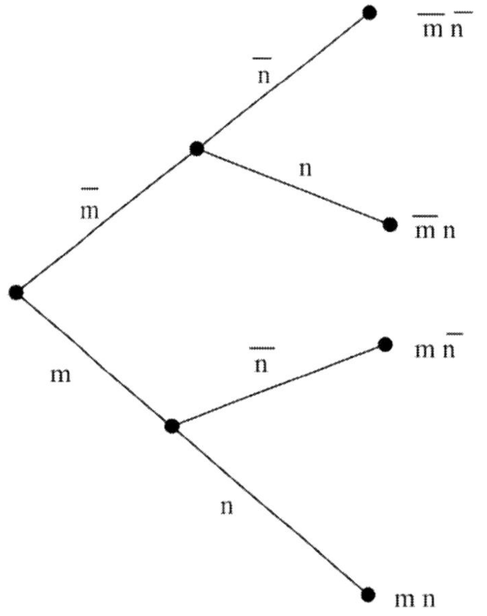

Fig. 3 Probability tree for the maintenance person performing tasks m and n

By examining the Fig. 3 diagram, it can be noted that there are three distinct possibilities (i.e., $\bar{m}\bar{n}, \bar{m}n, and\ m\bar{n}$) for not successfully accomplishing the overall mission by the maintenance person. Thus, the probability of not successfully accomplishing the overall mission by the maintenance person is given by

$$\begin{aligned} P_f &= P(\bar{m}\bar{n} + \bar{m}n + m\bar{n}) \\ &= P_{\bar{m}}P_{\bar{n}} + P_{\bar{m}}P_n + P_m P_{\bar{n}} \end{aligned} \quad (1)$$

where

P_f is the probability of not successfully accomplishing the overall mission by the maintenance person.

P_m is the probability of performing task m correctly by the maintenance person.

P_n is the probability of performing task n correctly by the maintenance person.

$P_{\bar{m}}$ is the probability of performing task m incorrectly by the maintenance person.

$P_{\bar{n}}$ is the probability of performing task n incorrectly by the maintenance person.

Since $P_{\bar{m}} = 1 - P_m$ and $P_{\bar{n}} = 1 - P_n$, Equation (1) reduces to

$$\begin{aligned} P_f &= (1-P_m)(1-P_n) + (1-P_m)P_n + P_m(1-P_n) \\ &= 1 - P_n P_m \end{aligned} \quad \textbf{(2)}$$

Similarly, by examining the Fig. 3 diagram, it can be noted that there is only one possibility (i.e., mn) for successfully accomplishing the overall mission by the maintenance person. Thus, the probability of successfully accomplishing the overall mission by the maintenance person is given by

$$\begin{aligned} P_s &= P(mn) \\ &= P_m P_n \end{aligned} \quad \textbf{(3)}$$

where

P_s is the probability of successfully accomplishing the overall mission by the maintenance person.

Example 2

Assume that in Example 1, the probabilities of the maintenance person performing tasks m and n correctly are 0.9 and 0.95, respectively. Calculate the probability of not successfully accomplishing the overall mission by the maintenance person.

By substituting the given data values into Equation (2), we get

$$\begin{aligned} P_f &= 1 - (0.95)(0.9) \\ &= 0.855 \end{aligned}$$

Thus, the probability of not successfully accomplishing the overall mission by the maintenance person is 0.855.

8.2. Pontecorvo Method

This is a quite useful method that can be used to obtain reliability estimates of task performance by a maintenance person. The method first obtains reliability estimates for separate and discrete subtasks having no correct reliability figures, and then it combines these estimates to obtain the total task reliability. Usually, Pontecorvo approach is applied during initial design phases and is composed of six steps shown in Fig. 4 [13, 22].

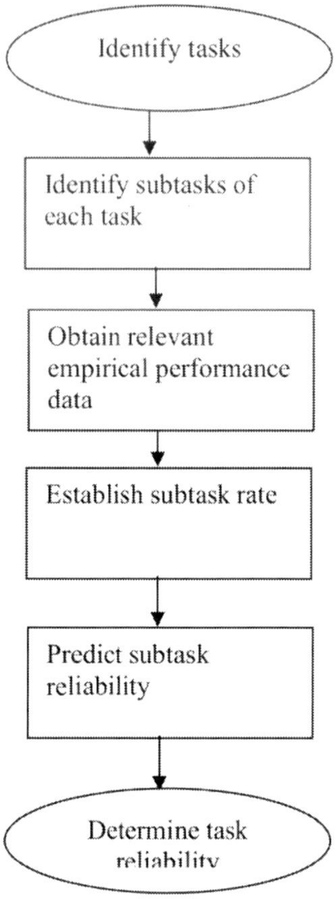

Fig. 4 Pontecorvo method steps

Step 1 is concerned with the identification of tasks to be performed. These tasks are to be identified at a gross level (i.e., each task is represented by one complete operation). Step 2 is concerned with the identification of those subtasks that are essential for task completion. Step 3 is concerned with collecting data from sources such as in-house operations and experimental literature.

Step 4 is concerned with rating each subtask according to its potential for error or level of difficulty. Normally, a 10-point scale is used to judge the appropriate subtask rate. The scale varies form least error to most error. Step 5 is concerned with predicting the subtask reliability and is accomplished by expressing the judged ratings of the data and the empirical data in the form of a straight line. The regression line is tested for goodness of fit.

Finally, Step 6 is concerned with determining the task reliability. The task reliability is obtained by multiplying reliabilities of all the subtasks.

It is to be noted that the above approach is used to estimate the performance of a single individual acting alone. However, when a backup person is available, the probability of the task being performed correctly (i.e., the task reliability) improves. Nonetheless, the backup individual may not be available all of the time. Under such scenario, the overall reliability of two individuals working together to accomplish a specified task can be estimated by utilizing the following expression [13, 22]:

$$R_O = \left[\left\{1-(1-R_s)^2\right\}PT_1 + R_s\, PT_2\right] / (PT_1 + PT_2) \quad (4)$$

where

R_s denotes the single person reliability.
PT_1 denotes the percentage of time the backup person is available.
PT_2 denotes the percentage of time the backup person is unavailable.

Example 3

Two maintenance workers are working independently together to carry out a maintenance-related task. The reliability of each worker is 0.90, and only 40% of time is the backup worker available. Calculate the reliability of performing the maintenance task correctly.

Thus, as per the specified data value, the percentage of time the backup maintenance worker unavailable is given by

$$\begin{aligned} PT_2 &= 1 - PT_1 \\ &= 1 - 0.40 \\ &= 0.60\, or\ 60\% \end{aligned}$$

Using the above calculated value and the given data values in Equation (4), we get

$$\begin{aligned} R_O &= \left[\left\{1-(1-0.9)^2\right\}0.4 + (0.9)(0.6)\right] / (0.4+0.6) \\ &= 0.936 \end{aligned}$$

Thus, the reliability of carrying out the maintenance task correctly is 0.936.

8.3. Pareto Analysis

The method is named after Vilfredo Pareto (1848 – 1923), an Italian economist and it is a quite useful method that can be used to separate the important causes of maintenance error-related problems from the trivial ones.

Thus, the method is considered a powerful tool to identify areas for a concerted effort to minimize or eliminate the occurrence of maintenance errors. The method is composed of the six steps listed below [23, 24].

- **Step 1:** List causes in tabular form and count their occurrences.
- **Step 2:** Arrange the causes in descending order.
- **Step 3:** Calculate the total for the entire list.
- **Step 4:** Determine the percentage of the total for each cause.
- **Step 5:** Develop a Pareto diagram that shows percentages vertically and their corresponding causes horizontally.
- **Step 6:** Conclude from the final results.

Additional information on Pareto analysis is available in Refs. [23, 24].

8.4. Markov Method

This is a widely used tool to perform various types of reliability analysis and it can be sued to perform human error analysis in maintenance work. The method is described in Ref.[19]. Its application in the area of maintenance, is demonstrated through the following mathematical model.

Mathematical Model

This mathematical model represents a maintenance person performing a maintenance task. He/She can make and self-correct an error. The state space diagram of the model is shown in Fig. 5 [24]. Numerals in boxes denote system states.

The model is subject to the following assumptions:

- The maintenance person's error and self-error-correction rates are constant.
- The maintenance person can self correct his/her errors.
- After the error correction the maintenance person's performance remains normal.

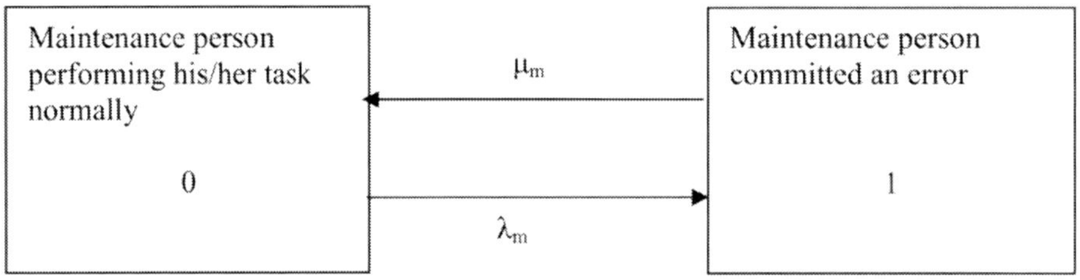

Fig. 5 State space diagram for the maintenance person

The following symbols are associated with the model:

i	is the maintenance person's state; for i = 1 (maintenance person performing his/her task normally), i = 1 (maintenance person committed an error).
$P_i(t)$	is the probability that the maintenance person is in state i at time t; for i = 0,1.
λ_m	is constant error rate of the maintenance person.
μ_m	is constant self-error-correction rate of the maintenance person.

With the aid of the Markov method, we write down the following equations for the Fig. 5 diagram:

$$\frac{dP_0(t)}{dt} + \lambda_m P_0(t) = P_1(t)\mu_m \quad (5)$$

$$\frac{dP_1(t)}{dt} + \mu_m P_1(t) = P_0(t)\lambda_m \quad (6)$$

At time t = 0, $P_0(0) = 1$ and $P_1(0) = 0$.

Solving Equations (5) – (6), we get

$$P_0(t) = \frac{\mu_m}{(\lambda_m + \mu_m)} + \frac{\lambda_m}{(\lambda_m + \mu_m)} e^{-(\lambda_m + \mu_m)t} \quad (7)$$

$$P_1(t) = \frac{\lambda_m}{(\lambda_m + \mu_m)} - \frac{\lambda_m}{(\lambda_m + \mu_m)} e^{-(\lambda_m + \mu_m)t} \quad (8)$$

As time t becomes very large, we get the following steady state probability equations from Equations (7) – (8), respectively:

$$P_0 = \frac{\mu_m}{\lambda_m + \mu_m} \quad \textbf{(9)}$$

$$P_1 = \frac{\lambda_m}{\lambda_m + \mu_m} \quad \textbf{(10)}$$

where

P_0 and P_1 are the steady state probabilities of the maintenance person being in states 0 and 1, respectively.

Example 4

A maintenance person is performing a maintenance task and his/her error and self-error-correction rates are 0.0003 errors/hour and 0.0001 errors/hour, respectively. Calculate the maintenance person's probability of correctly performing his/her task during an 8 hour period.

By substituting the specified data values into Equation (7), we get

$$P_0(8) = \frac{0.0001}{(0.0003+0.0001)} + \frac{(0.0003)}{(0.0003+0.0001)} e^{-(0.0003+0.0001)(8)}$$
$$= 0.9976$$

Thus, the maintenance person's probability of performing his/her task correctly is 0.9976.

9. References

[1] Meister, D., The Problem of Human-initiated Failures, Proceedings of the 8th National Symposium on Reliability and Quality Control, 1962, pp. 234 – 239.

[2] Meister, D., Human Factors in Reliability, in Reliability Handbook, edited by W. G. Ireson, McGraw-Hill, New York, 1966, pp. 12.2 – 12.37.

[3] Hagen, E.W., Human Reliability Analysis, Nuclear Safety, Vol. 17, 1976, pp. 315 – 326.

[4] Mason, S., Improving Maintenance by Reducing Human Error, 2007. Available from Health Safety and Engineering Consultants Ltd., 70 Tamworth Road, Ashby-de-la-Zouch, Leicestershire, U.K.

[5] Dunn, S., Managing Human Error in Maintenance, 2007. Available from Assetivity Pty Ltd., P.O. Box 1315, Boorgoon, Washington 6154.

[6] AMCP 706 – 134, Maintainability Guide for Design, U.S. Army Material Command, Department of the Army, Washington, D.C., 1972.

[7] Report: Investigation Into the Clapham Junction Railway Accident, Department of Transport, Her Majesty's Stationery Office, London, U.K., 1989.

[8] Reason, J., Hobbs, A., Managing Maintenance Error: A Practical Guide, Ashgate Publishing Company, Aldershot, U.K., 2003.

[9] Circular 243-AN/151, Human Factors in Aircraft Maintenance and Inspection, International Civil Aviation Organization, Montreal, Canada, 1995.

[10] Human Factors in Airline Maintenance: A Study of Incident Reports, Bureau of Air Safety Investigation, Department of Transport and Regional Development, Canberra, Australia, 1997.

[11] Christensen, J.M., Howard, J.M., Field Experience in Maintenance, in Human Detection and Diagnosis of System Failures, edited by J. Rasmussen and W.B. Rouse, Plenum Press, New York, 1981, pp. 111 – 133.

[12] Sauer, D., Campbell, W.B., Potter, N.R., Askern, W.B., Relationships Between Human Resource Factors and Performance on Nuclear Missile Handling Tasks, Report No. AFHRL-

TR-76-85/AFWL-TR-76-301, Air Force Human Resources Laboratory/Air Force Weapons Laboratory, Wright-Paterson Air Force Base, Ohio, 1976.

[13] Dhillon, B.S., Human Reliability: with Human Factors, Pergamon Press, New York, 1986.

[14] Rigby, L.V., The Sandia Human Error Rate Bank (SHERB), Report No. SC-R-67-1150, Sandia Labs., Albuquerque, New Mexico, 1967.

[15] Strauch, B., Investigating Human Error: Incidents, Accidents, and Complex Systems, Ashgate Publishing Limited, Aldershot, U.K., 2002.

[16] Ellis, H.D., The Effects of Cold on the Performance of Serial Choice Reaction Time and Various Discrete Tasks, Human Factors, Vol. 24, 1982, pp. 589 – 598.

[17] Van Orden, K.F., Benoit, S.L., Osga, G.A., Effects of Cold Air Stress on the Performance of a Command and Control Task, Human Factors, Vol. 38, 1996, pp. 130 – 141.

[18] Wyon, D.P., Wyon, I., Norin, F., Effects of Moderate Heat Stress on Driver Vigilance in Moving Vehicle, Ergonomics, Vol. 39, 1996, pp. 61 – 75.

[19] Dhillon, B.S., Engineering Maintenance: A Modern Approach, CRC Press, Boca Raton, Florida, 2002.

[20] Under, R.L., Conway, K., Impact of Maintainability Design on Injury Rates and Maintenance Costs for Underground Mining Equipment, in Improving Safety at Small Underground Mines, Compiled by R.H. Peters, Special Publication No. 18 – 94, Bureau of Mines, United States Department of the Interior, Washington, D.C., 1994.

[21] Dhillon, B.S., Singh, C., Engineering Reliability: New Techniques and Applications, John Wiley and Sons, New York, 1981.

[22] Pontecorvo, A.B., A Method of Predicting Human Reliability, Proceedings of the 4th Annual Reliability and Maintainability Conference, 1965, pp. 337 – 342.

[23] Kanji, G.K., Asher, M., 100 Methods for Total Quality Management, Sage Publications Ltd., London, 1996.

[24] Dhillon, B.S., Design Reliability: Fundamentals and Applications, CRC Press, Boca Raton, Florida, 1999.

Career Management and Employee Motivation in Low Skilled, Low Margin Environments.

S. Salman[1] **and A. Khan**[1,2,a]

[1] Faculty of Business and Management, University of Wales Trinity Saint David, Swansea , UK.

[2] Educational Excellence (e^2) Ltd, Pembrokeshire, Wales, UK.

Email[2,a]***: e2ukoffice@aol.com***

Abstract. Career management enhances the level of motivation, resilience, awareness and the direction of employees' careers goals relative to the existing job opportunities within and outside an organisation. The main aim of this paper is to evaluate how a low-skilled company environment can use career management as a tool for employee motivation. While using Company A (Food company) as its case study organisation, the research utilises an interpretivism philosophy, deductive research approach, case study research strategy, and random sampling method. Primary data is collected by way of interviews and questionnaires and analysed using coding method. The study establishes that Company A has a robust employee career management programme. This programme is based on both short-term and long-term goals and aims to meet both individual employees' and organisation-wide goals. Specifically, the study establishes that Company A employee career management programme helps employees to make wise career decisions, become aware of available career opportunities, set career goals, execute career goals, explore their careers, and experience career progression. The findings have implications for employee motivation in other low skilled work environments

1. Introduction

Employee motivation enhances organisational success [1, 2]. Motivation can be defined as the degree to which an individual wants and chooses to engage in certain manner [3]. According to [4], motivation is also the psychological process that gives behaviour purpose and direction. One of the greatest challenges for any organization is to improve its level of efficiency and productivity [2]. However, organisations with motivated and enthusiastic employees that apply new techniques and technologies attain high levels of productivity.

Company A has a sophisticated career management programme that guarantees it a steady supply of employees. This is after realisation that there are limited opportunities for career development and hence there is need to develop it uniformly among employees. The employees own their career development goals with the understanding that there are established career lines available to them. Career management in Company A is more than just a promotion – it is about learning and career enrichment. This proactive career management process has a strong motivating aspect and it has great impact on the attitudes of the employees. Career management, no doubt enables an organization to attract the best employees, motivates them and to develop and retain the best talent.

Employee retention strategies are also very important to the growth of Company A. Employees often want to be appreciated by the management and they want to be motivated, it is therefore effective to ask the employees what they want the management to do for them. Thus the main aim of this paper is to evaluate the different strategies of employee motivation and career management at Company A (and its two branches) through different motivational theories with a view to improving both aspects, and for lessons to be learnt for other such company (low-skilled, low-margin) environments.

2. Literature Review of Employee Motivation

Employee motivation is a complex process as it has its influence in all areas including organizational development, resource management, internal project and upcoming programs. Motivation is the set of psychological processes that cause the arousal, direction, and persistence of individual's behaviours toward attaining a goal. As different motivation techniques are applied for different department of the same organization. Choosing which type of motivation for which department plays important role and within the same department, for different task different type of motivation applied [2].

There are two types of motivation namely intrinsic and extrinsic [5]. According to [6] motivation has to do with either internal or external factors that influence a persons' choice of action. Ryan and Deci [7] describe intrinsic motivation as the conduct of an activity for its inherent satisfactions rather than a separable consequence. Intrinsic motivation refers to motivation that is induced internally by a person. The reward of this source of motivation comes from conducting the task rather than as a consequence of the task. Intrinsic motivators satisfy an individual's direct needs and this provides satisfaction and pleasure [5]. People feel motivated when their values are addressed

Nevertheless, not all individuals are motivated by internal factors. According to [7] external factors have an impact on an individual's level of motivation. Pleasure derived from extrinsic factors comes about from something that a task induces. Examples of extrinsic factors include salary, work benefits, job security, work punishments, work conditions [5]. On a general note, extrinsic motivators such as determinate rewards, for example fixed salary increase at specific periods of the year, may reduce the level of motivation because individuals will quickly lose interest on a particular motivator [6].

Overall, intrinsic motivation is the most important motivation as it develops a person's social, cognitive and physical abilities [7]. Intrinsic motivation can only originate from within an individual. It is only the individual who can drive himself / herself to perform tasks zealously in an organizational setting. Stimulation to perform well can only stem from an individual and is considered to be a powerful force that emanate from an individual's inner-self [8]. Examples of intrinsic motivators include challenging tasks at the work place, recognition at work, personal growth and development, quality of work, achievement, and responsibility.

3. Research Methodology

There are several research strategies that a social science project can adopt. According to [9], these include case study, experiment, survey, grounded theory, ethnography, and action study. Each of these strategies has its own strengths and weaknesses. A survey strategy, for example, is broader and its findings can be generalised in a number of scenarios [10].

This research adopted a case study approach. The rationale for adopting a case study strategy was based on the notion that the researcher understands the case study organisation well since he has been working there for some time. To this effect, the researcher was most likely able to decipher hidden meanings of the various responses given by the study participants.

This study sample was made up of both managers and crew members working in the Chelmsford 1 store and Chelmsford 2 store. From these two stores, a total of 10 managers and 30 crew members (40 participants) were sampled. The decision to sample only 40 participants was based on time and monetary constraints. To this end, it was reasoned that 40 participants were sufficiently representative of the staff working at the Chelmsford 1 and Chelmsford 2 Company A outlet.

The study used random non-probability sampling method. This sampling method entails identifying a group of potential participants, seeking their consent to take part in the study and then subjecting them to a random selection process. The researcher used the 40 and 15 consenting potential participants to make two alphabetical lists, one for the managers and the other for the crew members were prepared.
This study utilised both secondary and primary data. The rationale behind the decision to include both primary and secondary was based on the notion that the findings of a credible research study should be based on past studies as well as case-specific data collected from human subjects or from still observations [11].

A large amount of the collected data was qualitative. The reason for collecting qualitative data as opposed to quantitative data was advised by the study topic, study objectives as well as study questions. Moreover, the study adopted an interpretive research philosophy which does not involve testing of hypotheses, it is only concerned in drawing broad patterns and combining them to create broad generalisation pertaining to the influence of career management on employee motivation [10].

4. Research Findings

This study sample was made of 40 participants. This sample comprised of 10 managers and 30 crew members working for Company As' Chelmsford 1 and Chelmsford 2 stores. Each store contributed 5 managers and 15 crew members for the study sample. The interviews were administered to the managers on a face-to-face basis, while questionnaires were conducted on the crew members. To this end, managers were considered more insightful than the crew members in discerning the influence of Company A's career management programme on employee motivation.

The use of questionnaires was done in the case of crew members, because of the fear by the subordinates to be reprimanded by their seniors. It was also done because it enhances a level of anonymity. The results of the questionnaires conducted on the crew members, were divided into **three sections**. The first section was concerned about the length of time that the crew members were working at Company As. The second section was concerned about the career aspirations of the crew members. The third section was concerned about Company A's career management program.

The only demographic detail collected for this research is the length of service possessed by the respondents. Figure 1 shows the distribution of the respondents according to their year of work.

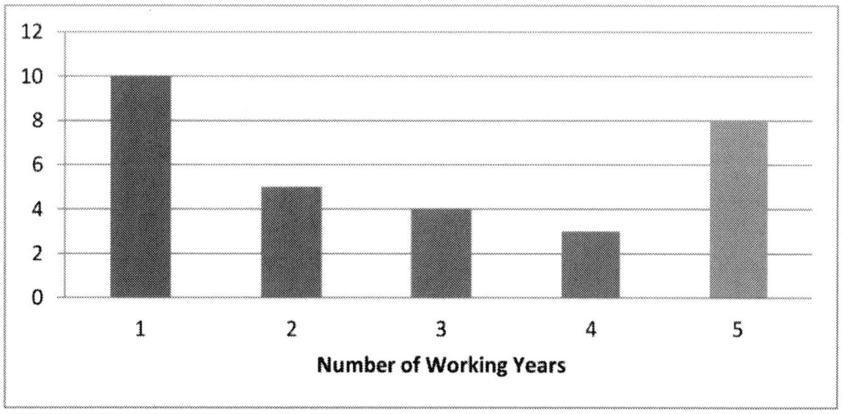

Figure 1: Respondents number of working years.

Respondents with one (1) year of work were discovered to have the most number of respondents. Respondents with 5 years of experience closely followed with 8. The dominance by employee with the least number of working years can be based on continuous changing of staffs like servers, drivers and others. Also, the respondents might just be working for Company As with most of them having experience in other organisations. The fact that 10 managers were involved might contribute to the high number of respondents with 8 years of working experience.
Findings in Figure 2 showed Chelmsford 1 employees to be more interested in becoming crew trainers. With 9 respondent positively disposed, it was discovered that most of the respondents understand that this post increases the remunerations and further increase their level within the organisation.

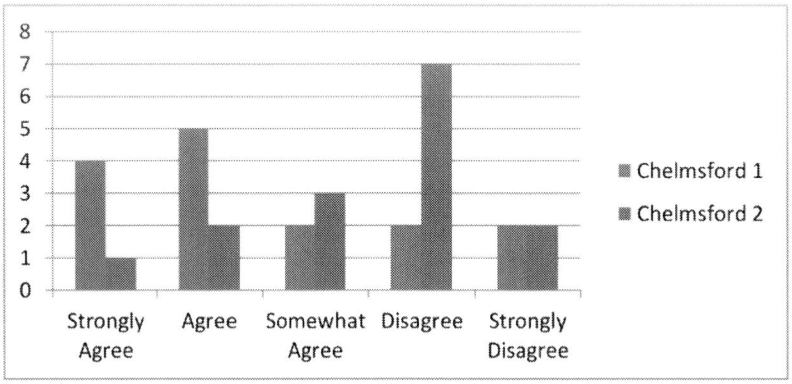

Figure 2: Crew Trainer Willingness Distribution.

Chelmsford 2 respondents disagreed suggesting the prevalence of employees with limited working years. It can be viewed that employees in Chelmsford 2 store are more concerned about part-time opportunities and do not see the crew trainer as very important to their career progression. In essence, the two stores have an employee base that has different perspectives concerning this question. Employees at Chelmsford 2 consist more of students that are interested in increased hourly pay instead of holding a permanent paying jobs such as crew trainer. In Chelmsford 1, more employees desire permanent employment with being crew manager giving them the opportunity to rise within the organisation.

The difference in the perspective of employees in the two stores was exposed in the responses gathered in for this question. Figure 3 showed that crew members at Chelmsford 2 were more confident and believed they can become senior executive at the organisation. However, employees at Chelmsford 1 believed it was nearly impossible to become senior executives. It seems Company A does not apply a similar career management programmes for all its stores. Unlike Store 1, Store 2 engaged employees with more qualification or invested in enhancing their employees' career skills. In addition, it is arguable that junior employees in Store 2 are in the process of advancing their academic and professional qualifications to occupy executive positions that emerge within the organisation. This argument is based on the conventional notion that senior management positions require higher academic and professional credentials.

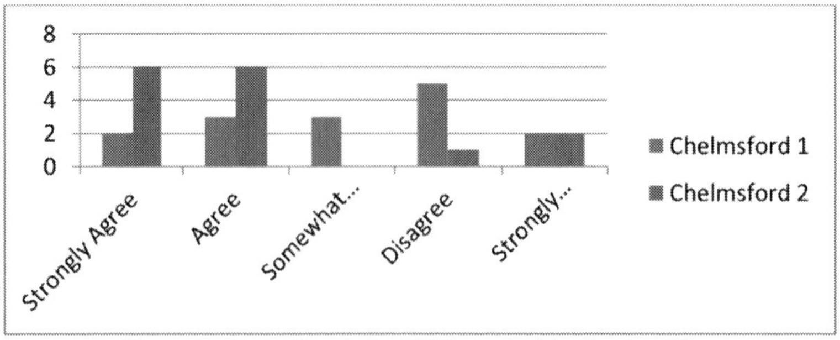

Figure 3: Senior Executive Willingness Distribution Chart

Three sets of questions were asked about the Company A's Career Management Programme (CMP):

(a) CMP is based on short-term targets – between 1 and 2 years:
The distribution in response to the question showed a positive perspective. A total of 15 (see Table 1) respondents indicated that Company A's CMP is based on short-term goals. It can be deduced that as much as the company wants expansion and great long term improvement sales, current sales and customer satisfaction is an important object of the company's employee career management.

	Chelmsford 1	Chelmsford 2
Strongly Agree	2	7
Agree	3	3
Somewhat Agree	5	1
Disagree	4	3
Strongly Disagree	1	1

Table 1: Respondent Perspective on short term targets of Company A's CMP

Furthermore, daily sales and performance is the most important aspect of employee career. The store is more concerned about what you do in the nearest future irrespective of their focus on future career improvement. A notable discovery is the presence of 6 respondents that are indifferent across the two stores showing that the direction in the short term of CMP implementation is unclear. Also, 9 respondents in the negative were captured. It can be deduced that the future is more important as this company continues to provide long term and strategic innovations for long term implementation of CMP.

(b) CMP is based on intermediate-term targets – between 3 and 7 years:
Table 2 below showed that the average of respondents were sure of the intermediate targets. This showed the vagueness which the CMP implementation is based in the intermediate sphere in the perspective of employees. Respondents that were not sure are 7 showing that the intermediate targets in the CMP implementation is not clear and employees are finding it difficult to understand.

It can be viewed that the intermediate targets is somehow silent as the most prominent reason for CMP is usually the short term and the long term" Respondents are of the opinion that intermediate term usually become the short term in a couple of years making it to be less important in the reason for CMP implementation

	Chelmsford 1	Chelmsford 2
Strongly Agree	3	5
Agree	4	2
Somewhat Agree	3	4
Disagree	2	1
Strongly Disagree	3	3

Table 2: Respondent Perspective on intermediate targets of Company A's CMP

(c) CMP is based on long-term targets – over 10 years:
Chelmsford 1 respondents were more positive than Chelmsford 2 in this question (Table 3). Employees in Chelmsford 1 see the long-term nature of CMP implementation as a lot of them has been with Company As

IOP Conf. Series: Materials Science and Engineering **65** (2014) 012032 doi:10.1088/1757-899X/65/1/012032

for over 3 years. The employees have experienced CMP implementation and understood the focus better than employees at Chelmsford 1 that are mostly part time.

Chelmsford 2 employees were influenced by their individual educational endeavours and do not plan the career solely on Company As. In this aspect, some of them are not sure the long term nature of the CMP implementation coupled and the fact that they can move to a better employment after completion. For Chelmsford 1 employees, CMP is based on in house career development while Chelmsford 2 employees are more disposed to individual career path development. It can be deduced that Company A's pursues mixed strategies for different employee cadres. The company seems to follow a long-term career management strategy especially for senior positions and short-term and intermediate strategies for junior and middle level employees.

	Chelmsford 1	Chelmsford 2
Strongly Agree	5	3
Agree	5	2
Somewhat Agree	1	6
Disagree	2	3
Strongly Disagree	2	1

Table 3: Respondent Perspective on long-term targets of Company A's CMP

The major distribution for Chelmsford 2 is the indifferent category with 6 respondents. This deduced that CMP implementation is directed towards improved performance of the company and not necessarily the long term growth. Managing the careers of employees is important for the organisation irrespective of targets.

(d) Company A's Career Management Programme (CMP) creates new, long-term opportunities for employees.

Figure 4 shows the distribution as it concerns the long term and new nature of opportunities created by CMP at Company As. Both stores have 10 respondents positively disposed showing a satisfactory distribution in the cumulative. These substantiated by stating that careers in Company As are planned in such a way that employees can maintain their loyalty and contribute to the organisation in a new way that might be needed.

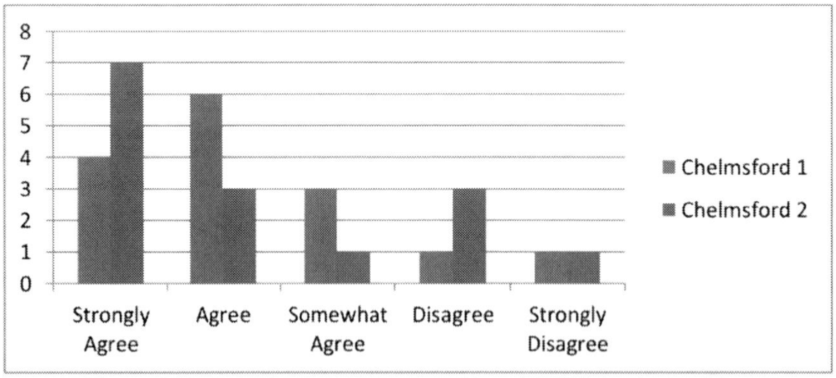

Figure 4: New and Long-term Opportunities of CMP Distribution Chart

Figure 5 is a display the distribution that is even but still skewed suggesting store opinions that are encouraging. On the cumulative, the two stores have 18 respondents that are agree in some degree that employee motivation enhancement is an aim that CMP implementation at Company As is focused.

However, Chelmsford 2 had the most encouraging distribution positive distribution with 10 respondents agreeing.

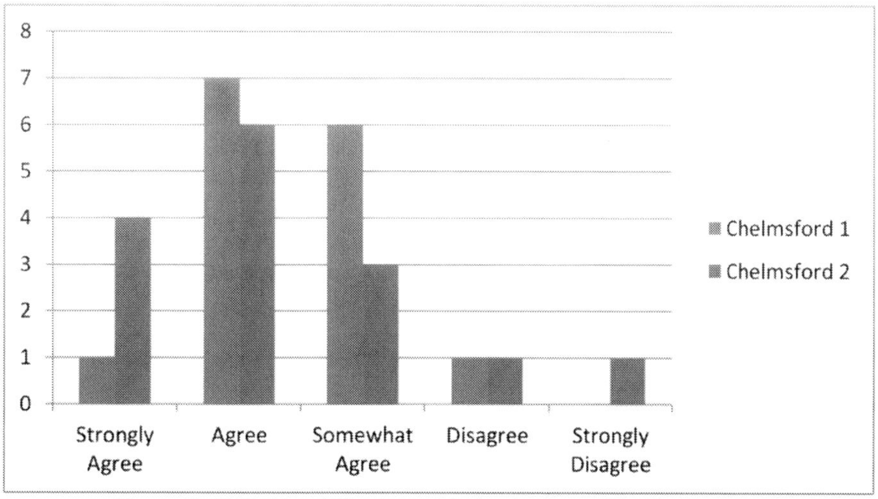

Figure 5: CMP Aim of Employee Motivation boost Distribution Chart

It is worthy of note that Chelmsford 2 does not have any respondent that is strongly disagreeing with the statement. However, Chelmsford 1 have sizeable amount (6) of respondents that are in the middle. This number is disturbing considering that it can swing to either side of the perspective. These perspectives are based on the fact that motivation is wide and CMP cannot be solely responsible for that. This distribution showed that each store has unique perspective within employee base about what motivation implies and if the aim of CMP is enough to effectively impact.

This section has presented and discussed the study findings. It has emerged that Company A's CMP is based on short-term, intermediate, and long-term goals. The company pursues these three different CMP approaches for different employee cadres and different situations depending on the underlying circumstances. However, it has emerged that Store 2 has a better CMP than Store 1 – comparatively, Store 2 experience low employee turnover (high motivation levels) than Store 1, Store 2 offers its employees a better CMP package that encourage them to desire to ascend through the company's structures to occupy high positions. In extension, Store 2 seems to experience high employee job engagement than Store 1 – most Store 2 employees believe that a better CMP will enhance their career opportunities in the form of getting promoted to better paying positions within the organisation. Store 2 CMP seems to satisfy industry, legal, and ethical standards, it is more effective in boosting employee job engagement, and job performance.

5. Research Conclusions

The case study analysed the results gathered from the two stores. Although the findings are limited by the case study being in one company (and two separate locations), there are still important lessons to be learnt (also supported by the published literature cited earlier) for these types of environments. What was evident during this study is that the levels of motivation in the two Company A's stores are high. This was mainly attributed to the career management programme instituted by Company As. These results showed that the career management programme instituted by the two stores had a significant impact on various factors within the Company A's organizational structure. Many of the respondents agreed that the programme had a significant impact on matters such as motivation, employee turnover, performance and the organizations future aspiration in terms of strategies goals and suitability of the organization.

Findings in the two stores in some areas of the research differs thereby suggesting different perspective that was filtered from the interview. The career aspiration from the two stores differs based on the fact that the employee bases in the stores have differing focus. Chelmsford 1 present its employee base as long term focused with detail towards intra organisational growth. This perspective was reflected in their targets for career development as most of the respondents from the store have long-term targets within the organisation. This psychology influence their career aspiration and CMP implementation. Chelmsford 2 has younger employee base with some taking the job from the store as part time. Their major motivation is to get extra pay and develop personally to occupy top managerial position. Philosophically, the difference in perspective in stores is based on the focus of the employee base in the stores. For example, Chelmsford 2 employees have high agreement to become shift management as it allows them to raise their basic pay in part time basis while Chelmsford 1 employee desired to be business manager which is more of a full time job. Another pattern was displayed as majority of the Chelmsford 2 employees desire to be senior executive.

This research has answered some intricate questions and shed more light in the subject area, which can be concluded as follows:

- Understanding the role of employee motivation and career management programmes are vital even in low-skilled low margin organisations.
- Discovering the best practices for career management and how to derive maximum value from its implementation is crucial.
- To be cognizant of the impact of organisational culture on the implementation of employee motivation.
- To link career management to strategic planning of the organization in order to ensure that the organization identify talent from different managerial levels of the organization.

Future work in this area should focus on choosing a greater number of companies to research the initial findings.

References:

[1] Greenberg, Jerald, and Baron, Robert ,A. (2003). *Behaviour in organizations*, 8th ed., Prentice-Hall.

[2] Robbins, S. P. and Judge, T. A., (2008). *Organizational Behaviour*. 13th ed. Prentice-Hall.

[3] Mullins, L. J., (2005). *Management and Organizational Behaviour*, 7th ed. Financial Times Prentice Hall, Harlow.

[4] Kreitner, R., (2007). *Management*. 10th ed. Cengage Learning .USA: Houghton Mifflin Company.

[5] Jones, G. R. and George, J. M., (2003). *Contemporary Management*. 3rd ed. Boston: McGraw-Hill.

[6] Mitchell, T. R., (1982). *Motivation: New directions for theory research and practice*. Academy of Management Review, 7 (1), 80-88.

[7] Ryan, R. M. and Deci, E. L., (2000). *Intrinsic and Extrinsic Motivations: Classic Definitions and New Directions*, Contemporary Educational Psychology 25, 54-67.

[8] Bruce, A. and Pepitone, J. S., (1998). *Motivating Employees, a Briefcase Book*. McGraw-Hill Professional.

[9] Creswell, J., (2009). *Qualitative inquiry and research design*. Thousand Oaks, CA: Sage Publications.

[10] Saunders, M., Lewis, P. and Thornhill, A., (2009). *Research methods for business students*, 5th ed. Harlow: Pearson.

[11] Cooper, D.R. and Schindler, P.S., (2011). *Business research methods.* 11th ed. Boston, MA: McGraw-Hill Publishers.

Can we (control) Engineer the degree learning process?

AS White[1], M Censlive[1], D. Neilsen[2]
[1]School of Science and Technology, Middlesex University, The Burroughs, Hendon NW44 BT
[2]Business Faculty, University of Greenwich, 30 Park Row, London SE10 9LS

E-mail: a.white@mdx.ac.uk, d.nielsen@gre.ac.uk

Abstract. This paper investigates how control theory could be applied to learning processes in engineering education. The initial point for the analysis is White's Double Loop learning model of human automation control modified for the education process where a set of governing principals is chosen, probably by the course designer. After initial training the student decides unknowingly on a mental map or model. After observing how the real world is behaving, a strategy to achieve the governing variables is chosen and a set of actions chosen. This may not be a conscious operation, it maybe completely instinctive. These actions will cause some consequences but not until a certain time delay. The current model is compared with the work of Hollenbeck on goal setting, Nelson's model of self-regulation and that of Abdulwahed, Nagy and Blanchard at Loughborough who investigated control methods applied to the learning process.

Keywords: Learning, control theoretic model, simulation, Kolb model

1. Introduction

The role of learning theory in Engineering Higher education in the UK has only been a topic of serious investigation since the 1970's following the expansion of higher education in the late 1960's, leading to the integrated use of the student centred project method of teaching. In the 1970's investment was put into the use of teaching machines (Skinner [1]) with few consistent results reported. This basic technology was replaced by the use of personal computers in the 1980's.

If we are expected to use the new online techniques embodied by "Massive Open Online Courses" (MOOC) then we need to have a much better understanding of the process of learning in engineering education. The accepted view in the UK is described in Houghton [2] which uses the Kolb [3] theory in terms of:

1. Concrete Experience
2. Observation and reflection
3. Abstract conceptualisation
4. Testing concepts in new situations

These learning concepts can be mapped to the SARTOR (Engineering Council [4]) three part scheme: Knowledge; Skills and Understanding

These components of learning are compatible with a control theoretic approach and the principle of feedback is a key component of all reflection activities. A Delay or knowledge "soak time" is acknowledged by most academics to be a normal component of the learning process.

2. Review

The idea that people rely on mental models of situations was first proposed by Kenneth Craik [5] and this process was used to construct small scale models of reality to underlie reason, to anticipate events and to help with explanation. Norman [6] describes Mental Models as follows; "In interacting with the environment, with others, and with the artefacts of technology, people form internal, mental models of themselves and other things with which they are interacting. These models provide predictive and explanatory power of understanding the interaction." This will mean that people will map the observations of the real world to the simplest model that fits the facts as observed or interpreted by that individual.

The uses of control theoretic models in psychology dates back to Erez's [7] work on feedback and goals. Campion and Lord [8] applied control theory in light to goal-setting. This work was developed by Klein [9] studying work motivation and Hollenbeck [10] looking at work environments. In engineering the model described by Kolb was adopted by a number of investigators because it seemed to fit the existing engineering pedagogy. Abdulwahed et.al. [11] Summarises the work undertaken at Loughborough to develop control models of learning using the Kolb theory of learning.

Seri [12] illustrates a model of learning (figure 1), showing the necessary feedback paths and recognition of difference and error which agrees with the models used by Abdulwahed et.al.

White [13] reviews the results of experiments of control and mental models combining all these observations together with existing double loop learning models to create an integrated model (figure 2). Here all the aspects of prior education, environment and poor judgement combine to make a given strategy and subsequent decisions.

This model shows how difficult it is to isolate the circumstances for a successful learning process. An et.al [14] created a System Dynamics/control theoretic model of service levels for IBM. One of the key aspects of System Dynamics models of project developments is the important contribution of time pressure operating within deadlines. This is included in the model shown in figure 3.

3. Control model of learning in higher education

In this section we propose a control model of a constrained learning process, the learning target for the model is gaining the knowledge and skill to manipulate a given computer package. The model shown in figure 3 represents the process of obtaining capability in using specialist software. It has been developed using the Kolb model of experiential learning and building on the control models of Abdulwahed et.al. This implementation incorporating a delay is supported by experimental data for the exponential model of delay to the learning process from the results of Towill [15].

The capability *Cap* has a desired value *Cdes* which is rated as 1 or 100%. The effort needed to reach this level of capability is *E* and the time delay between the effort being applied and the *Cap* value observed is represented by a time delay transfer function $\frac{KEC}{(T_L s+1)}$. The time constant T_L is the time to reach 66% of the final value.

Each learner has a measure of the value of the capability achieved *Cap*. This measured value is FE**Cap* and is different for each actor. The student recognises an error between the measured capability and the desired performance. He/she will then adjust the effort by a gain value *Keff*. This gain is in two parts, one fixed rate and an element caused by the pressure due to time remaining. These are indicated in figure 4. The effective gain increases as time progresses. This model uses a linear increase in gain but would probably be a greater rate e.g. parabolic. The model is described in equations 1 to 5

$$Cap = Cap_p + Cap_s + \frac{E\,KEC}{T_L s+1} \quad (1)$$

$$E = Keff(Cdes - FE\,Cap) \quad (2)$$

$$Cap_p = \frac{Kip}{T_p s+1}(Cdes - FE_p Cap) \quad (3)$$

$$Cap_s = \frac{Kis}{T_p s+1}(Cdes - FE_s Cap) \quad (4)$$

$$Keff = K_0 + K_t\left(\frac{1-e^{-Tms}}{s^2}\right) \quad (5)$$

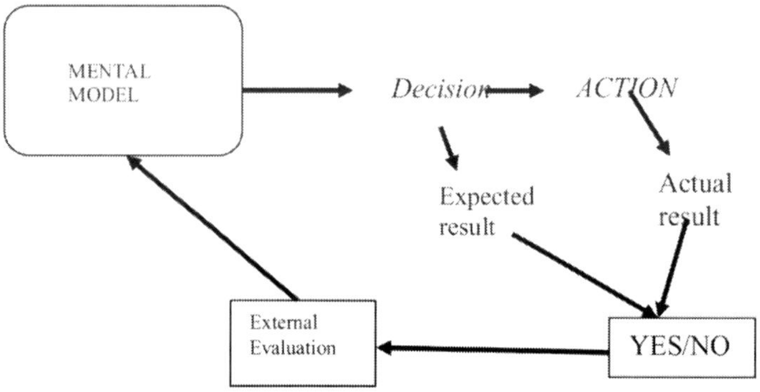

Figure 1. Individual loop of Learning adapted from Seri [13]

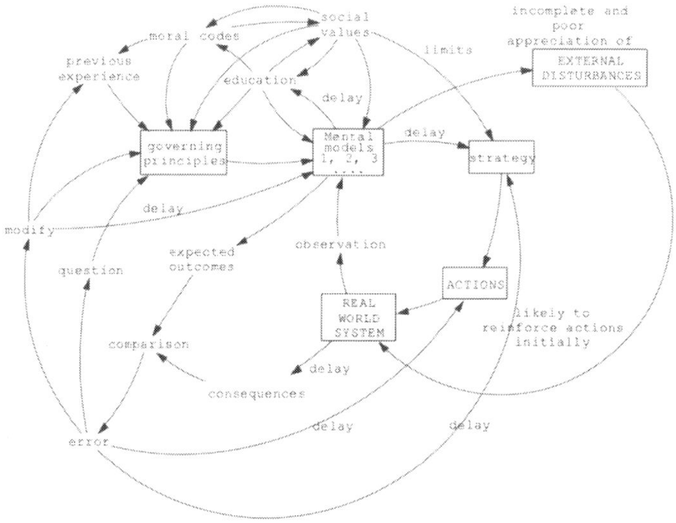

Figure 2. Model of Human control Learning Process from White [14]

The individual effort, whether by using electronic sources or by critical thought, is modified by two factors; one is the action of the supervisor or tutor who then intervenes when the student is struggling and the other due to the student gaining peer group help, which may be substantial. Both of these add to the final capability without learning (usually). Both of these are represented in the same way but with different feedback errors; gains and time delays. The capability acquisition through the module is shown in figure 3.

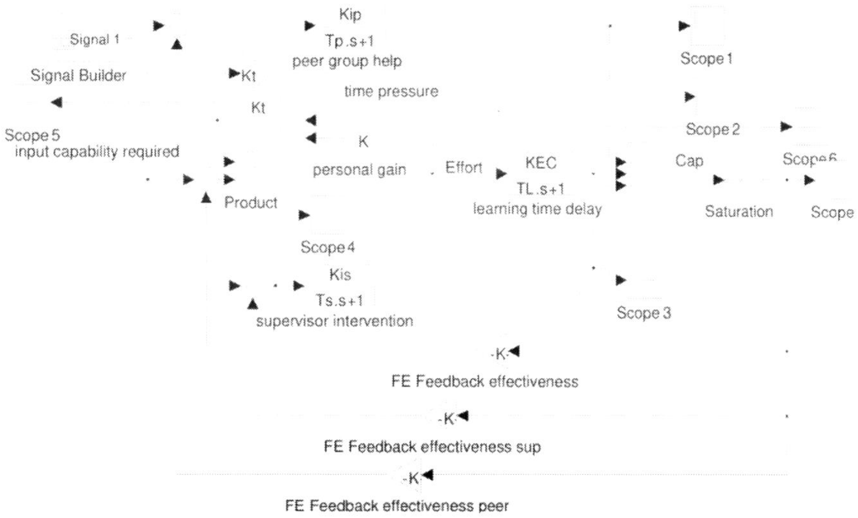

Figure 3. Control model of Learning

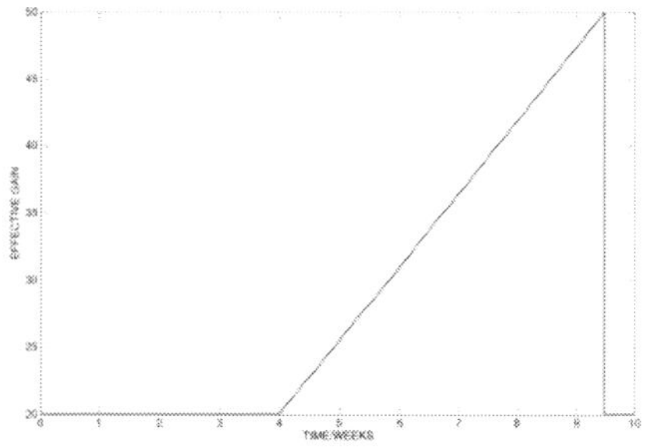

Figure 4. Effective gain

With the constants used in this first example half the final capability is due to the action of the peer group support and help from the tutor. Note that even then in this case the student does not achieve 100% capability. Figure 6 shows the sensitivity of the capability to reduced student performance or reduced peer group contribution. The data used is based on observations of teaching MATALB to two sets of students in 1999 and 2010. In the two cases the contributions of peer effects were greater in 2010 than in 1999.

4. MOOCs for Learning

The early stage assessments of MOOCs, based on the uptake by the number of students registered, indicated a very promising trend towards a new way of learning. Various studies have now raised questions about the success rates and effectiveness of MOOCs for student learning (Santos et al, [17]). Ramesh et al, [18] found a small clustering of student behaviour demonstrating engagement using these learning resources. Chen [19] argued that formal inclusion of regular feedback and interactions between the teacher and learner throughout the duration of the course is necessary for a successful and increased percentage of learning completion. In view of this and the applicable concept of value added feedback from supervisor intervention in the Control Model of Learning (figure 3) the design for

MOOCs courses must be reassessed to ensure the incorporation of these elements. Specific and clear pedagogical designs must be embedded to facilitate and encourage this model of learning.

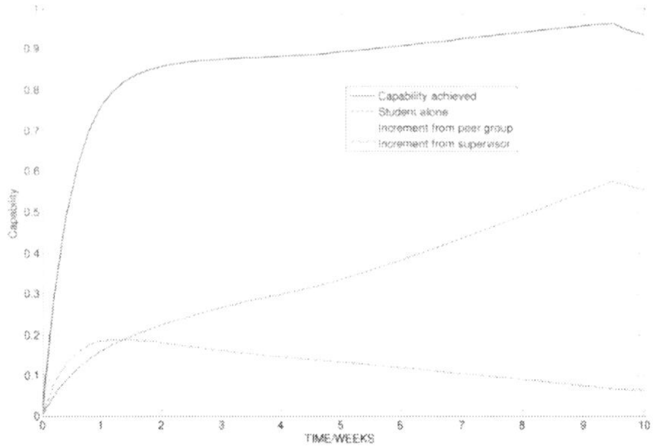

Figure 5 Capability performance

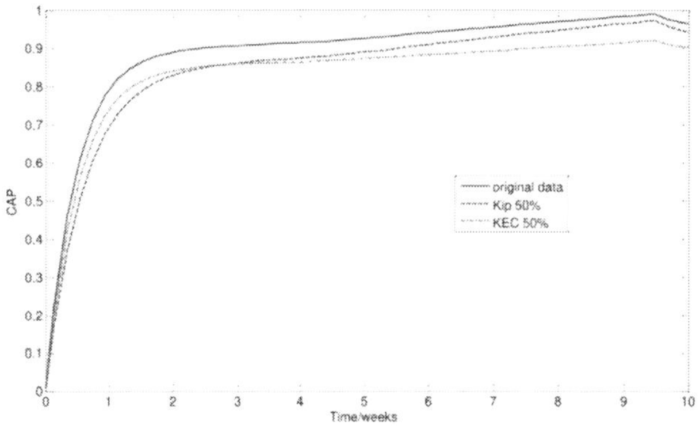

Figure 6 Overall performance sensitivity

5. Conclusions

A new control model has been devised building on the work of Abdulwahed et al. and White using the experiential model of Kolb

This model enables the effect of supervision and peer group interaction to be incorporated as well varying the effectiveness of the students own contribution.

Sample results show that the times of contribution and the amount of peer group contribution to be critical to reducing the time to learn a given set of operational skills for example the use of a given set of software applications.

References

[1] Skinner, B.F. (1958) Teaching Machines, *Science,* Vol 128, No3330, 969-977.
[2] Houghton, W, (2004) Learning and Teaching Theory for Engineering Academics, *Higher Education Academy*, Engineering subject centre.
[3] Kolb, D (1984) *Experiential Learning: experience as source of learning and development*, Prentice-Hall.
[4] Engineering Council (1985) Standards and Routes to Registration (SARTOR). EC 10 Maltravers Street, London WC2R 3ER, England.

[5] Craik, K. (1943) *Nature of Explanation*, CUP
[6] Norman, D. A. (1983). Some observations on mental models. In D. Gentner and A. L. Stevens (Eds.), *Mental models* (p. 7-14). Hillsdale, NJ: Lawrence Erlbaum.
[7] Erez, M. (1977). Feedback: A Necessary Condition for the Goal Setting Performance Relationship. *Journal of Applied Psychology*, Vol.62, pp.624-627.
[8] Campion, M. A., & Lord, R. G. (1982). A control systems conceptualization of the goal-setting and changing process. *Organizational Behavior and Human Performance*, Vol.30, pp.265-287.
[9] Klein, H. J. (1989). An integrated control theory model of work motivation. *Academy of Management*, Vol.14, pp.150-172.
[10] Hollenbeck, J. (1989). Control Theory and the Perception of Work Environments: The Effects of Focus and Attention on Affective and Behavioral Reactions at Work. *Organizational Behavior and Human Decision Processes*, Vol.43, pp.406-430
[11] Abdulwahed, M., Nagy, Z.K. and Blanchard, R.E., (2008). Constructivist project based learning design, a cybernetics approach. IN: Malpica, F. et al. (Eds). Proceedings of the 2nd International Multi-Conference on Society, Cybernetics and Informatics, (IMSCI 2008), June 29th - July 2nd, 2008
[12] Seri, P (2001) Losing areas and shared mental models: towards a definition of the cognitive obstacles to local development, *DRUID Winter Conf.* Copenhagen, 18-20 Jan.
[13] White, A. S. (2013) Human Supervision of Automated Systems and the Implications of Double Loop Learning, *International Journal of Information Technologies and Systems Approach*, Vol.6, No.2, pp.13-21, July-December
[14] An, L, Jeng, J and Lee Y. M, (2007). "Effective Workforce Lifecycle management via System Dynamics Modeling and Simulation", Proc. *2007 Winter Simulation Conf.*, IEEE, pp.2187-2195
[15] Towill, D (1977) Exponential smoothing of learning curve data, *Int.J.Prod. Res.*, Vol.15, No.1, pp.1-15.
[16] Santos, J.L., Klerkx, J., Duaval, E., Gago, D. and Rodriguez, L., (2014). Success, activity and drop-outs in MOOCs an exploratory study on the UNED COMA courses, Proc. 4th Int. Conf. on Learning Analytics And Knowledge, *ACM, NY,* ISBN: 978-1-4503-2664-3 doi>10.1145/2567574.2567627 pp. 98-102
[17] Ramesh, A., Goldwasser, D., Huang, B., Daume, H., III and Getoor, L. (2014) Uncovering hidden engagement patterns for predicting learner performance in MOOCs, *L@S '14: Proc.1st ACM conference on Learning @ scale conference,* pp. 157-158, ACM NY, ISBN: 978-1-4503-2669-8 doi>10.1145/2556325.2567857
[18] Chen, P. and Chen Y. (2014) Facilitating MOOCs learning through weekly meet-up: a case study in Taiwan, L@S '14 *Proc. 1st ACM conference on Learning @ scale conference*, pp. 183-184, NY, ISBN: 978-1-4503-2669-8 doi>10.1145/2556325.2567872

Symbols

Cap *Capability to perform computing task(0-100%)*
Cdes *Capability Desired*
E *Effort expended in person hours*
FE *Feedback effectiveness*
K *Personal gain, increment of effort per difference of perceived gap between own performance and desired value*
K_0 *Initial value of personal gain*
K_t *effect of time pressure on personal gain*
KEC *Personal capability increment per unit effort*
Keff *Effective personal gain allowing for time pressure*
Kip *Peer group gain increment per unit time*
Kis *Supervisor gain increment per unit time*
Kt *Part of personal gain due to time pressure*
T_L *Learning time delay for the individual*
Tm *Time to realise time pressure*
T_P *Time delay for effective peer group help*
T_S *Time delay for effective supervision.*

IOP Conf. Series: Materials Science and Engineering **65** (2014) 012034 doi:10.1088/1757-899X/65/1/012034

FROM E-MANUFACTURING TO INTERNET PRODUCT PROCESS DEVELOPMENT (IPPD) THROUGH REMOTE - LABS

Ernesto Córdoba Nieto

Permanent Professor, Mechanical and mechatronic Department, Universidad Nacional de Colombia
ecordoban@unal.edu.co

Paulo Andres Cifuentes Parra

M.Eng student at computer sciences, B.E. Mechatronic engineering, Universidad Nacional de Colombia
pacifuentesp@unal.edu.co

Juan Camilo Parra Díaz

B.E. Mechatronic Engineering, Universidad Nacional de Colombia,
jcparrad@unal.edu.co

Abstract. This paper presents the research developed at Universidad Nacional de Colombia about the e-Manufacturing platform that is being developed and implemented at LabFabEx (acronym in Spanish as "Laboratorio Fabrica Experimental"). This platform besides has an approach to virtual-remote labs that have been tested by several students and engineers of different industrial fields. At this paper it is shown the physical and communication experimental platform, the general scope and characteristics of this e-Manufacturing platform and the virtual lab approach. This research project is funded by COLCIENCIAS (Administrative Department of science, technology and innovation in Colombia) and the enterprise IMOCOM S.A.

1. Introduction

Nowadays, technological advantages are important factors for improving competitiveness in almost every economic activity in our global society [8]. Among these economical activities, Manufacturing has been one of the most affected by those new technological advances [4] such as communication technologies, software engineering, algorithms applied to manufacturing, virtual realities, web services and web presentation technologies. e-Manufacturing is becoming the new concept that is currently being applied in today's industrialized world [1]. This is the main reason why At LabFabEx (in Spanish as "Laboratorio Fabrica Experimental" and in English as Experimental Fabric laboratory) in Universidad Nacional de Colombia at the Cam-Mechatronics room Lab, an e-Manufacturing Platform is being developed and implemented. This platform has industrial and experimental machines that are grouped in three manufacturing cells, a communication platform (routers, servers, web cams) and software that supports and manages the information system for centralization of information and for remote operation of these machines.

This platform will provide several facilities regarding to:

- Centralize and synchronize machine performance process information.
- The final user will be able to set machine parameters. This configuration process will be provided through web-services.
- Remote operation of machines, through web VMM[1] interfaces from which the final user can develop product processes.
- Support and implement a collaborative robotic network for performing flexible manufacturing processes.

Experimental tests have been performed with the academic field (professors and students) and with the industrial field (industrial managers and engineers). Although this platform is currently being developed, these tests have shown advantages about remote operation. In fact, the virtual environments or VMM (virtual manufacturing machines) are components that eases the way for operate remotely these machines. The e-Manufacturing platform has been successfully tested through Internet from different parts of the world such as Colombia, Germany, France, Mexico and USA.

LabFabEx project will impact Colombian enterprises by making research about their industrial process and academic and teaching activities by allowing virtual remote laboratories through Internet. LabFabEx is part of the project RC-723 "FITTING OUT AN EXPERIMENTAL FABRIC LABORATORY (LabFabEx) BASED ON E-MANUFACTURING TECHNOLOGY FOR INTEGRATED DEVELOPMENT OF PRODUCTS AND PROCESS AND ITS TRANSFERENCE TO THE PRODUCTIVE SECTOR". COLCIENCIAS, IMOCOM S.A and Universidad Nacional de Colombia have funded this project.

2. LabFabEx (Experimental Factory Laboratory)

LabFabEx, (for its acronym in Spanish "Laboratorio Fabrica Experimental") has several industrial and experimental manufacturing machines and equipment. These machines are grouped into three manufacturing cells. The first one contains 6 industrial machines and it is named the Industrial Manufacturing Cell (IMC). The second one has 5 experimental manufacturing machines so that it is named the Experimental Manufacturing Cell (EMC). This EMC's machines have been designed and developed at Universidad Nacional de Colombia by the Mechanics and Mechatronics Department. The rapid prototyping cell has one industrial 3d printing Machine, one experimental 3D printing machine and one experimental parallel kinematic robot (Orthoglide) for CNC machining. There are two experimental AGVs (Automatic Guided Vehicle) that are being developed at the lab. These two vehicles will interconnect physically all the machines at the laboratory (IMC's machines, EMC's machines and rapid prototyping cell's machines) for several types of industrial processes. Figure 1 shows the IMC and Figure 2 shows the EMC. Besides, this e-manufacturing platform will be able to integrate another experimental and industrial machines.

---

[1] VMM (Virtual Manufacturing Machine) refers to a virtual 3D environment that represents the geometric features of the real machine at the lab.

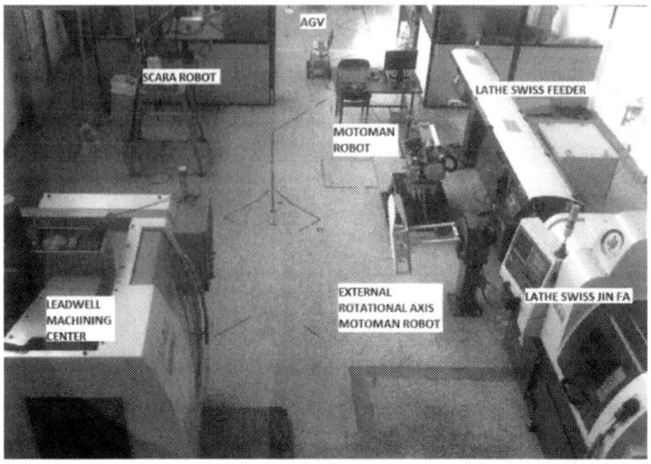

Figure 1. *Industrial Manufacturing Cell's Machines and their physical disposition. Universidad Nacional de Colombia. Mechanics and Mechatronics Department. Cam-Mechatronics room Lab*

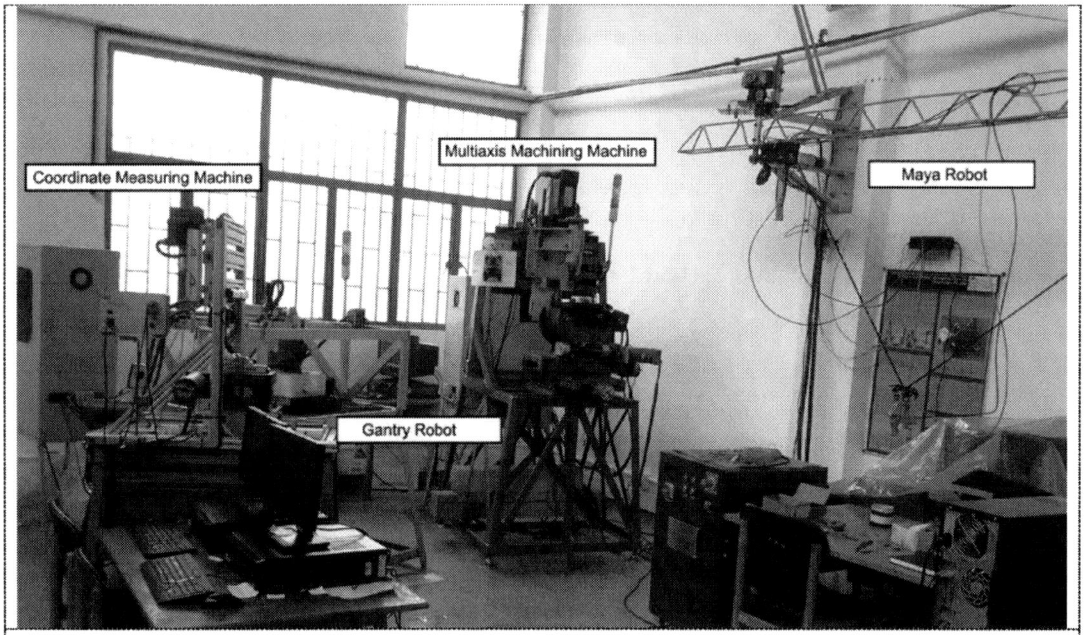

Figure 2. *Experimental Manufacturing Cell's Machines and their physical disposition. Universidad Nacional de Colombia. Mechanics and Mechatronics Department. Cam-Mechatronics room Lab*

The Industrial Manufacturing Cell's machines are listed below:

1. Motoman Robot MH6-DX100: 6 degrees of freedom and one external rotational axis and one external linear axis.
2. Leadwell v20 Machining center: This is a CNC machine that has four axes.
3. Scara Robot: This robot has four DOF and its main function is to pick and place.
4. Lathe Swiss Jin Fa. This machine has seven degrees of freedom (JSL SERIE 32AB)
5. AGV mobile robot: Automatic Guided Vehicle.

6. Presetting machine Zoller smile 400.

The Experimental Manufacturing Cell's machines are listed below:

1. Maya crane Robot: pick and place material around the Experimental Manufacturing Cell.
2. Gantry parallel Robot: three degrees of freedom.
3. Multi Axis Machining Machine: 5 degrees of freedom
4. AGV mobile robot: Automatic Guided Vehicle.
5. Coordinate Measuring Machine (CMM): four degrees of freedom.

3. Experimental Communication Platform

The Communication Platform of LabFabEx has the structure shown in the figure 3. This platform integrates each machine at the lab in a common communication infrastructure.

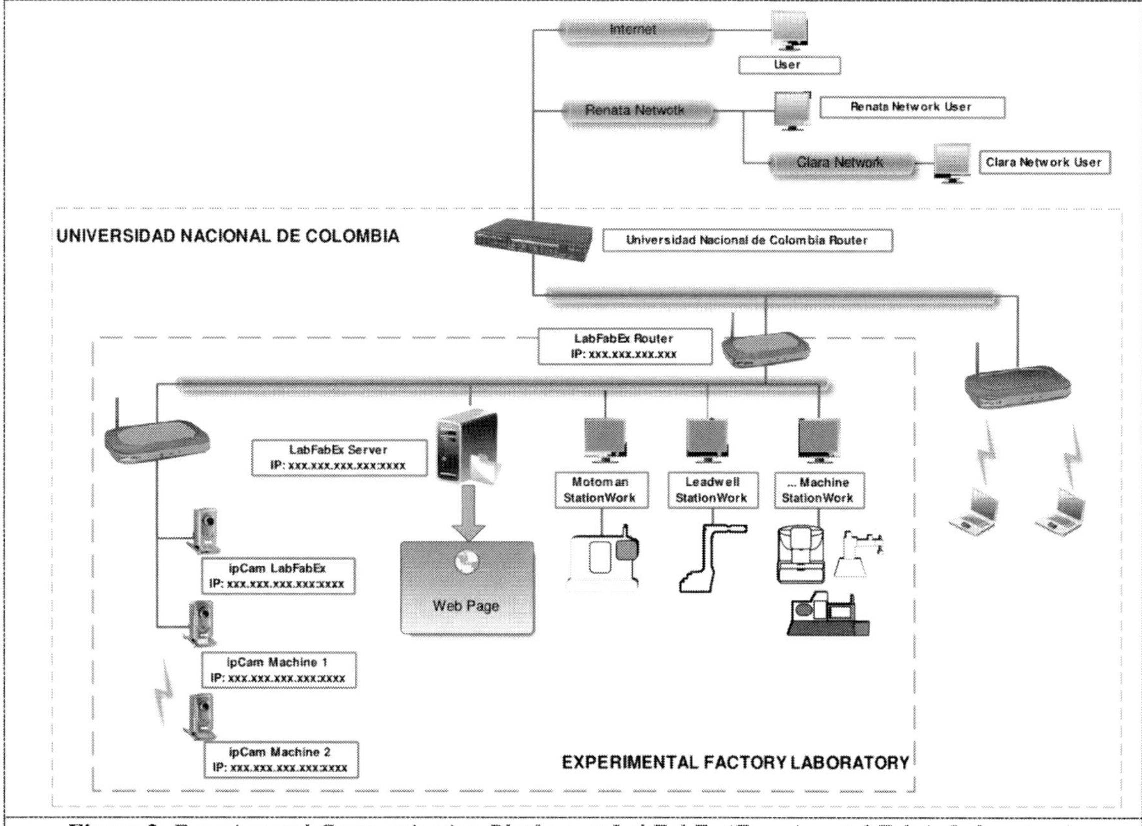

Figure 3. *Experimental Communication Platform at LabFabEx (Experimental Fabric Laboratory). Universidad Nacional de Colombia, Mechanics and Mechatronics Department. Cam-Mechatronics room Lab*

4. E-manufacturing platform

e-Manufacturing is a new approach that integrates the manufacturing infrastructure with diverse elements of business systems such as suppliers and customer network systems (refers to [1]). This process of integration is based on synchronization between the transformation system and the business system [1].

IOP Conf. Series: Materials Science and Engineering **65** (2014) 012034 doi:10.1088/1757-899X/65/1/012034

At the Cam-Mechatronics room Lab, at Universidad Nacional de Colombia, there is an e-Manufacturing platform that is under construction. Figure 4 shows the e-manufacturing platform's general architecture scheme with its operation characteristics, the machines located at the lab (and how get access to their controllers) and how the information and interfaces through Internet are presented.

The main idea of this approach is to centralize and synchronize machine performance process information. This information is presented through Internet using graphical computer technologies such as OpenGL (WebGL and processing), Client-Server communication technologies (webSockets, ajax, javaNIO, javaIO) and web presentation technologies (applets, html5, javascript).

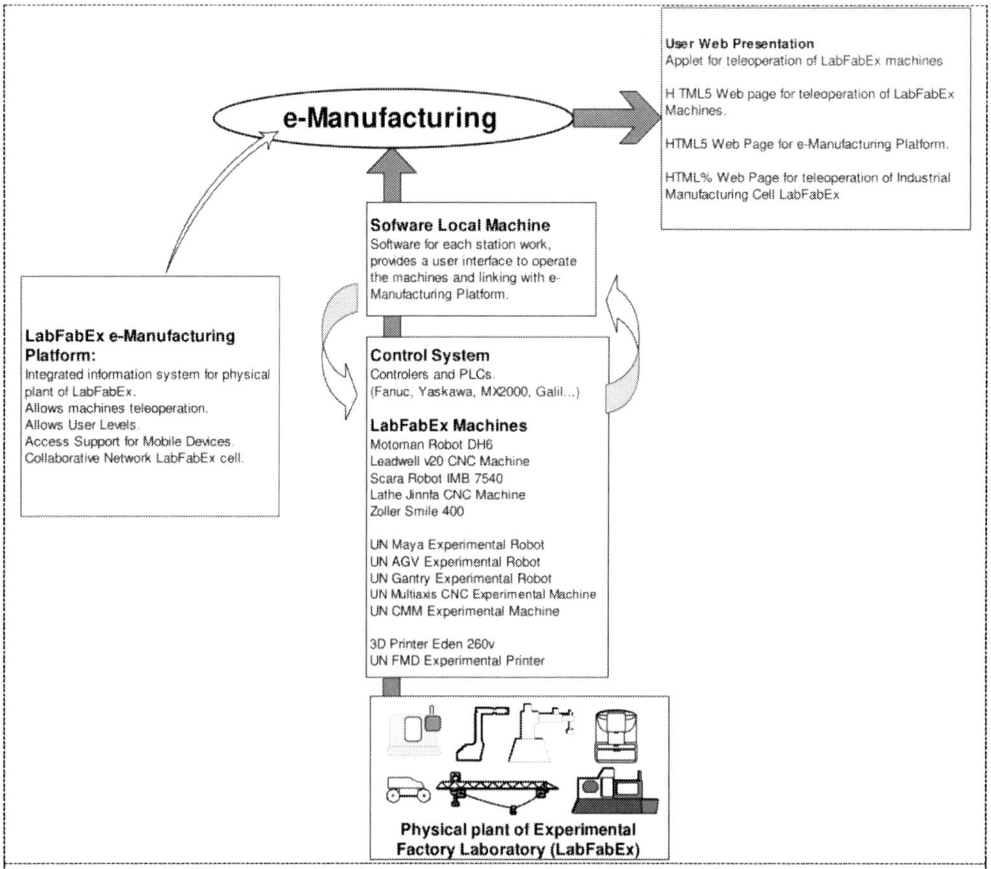

Figure 4. *LabFabEx e-Manufacturing platform architecture*

The virtual lab approach of this platform provides students with technological tools through Internet with which they will be able to access and know how works each machine located at the physical lab.

This e-Manufacturing platform has three modules (see figure 7): 1) Client module for remote access (User Web Presentation), 2) Server module for information management (LabFabEx e-Manufacturing Platform) and 3) station working module for each laboratory's machine (Software local Machine). Each one of these three interdependent modules is designed based on the axiomatic design methodology for software development.

Figure 5 shows the protocols used at each machine for getting communication with the e-manufacturing platform. Machines at LabFabEx have their respective VMM (virtual manufacturing machine that refers to the 3D virtualization of each machine and its environment) and web communication technologies.

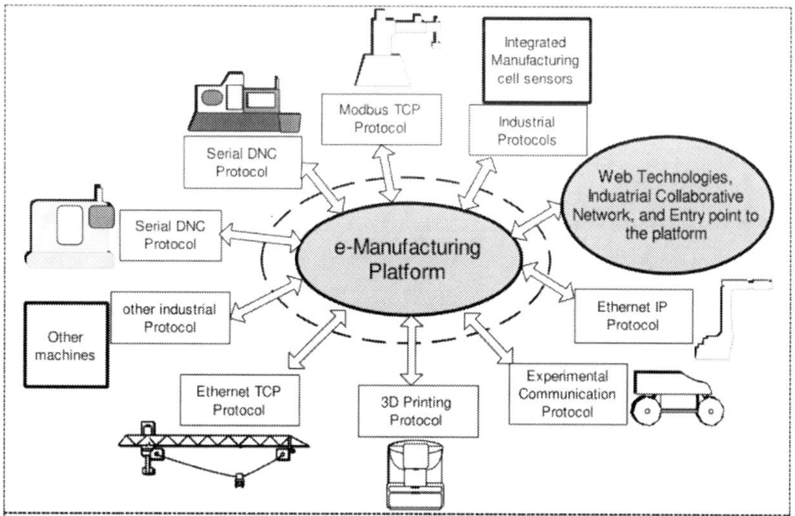

Figure 5. *Protocol used at each machine for getting communication with the e-Manufacturing platform*

The main functions of this e-Manufacturing platform are:

- **Centralize information of each machine:** the current status information of each machine is got, processed and presented to the final user
- **Teleoperation of each machine through internet:** each machine can be teleported through internet, from a typical Computer or from a mobile device (e.g. tablets or smart phones)
- **User access levels:** administrator (Super-user), guest user and operator (see figure 6). Super-User can be Admin User (server Manager that can enable Web Service) or Machine User (Machine manager that can get access to these machines without restrictions). Guest User only had video streaming from the LabFabEx. Operator user has three levels. As beginner user can only move the axis of each machine. As Moderated User can move the axis and send Machine commands. As Advanced User can also load and download Machine programs
- **Industrial collaborative robotic network (ICRN):** The ICRN will integrate three robotic platforms (Motoman Robot, Scara Robot, AGV mobile robot and Maya crane robot) with two machining machines (Leadwell Machining center and Lathe Swiss Jin-Fa) for getting a collaborative robotic network in which automatic industrial processes can be performed (machining, assembly, pick and place and raw material transportation).

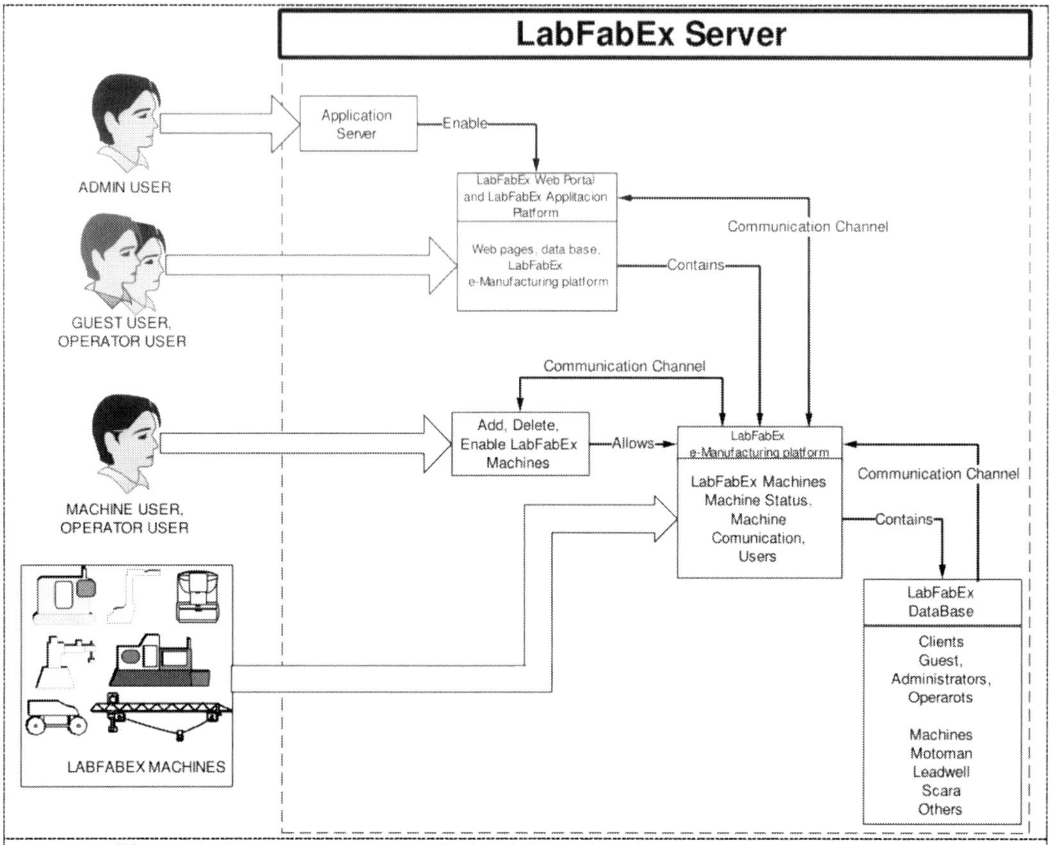

Figure 6. *User access levels at LabFabEx: Administrator, guest user and operator*

The e-Manufacturing platform has three main modules: 1) Client module for remote access; 2) Server module for information management; 3) Work station module for each machine. These three interdependent modules follow the axiomatic design methodology for software development (see figure 7).

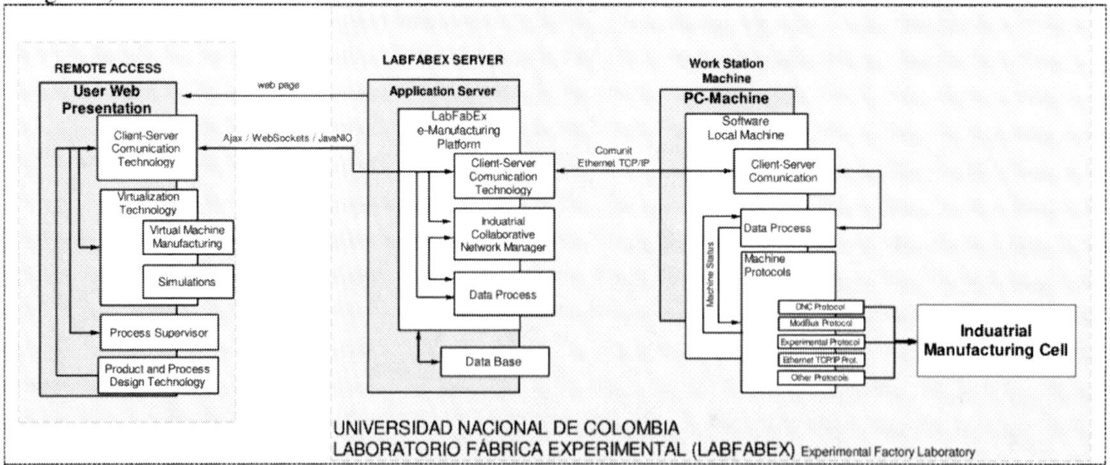

Figure 7. *Methodical diagram. e-Manufacturing platform's modules: 1) Client module for remote access; 2) Server module for information management; 3) Work station module for each machine.*

Through this e-manufacturing platform the final user can develop industrial processes for specific products. The user can operate remotely each machine through the web presentation layer that has several operation features regarding to the industrial process that is going to be perform at the lab

5. Virtual platforms

The physical machines have been mapped to VMM's (Virtual Manufacturing Machine) for developing a virtual laboratory that can be accessed through Internet. Some of these machines belong to the old teleoperation platform (Renata[2] V 1.0) and the other ones have virtual interfaces developed by using new computational graphics technologies. The new interfaces have several features such as programmed task simulations, the VMM (Virtual manufacturing machine), web-chat module for getting an efficient communication between the operator at the lab and the final user, Control panel with commands for teleoperation, Status-Console panel for showing Input/output events. This interface implements APIS of Processing (Programming language for computer graphics), proscene[3], ControlP5 and OBJLoader and manages connection exceptions, data and internal processes.

5.1. Motoman MH6-DX100 Robot

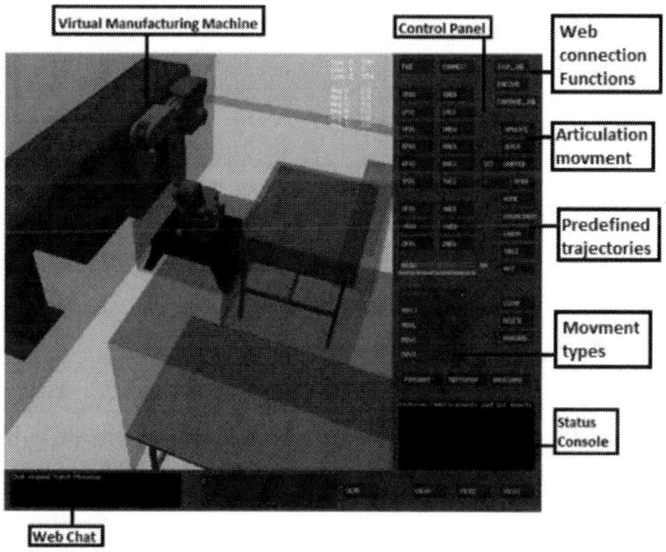

Figure 8. *Virtual interface[4] for operate remotely the Robot Motoman MH6-DX100 trough Internet. This interface shows the virtual environment and its facilities for programming this industrial robot. Universidad Nacional de Colombia. Mechanics and Mechatronics Department. Cam-Mechatronics room Lab*

[2] Renata is the academic network advanced technology in Colombia. Through this technological communication infrastructure, several projects has been developed for getting remote access to the machines at the Cam-Mechatronics room Lab at Universidad Nacional de Colombia. (See [3])

[3] Processing library developed by Prof. Dr. Jean Pierre Charalambos

[4] This virtual interface has been registered in Colombia as "Plataforma virtual para la teleoperación del robot Motoman MH6-DX100 en Internet". Into English it is named "Virtual platform for tele-operating the Motoman Robot MH6-DX100 in internet"

Figure 8 shows the 3D virtual robot and its environment. This Robot has 8 degrees of freedom (DOF). The first six DOF's refer to the robotic arm (nomenclature: S, L, U, R, B, T) and the last ones are two external axis (one rotational and one linear external axis).

The mathematical modeling (inverse and forward kinematics) has been implemented in order to perform accurate trajectory simulations that can be lineal, circular or spline as it is shown in figure 9. (See [2]).

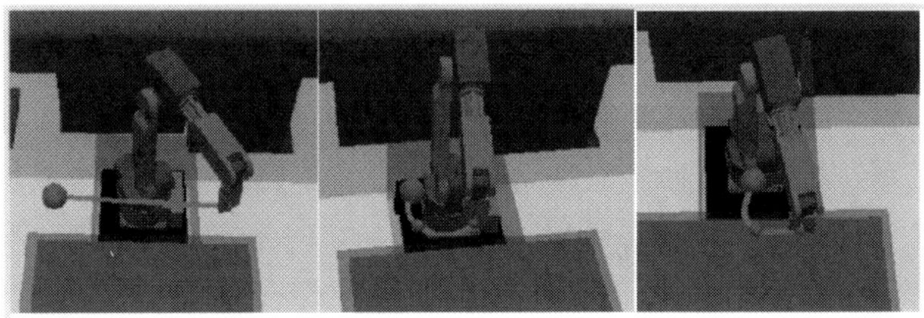

Figure 9. *Trajectory simulations (linear, circular and spline trajectories) based on the mathematical modeling of inverse and forward kinematics of this industrial robotic arm.*

While performing the simulation, if any shock arises, the user must change the trajectory before executing the programed task. If the user does not do that, the 3D virtual interface will not allow the user to execute the trajectory programed. Figure 10 shows a shock between the robot and a security plane.

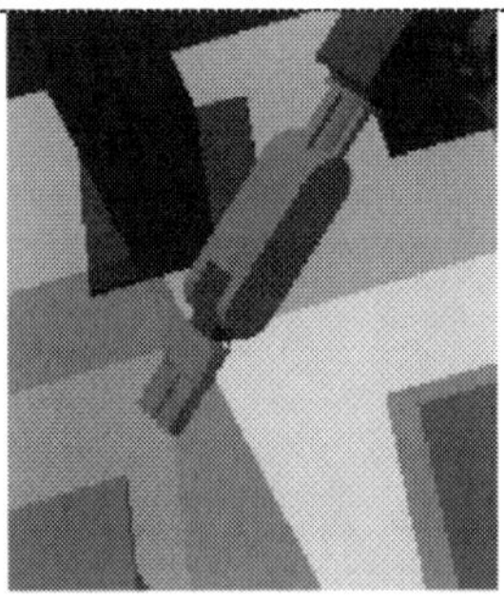

Figure 10. *Security planes get red if there is any shock or mechanical position out of range. A trajectory simulation can be executed only if and only if there is not any chock between the robot and its environment (security planes) or with itself.*

5.2. Scara Robot 7540

This robot has 4 degrees of freedom. Figure 11 shows its general operation diagram.

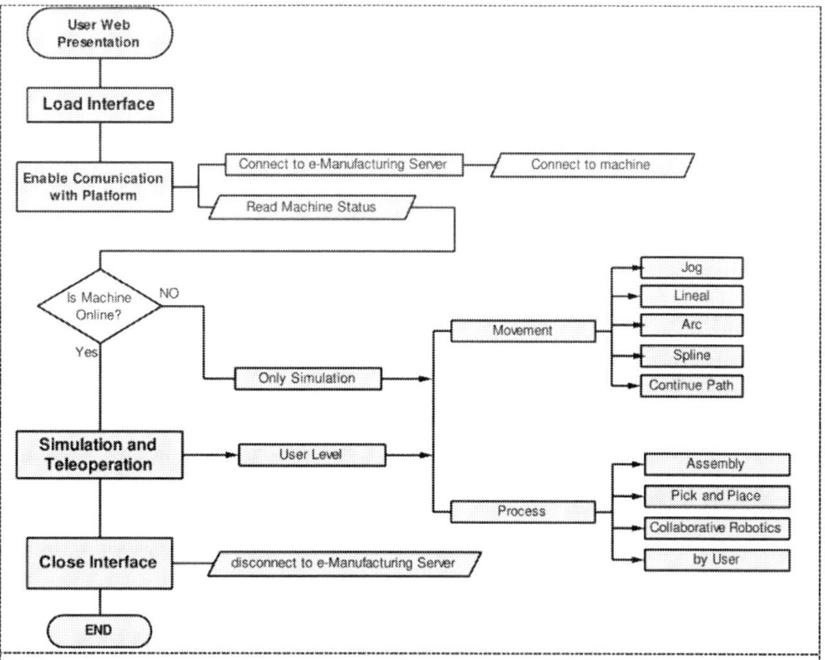

Figure 11. *General operation diagram for Scara Robot 75-40. Universidad Nacional de Colombia. Mechanics and Mechatronics Department*

The figure 12 shows the VMM (Virtual Manufacturing Machine) for operating this machine.

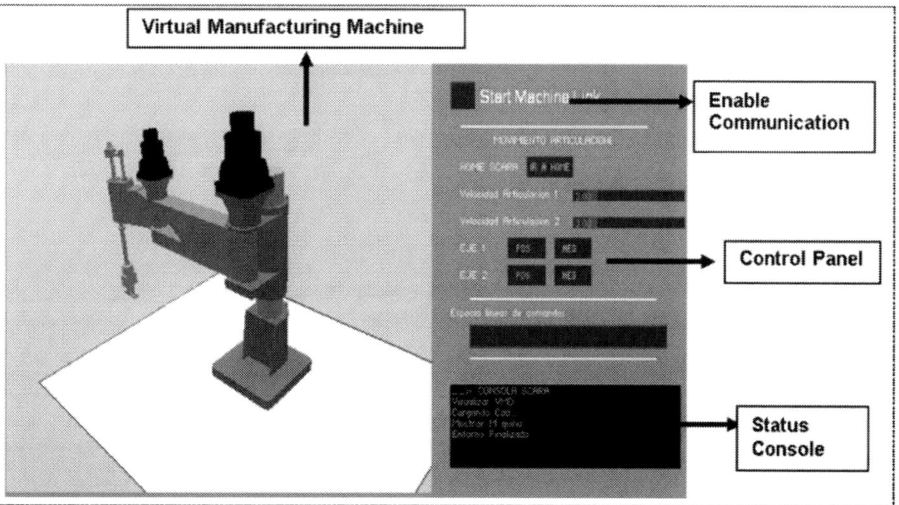

Figure 12. *Virtual interface for operating remotely the Robot Scara 7540 trough Internet. This interface shows its 3D virtual environment and its control panel. Universidad Nacional de Colombia. Mechanics and Mechatronics Department. Cam-Mechatronics room Lab*

5.3. Leadwell V20i Machining Center

This is a CNC Machining center with 4 degrees of freedom. The figure 13 shows the operation diagram of this Machine in the e-Manufacturing Platform.

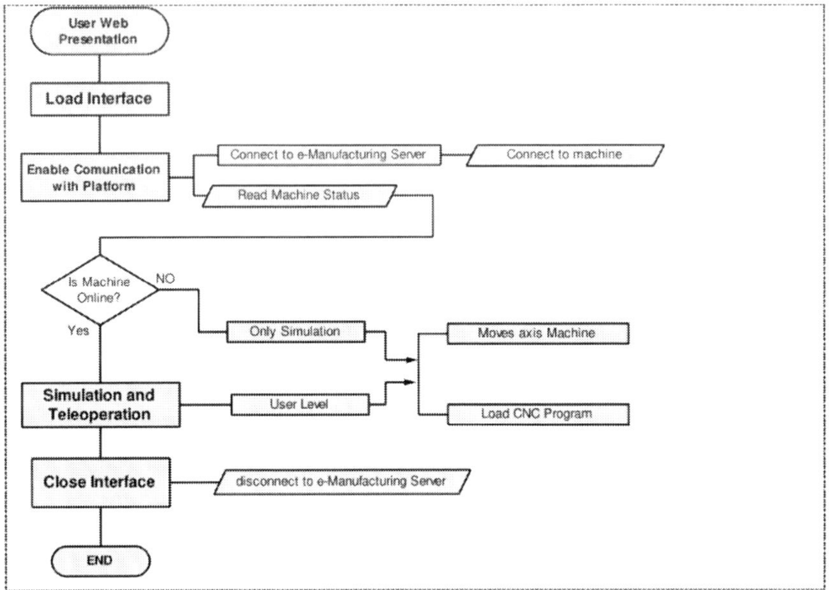

Figure 13. *Operation Diagram of the Leadwell V20i CNC Machining Center. Universidad Nacional de Colombia. Mechanics and Mechatronics Department*

The figure 14 shows the HMI for operating the Leadwell Machining Center through Internet.

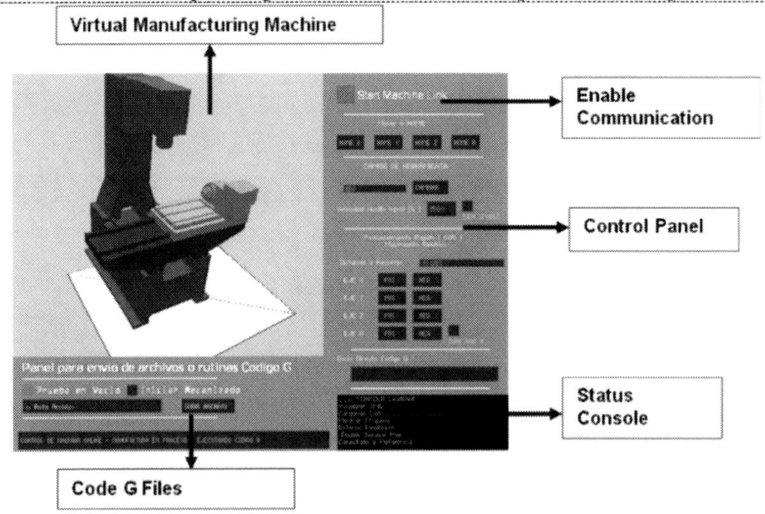

Figure 13. *Virtual interface for operating remotely the Leadwell V20i Machining center through Internet. This interface shows the virtual environment and its facilities with which the final user can program this industrial Machining center. Universidad Nacional de Colombia. Mechanics and Mechatronics Department. Cam-Mechatronics room Lab*

IOP Conf. Series: Materials Science and Engineering **65** (2014) 012034 doi:10.1088/1757-899X/65/1/012034

6. Our attempt to cloud manufacturing

While developing the e-Manufacturing platform and after making some validation tests with students and engineers from diverse academic and industrial fields, the platform performed an approach to cloud Manufacturing because product process are developed through internet. The virtual work-stations for each machine and practical exercises implemented by several students have lead to platform model based on the concept of cloud Manufacturing for flexible Manufacturing cells

The Cloud-Manufacturing concept, that came up first on 2010 [5], describes a manufacturing system in the cloud, based on the philosophy of "design Anywhere, Manufacture Anywhere". This platform centralizes and manages manufacturing enterprise's distributed resources for offering support and services in the cloud, based on a model that integrates supply system, business domain and user domain. This model is represented in several layers as shown in the figure 14.

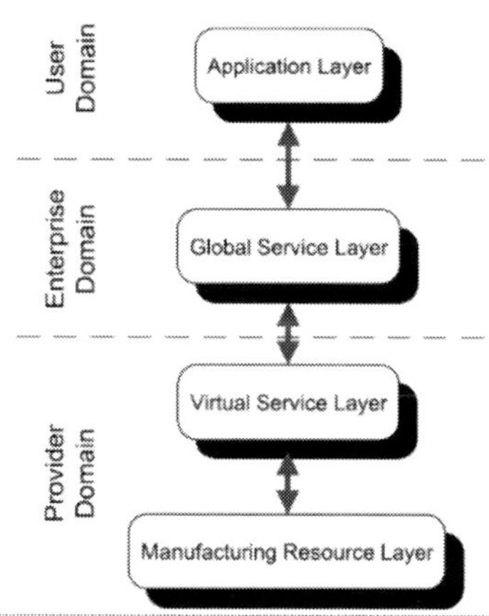

Figure 14. Cloud manufacturing architecture layers [6]

RESULTS

This e-manufacturing platform integrates industrial machines in a 3D virtual environment. The final user can access this 3D virtual platform for operating remotely each machine through the VMM (Virtual Manufacturing Machine) and also for getting centralized information of each machine. This platform has several features such as trajectory simulations, CNC G code compiling, current status information through Internet and software flexible frameworks that can be easily implemented to new machines that are being added to this e-manufacturing platform. This e-manufacturing platform has been tested from several parts of the world, such as Colombia, Germany, France, U.S.A and Australia. Through this testing, several improvements, regarding to virtual interface and robust communication, have been implemented in a continuous way.

DISCUSSION

This platform was initially developed based on only java technologies. However, there are new emergent web communications technologies that have several advantages. Java-applets are technologies that cannot be accessed through mobile devices and have some security restrictions. This has been one of the main reasons for implementing new web communication and computer graphics technologies on the e-manufacturing platform's client side, such as HTML5, javascript, webGL, webSockets, ajax, GWT (these technologies are standardized technologies by W3C[5]). Using these new technologies, the e-manufacturing platform can be accessed through mobile devices and this would be one the most important improvements on this platform

References

[1] Koç, M., Ni, J., Lee, J., & Bandyopadhyay, P. (2004). The Industrial Information Technology Handbook - Introduction to e-Manufacturing. (R. Zurawski, Ed.) CRC Press.
[2] Juan Parra, Julio Garcia, Ernesto Córdoba, Gabriel Mañana, Virtual teleoperation monitoring platform for Motoman Robot MH6-DX100 through Internet
[3] D. Escobar (2010): Accesibilidad a la Celda de Manufactura Flexible del Laboratorio de Mecatrónica a través de la Red Académica de Tecnología Avanzada Colombia (RENATA). Universidad Nacional de Colombia, Universidad de los Andes, ITESM-México, Universidad Autónoma de Occidente, Colciencias, Robótica ID.
[4] Cheng, F.-T., Tsai, W.-H., Wang, T.-L., Yung-Cheng, J. C., & Su, Y.-C. (2010). Advanced e-Manufacturing Model. IEEE Robotics & Automation Magazine.
[5] Li BH et al., "Cloud manufacturing: a new service-oriented networked manufacturing model," Computer Integrated Manufacturing Systems , vol. 16, pp. 1–7, 2010.
[6] Xun Xu, "From Cloud Computing to Cloud Manufacturing," Robotics and Computer-Integrated Manufacturing, vol. 28, pp. 75-86, 2012.
[7] García, J. C., & Parra, J. C. (2012). Monitoreo, supervisión y teleoperación para el robot motoman mh6 a través s de internet. Bogotá: Tesis de Grado Universidad Nacional de Colombia.
[8] Lee, J. (2003). E-manufacturing—fundamental, tools, and transformation. En Robotics and Computer Integrated Manufacturing (Vol. vol 19, págs. 501–507). Pergamon

[5] World Wide Web Consortium

IOP Conf. Series: Materials Science and Engineering 65 (2014) 012035 doi:10.1088/1757-899X/65/1/012035

Development the conceptual design of Knowledge Based System for Integrated Maintenance Strategy and Operation

Milana, M K Khan and J E Munive

School of Engineering and Informatics, University of Bradford, BD7 1DP UK

E-mail: m.milana@student.bradford.ac.uk, m.k.khan@bradford.ac.uk, j.e.munive@bradford.ac.uk

Abstract. The importance of maintenance has escalated significantly by the increasing of automation in manufacturing process. This condition switches traditional maintenance perspective of inevitable cost into the business competitive driver. Consequently, maintenance strategy and operation decision needs to be synchronized to business and manufacturing concerns. This paper shows the development of conceptual design of Knowledge Based System for Integrated Maintenance Strategy and Operation (KBIMSO). The framework of KBIMSO is elaborated to show the process of how the KBIMSO works to reach the maintenance decision. By considering the multi-criteria of maintenance decision making, the KB system embedded with GAP and AHP to support integrated maintenance strategy and operation which is novel in this area. The KBIMSO is useful to review the existing maintenance system and give reasonable recommendation of maintenance decisions in respect to business and manufacturing perspective.

1. Introduction

For a long time manufacturing has been pointed as one of the drivers of business achievement [1], but not maintenance. It was caused by the way of maintenance being considered. It was not counted as a substantial function as long as the manufacturing works in the expected performance. Instead, when the manufacturing process deteriorated, the quality of product decreased, the production target was not achieved, and the corrective action ate up the production budget for repairing action, then maintenance was considered as the inevitable cost.

The rapid development of automation in manufacturing process leads the significant changes to maintenance perspective recently. The implementations of advanced technology and lean manufacturing philosophy have increased the complexity of maintenance [2]. By providing the logistic support for manufacturing, today's maintenance perspective is considered affecting manufacturing process more than ever before. The automation could reduce cost of manufacturing process but increase budget allocation for maintenance purpose instead. It has been proved by which the maintenance cost could reach up to 70 % of total production cost [3]. Furthermore, the activities of maintenance to keep manufacturing equipments working on the expected performance are considered as inseparable part of the overall business competitiveness [4]. Hence, every decision for maintenance strategy and operation should always integrate to business and manufacturing perspective. This paper aims to develop the conceptual design of Knowledge Based System for Integrated Maintenance Strategy and Operation (KBIMSO). It elaborates the indicators identified in KBIMSO framework and shows how those indicators are collaborated to obtain maintenance decisions.

2. Relationship of Maintenance with Manufacturing and Business Perspectives

The relationship between maintenance and manufacturing is firmly visible within the manufacturing plant as to keep manufacturing assets works on their expected performance. Maintenance is defined as the combination of technical, administrative and managerial actions during the life cycle of equipment to keep it in, or restore it to its designed function [5,6]. It clearly means that maintenance is working based on demand from manufacturing function by referring to organization goal, as depicted in Figure 1. Maintenance and manufacturing function as an individual system within the organization have their own elements which work collaboratively to gain the function's objective. As interrelated functions, the main role of maintenance is fulfilling demands of manufacturing to retain its production capacity by maintaining the manufacturing equipments to work on their expected function. The contribution of maintenance to manufacturing performance, such as product quality and quantity, lead time and safety makes maintenance is considered as one of competitive drivers which affect overall business competitiveness. Therefore, maintenance strategy and operation need to be referred to business strategy in order to achieve organization goal.

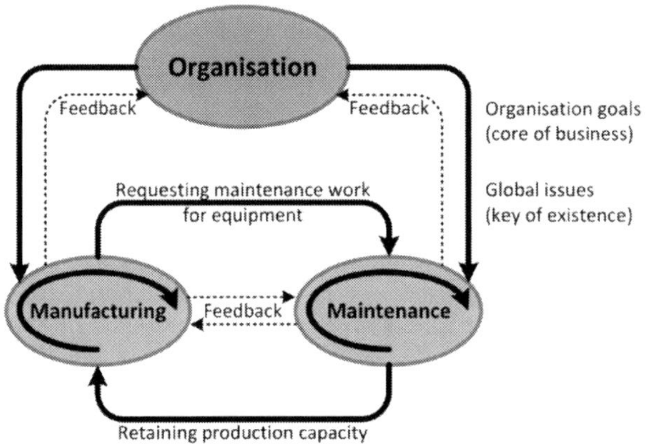

Figure 1. Relationship between organisation, manufacturing and maintenance [7]

The interaction of maintenance, manufacturing and business has been investigated on many studies. The survey conducted by Pinjala et al.[2] argue that a pro-active maintenance and better planning and control of manufacturing system are implemented by a quality business competitor. Meanwhile, Swanson [8] confirms the relationship between maintenance with manufacturing and business performance in term of product quality, equipment availability, and production cost. Gulati [9] emphasizes that implementation of best practice maintenance can improve performance, competitiveness and market share. Following these evidences, this paper is intended to configure the methodology required to take maintenance decision by referring to all contributed elements in maintenance itself, manufacturing, and business perspectives, which is called Knowledge Based System for Integrated Maintenance Strategy and Operation (KBIMSO).

3. Conceptual Model of Knowledge Based System for Integrated Maintenance Strategy and Operation (KBIMSO)

The KBIMSO conceptual model is generated from the KBIMSO framework [7] which is rearranged and emphasized to present the structure of KBIMSO model, as presented in Figure 2. Every single element on business perspective, manufacturing perspective and maintenance perspective is treated as a module and explored in detail into sub-modules to present the key factors contributed into KBIMSO. The business and manufacturing perspectives in KBIMSO framework are pointed as the strategic level in KBIMSO structure, while the maintenance perspective is presented through the operational level.

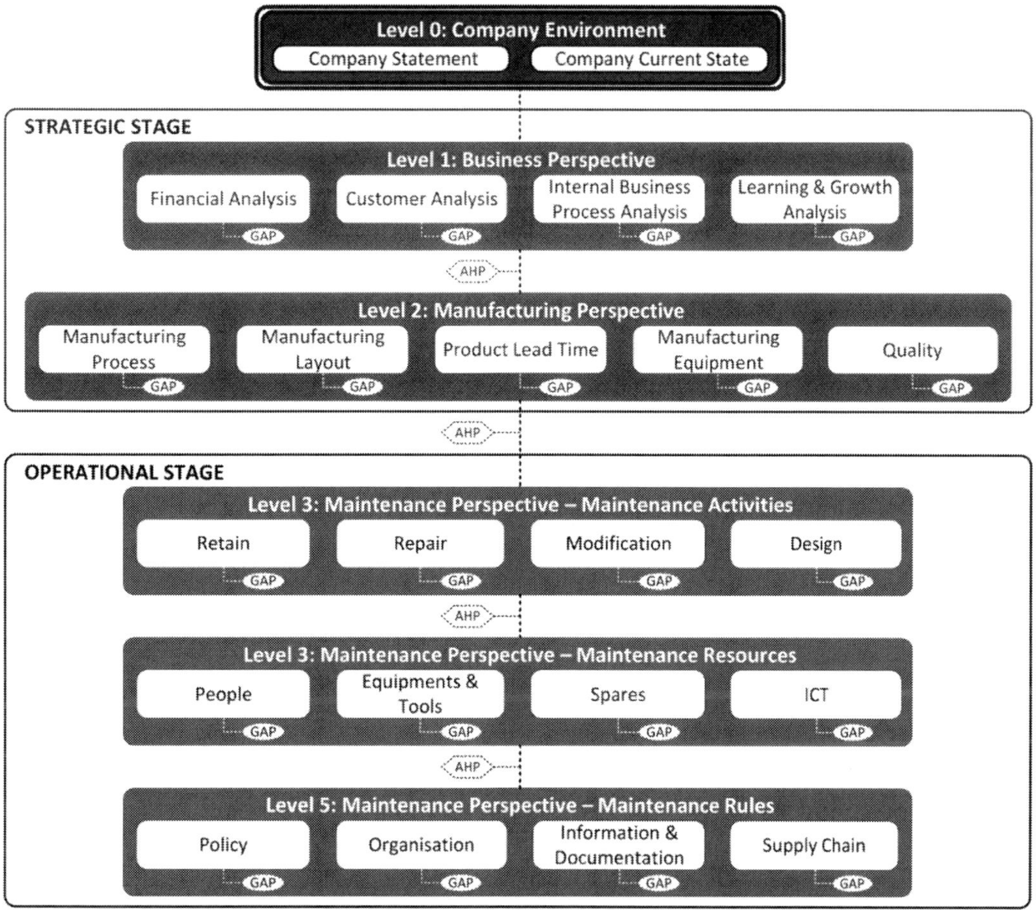

Figure 2. Structure of KBIMSO model

3.1. *Level 0 – Company Environment*

Level 0 is a top level of KBIMSO model which is intended to describe the company environment through identifying company statement and company current state. Company statement includes vision, mission and objective of company to compete in business. The company statement is required to generate the suitable strategy and operation for the lower level of function. To accommodate the uniqueness of the company, it is important to consider the current state in which the company is running. The point of this identification process is to consider each company as a customized system and to classify any particular company into different characteristic which enable to arrange the suitable requirement on the development of KBIMSO model.

3.2. *Level 1 – Business Perspective*

The first level in strategic stage is *Business Perspective*. This perspective adopts Balanced Scorecard (BSC) approach to provide the overall business performance measurement through four perspectives; financial, customer, internal business process, and learning and growth. These four perspectives are noted to be able to effectively represent existing business performance and help to identify the prospective achievement in the future [10]. The ability of BSC to translate company statements into functional strategies points it as the main techniques to identify the business requirement towards manufacturing and maintenance in the business perspective of KBIMSO.

3.3. Level 2 – Manufacturing Perspective
Another factor examined in the strategic stage is *Manufacturing Perspective*. Maintenance as the manufacturing's logistic function treats manufacturing to take part to influence the maintenance decision making [11]. This level is corresponded to some factors of manufacturing which have high interaction with maintenance. Its modules consist of the manufacturing equipment, manufacturing process, manufacturing layout, product lead time, and quality. The chosen specifications of those aspects then will be sent to maintenance function to assist describing the expected maintenance performance required to support manufacturing function.

3.4. Level 3, 4 and 5 – Maintenance Perspective
The operational stage totally reflects the elements of maintenance system, which are *Maintenance Activities*, *Maintenance Resources*, and *Maintenance Rules*. It is started from identifying the required maintenance activities to keep manufacturing equipment's reliability in acceptable business recommendation. To execute maintenance activities, the maintenance resources are obviously required. Hayes et al. [12] emphasize trade-offs as an option to deal with resources limitation and resources allocation in respect to business strategy. Furthermore, the discussion about how those maintenance resources are collaborated and managed in a particular manner to gain the optimal result should also be taken into account. Thus, after the hardware of resources to execute the maintenance actions, maintenance rules are required as the software to manage and control it. These infrastructure elements cannot be separated from the structural and physical elements to gain successfulness of maintenance strategy and operation [2]. It examines the required rules and control facilities to manage maintenance activities and maintenance resources.

Maintenance perspectives on the operational level are not only describing the current performance of maintenance, but intended to specify the best combination of maintenance activities, resources and rules to fulfil the demand of manufacturing function and to achieve the aim of maintenance as a driver of business competitiveness.

4. Methodology of Knowledge Based System for Integrated Maintenance Strategy and Operation (KBIMSO)

Regarding the contributed elements in maintenance strategy and operation, the challenge to take maintenance decision can be figured out based on three points. First, determine the targets and prerequisite conditions to be reached, and then identify the gaps between the current condition and the prerequisite condition. Second, decide priority of contributed elements to achieve the optimal result as there are many elements which should be considered at the same time. Third, enable the computer programme to support the complicated calculation and decision making process through Knowledge Based (KB) system.

4.1. Gauging Absences of Prerequisites (GAP)
GAP is a benchmarking technique used to identify and assess the gap between a prerequisite state and an existing state in a particular system [13]. It analyses the responses of users toward some questions provided regarding the existing condition of organization. The responses are divided into two categories. The availability of prerequisite condition is identified as Good Point (GP). Instead, the absence of any prerequisite condition is described hierarchically from the most important one, so called Problem Category (PC-1) to the less important of that does not have any effect at all to the system sustainability, so called PC-5 [14].

4.2. Analytic Hierarchy Process (AHP)
AHP is performed by breaking down the problem, focussing on the goal, identifying the contributed criteria and structuring them into a hierarchy [15]. The process is continued by comparing criteria with pair-wise comparison manner. The weight of each criterion represents the level of importance of

them in respect to the aim of decision making. In KBIMSO, level of importance of criteria is referred to GAP analysis as the preliminary step.

Although AHP plays a satisfying method to tackle multi criteria problems, it is considered that the calculation of KBIMSO might be tedious in practices. For this reason, the collaboration of GAP and AHP in KB system surely will provide the valid and consistent recommendation within a shorter time.

4.3. Knowledge Based (KB) System

Decision making is usually taken by some experts within the organisation who acquire knowledge on that particular area. Unfortunately, such knowledge might be disappeared when they leave the organization [16]. Therefore, the computerized system of KB system is intended to imitate the thinking process of expert to generate the reasonable conclusion. KB system consists of facts and rules. Expressing facts in rules could be done on a simple way called production rules by using IF, AND, THEN, and OR statements.

To provide the illustration of KB rules in KBIMSO, the example of KB rules for sub-module *Repair* in module *Maintenance Activities* is presented as the following: (only some rules are presented)

IF *the equipment is repaired after its expected life time (Yes: GP; No: PC-1)*
AND *the equipment is repaired before the functional failure (Yes: GP; No: PC-1)*
AND *the deterioration of equipment has been detected (Yes: GP; No: PC-1)*
AND *the equipment is repaired due to run-to-failure approach (Yes: GP; No: PC-1)*
AND *the corrective action is done on the planned maintenance schedule (Yes: GP; No: PC-1)*
AND *the corrective action is done by the planned maintenance resources (Yes: GP; No: PC-1)*
AND *the corrective action is preferable than replacement (Yes: GP; No: PC-1)*
THEN *the maintenance activity of repair is applied effectively*
OR *the maintenance activity of repair is not applied effectively*

The KB rules above indicate the prerequisite condition of effective maintenance activity of repair. The fulfilment of such statements mean the corrective action is natural, under schedule and can be managed appropriately. Therefore, any positive response is considered as Good Point (GP). Instead, the negative response is classified as Problem Category (PC)-1 based on GAP description, which imply a very serious problem to overall system performance. The rest of KBIMSO modules and sub-modules will have the same pattern as KB rules above.

4.4. Typical Output of KBIMSO

The KB rules are presented to the users to get the responses of the most representative current state. The results obtained are then tabulated, as shown in Table 1, to find out the aspects that have been implemented on its good practice already and the aspects that have not reached its prerequisite condition, namely Problem Category (PC).

Table 1 Typical output of KBIMSO for *Maintenance Activities* module

Sub-Module	Dimension	Number of Questions	Good Point	Problem Category (PC)				
				PC-1	PC-2	PC-3	PC-4	PC-5
Retain	Work Planning	38	21	17	0	0	0	0
	Work Scheduling	21	13	8	0	0	0	0
	Resources Availability	25	16	5	4	0	0	0
Repair	Corrective Procedures	24	15	9	0	0	0	0
	Economic Analysis	26	19	5	2	0	0	0
	Risk and Safety Analysis	12	9	3	0	0	0	0
Total		146	93	47	6	0	0	0

As illustrated in Table 1, there are only two sub-modules presented in *Maintenance Perspective* module out of four sub-modules, for sake of brevity. The total number of rules contained in both sub-modules is 146, with 93 rules shows good practices (Good Point) and 53 rules identify problems.

From those problem, there are 47 problems are categorised as PC-1 and 6 problems are categorised as PC-2 based on GAP description. Later on, all identified problems will be weighted within AHP technique to get prioritisation for improvement of maintenance strategy and operation.

5. Conclusion

KBIMSO is developed to synchronize maintenance decision with business and manufacturing concern. KBIMSO conceptual design elaborates KBIMSO framework onto two main stages, strategic and operational. Strategic stage discusses the contributed elements of business and manufacturing perspectives toward maintenance, while operational stage focuses on maintenance elements to execute maintenance strategy and operation. As the methodology, KBIMSO employs GAP and AHP into the KB system. GAP works by identifying the gap between the existing and expected condition within maintenance environment. Then AHP works by structuring and weighting the contributed elements in order to provide recommendation for maintenance decision.

The verification and validation of this KBIMSO conceptual design will be done in the automotive industries due to its high automation equipments, advanced technology implementation, and flexibility. Nevertheless, with its generic design, KBIMSO can be applied on most manufacturing industries to review current maintenance practice as well as to give recommendation for optimal maintenance strategy and operation.

Acknowledgement

Support for this study is provided by the Directorate of Higher Education, Ministry of National Education, Republic of Indonesia and the University of Bradford, the United Kingdom.

References

[1] Skinner W 1969 *Harvard Business Review* Manufacturing–Missing Link in Corporate Strategy May ed
[2] Pinjala S K., Pintelon L and Vereecke A 2006 *International Journal of Production Economics* **104** 214-229
[3] Bevilacqua M and Braglia M 2000 *Reliability Engineering & System Safety* **70** 71-83
[4] Parida A and Kumar U 2006 *Journal of Quality in Maintenance Engineering* **12** 239-251
[5] BSI Standards Publication 2010 *BS EN 13306:2010* Maintenance Terminology (United Kingdom: BSI)
[6] Dhillon B S 2002 *Engineering Maintenance: A Modern Approach* (Boca Raton: CRC Press)
[7] Milana, Khan M K and Munive J E 2013 *International Conference on Advances in Mechanical and Manufacturing Engineering (ICAM2E)* (Kuala Lumpur)
[8] Swanson L 2001 *International Journal of Production Economics* **70** 237-244
[9] Gulati R 2013 *Maintenance and Reliability Best Practices* (New York: Industrial Press)
[10] Kaplan R S and Norton D P 1996 *The Balanced Scorecard: Translating Strategy into Action* (Boston: Harvard Business School Press)
[11] Simões J M, Gomes C F and Yasin M M 2011 *Journal of Quality in Maintenance Engineering* **17** 116-137
[12] Hayes R, Pisano G, Upton D and Wheelwright S 2005 *Operations, Strategy, and Technology: Pursuing the Competitive Edge (*New Jersey: John Wiley & Sons)
[13] Nawawi M K M, Khan M K, Hussain K 2008 *Journal of KONBiN* **5** 145-156
[14] Khan M K and Wibisono D 2008 *Business Process Management Journal* **14** 129-146
[15] Saaty T L 2008 *International Journal of Services Sciences* **1** 83-98
[16] Turban E, Volonino L, McLean E and Wetherbe J 2008 *Information Technology for Management : Transforming organizations in the digital economy* (New York: Wiley)

IOP Conf. Series: Materials Science and Engineering **65** (2014) 012036 doi:10.1088/1757-899X/65/1/012036

Latest German-English Terminology in CADCAM & Robotics

S M Hussain
Dy. General Manager, HRD, Bharat Heavy Electricals Ltd., Hyderabad, India

E-mail: katzmann1@yahoo.de

Abstract: As a technical translator who speaks German fluently, the author would like to study and display the latest bilingual technical terminology in German being used in hardware, software and process technology involved in CADCAM & ROBOTICS. This will greatly help technical translators to correctly translate difficult texts from this area of technology.

AAL Roboter= Ambient Assisted Living Robot
Absolutgenauigkeit=absolute accuracy
abstrahieren=to abstract
Achsbewegung=movement of axis
Achsbezeichnung=name of axis
Achse, lineare=linear axis
Achse, rotatorische =rotational axis
Achse, translatorische =translational axis
Achse, zwangsgekoppelte =force-coupled axis
Achse=axis
Achsenanordnung=axis orientation
Achsenanzahl=number of axes
Achsposition=axis position
Achssteuerung=axis control
Adapterplatte, montierbare=mounting adapter plate (MAP)
Agieren=reaction
Amphibienroboter=amphibious robot
Anbauflansch=mounting flange
Angriffspunkt des Effektors= Tool Center Point (TCP)
Anhänger, programmierbarer=teach pendant
Annäherungssensor= proximity sensor
Anschlusskupplung=utility couple
Ansteuerung=control
Antrieb, harmonischer=harmonic drve
Antriebsteuerung=drive control
Anwesenheitssensor=presence sensor
Arbeitsbereich, maximaler=maximum envelope
Arbeitsraum, betrieblicher=operating envelop
Arbeitsraum, eingeschränkter=restricted envelop
Arbeitsraum, maximaler=maximum envelop
Arbeitsraum=work envelop
Arbeitszelle=work cell
Arm, gefalteter=folded arm
Arm, gestreckter=outstretched arm
Armlänge=arm length

Assistenzroboter=secondary robot
Astmessbereich=branch measuring range
Aufbauroboter=assembly robot
Aufspüren=recognition
Aufstellungsort= place of erection
Ausbildungsroboter=teaching robot
ausgewuchtet, dynamisch=dynamically balanced
Ausgleicher=compensator
Ausleger, mikrorobotischer= micro robotic arm
Ausleger=arm
Ausrichtung, kolineare=collinear alignment
Ausrichtung, kontrollierte=controlled orientation
Aussenquellen= external libraries
Autonomie, niedrige= low level autonomy
Bahngenauigkeit=track accuracy
Bahnkurve=trajectory
Basis-KOS=base coordinate system
Bearbeitungsaufgabe=processing job
bedienergeführt=operator guided
Benehmen, selbstständiges= proactive behaviour
Berührung=contact
Berührungssensor== tactile sensor
Beschleunigungsniveau=acceleration level
Betätigungselement=actuator head
Betrieb, kontinuierlicher, bemannter=attended continuous operation
Betrieb, kontinuierlicher, mit Überwachung=attended continuous operation
Betrieb, vollautomatischer=fully automatic operation
Bewegung, multimodale=multimodal locomotion
Bewegung, autonome = autonomous movement
Bewegung, koordinierte, geradlinige=coordinated straight line motion
Bewegungsablauf=movement sequence
Bewegungsachse=movement axis
Bewegungseinschränkung=limiting device
Bewegungsfreiheitsgrad=degree of movement allowed
Bewegungsmöglichkeit=possibility of movement
Bewegungsplanung== motion planning
Bezugskoordinatensystem=reference coordinate system
Bezugs-KOS=reference coordinate system
Bockkranroboter= gantry robot
Bodenfläche=ground or base
Bodenkarte= terrain height map, relief map
Bodenreaktionskontrolle= ground reaction force control
Bodenstörung= terrain perturbation
Bündigfräsen= flush milling
Chirurgie, robotergestützte= robotic surgery
Code-Einsatz=code deployment
Datenerfassungseinrichtung=data acquisition device
Datenträger=data carrier
Digitalwertumwandlung=quantization

Drehachse=rotary axis
Drehbewegung=rotational movement
Drehdurchführung=rotary penetration passes
Drehgelenk= rotational joint
Drehmoment=torque
Drehmomentsensor=torque sensor
Drehschubgelenk=rotational displacement joint
Dreick schiefwinklige=acute angled triangle
Druckerhaltungsventil=pressure retention valve
Drucksensor=pressure sensor
Durchführbarkeit=operability
Effektorstellung=effector position
Eigengewicht=intrinsic weight
Eigenträgheit=intrinsic inertia
Einflusskoeffizient, kinematischer=kinematic influence coefficient
Einhaltung=compliance
Einheitsvektor=unity vector
Einstellparameter= setting parameters
Einzelachse=individual axis
Einzelpunktkontrolle=single point control
Elektrospindel= high speed cutting spindle
Endeffektor=end effector
Entgraten, robotgestütztes=robotic deburring
entschärfen=attenuate
Erkennungsfähigkeit= cognitive capability
Erkundungsroboter=reconnaissance robot
Erweiterbarkeit=expandability
Exzentrizität=eccentricity
Fahrroboter=traversing robot
Feinfräsen= fine milling
Feinjustieren=fine adjustment/fine tuning
Feldroboter= field robots
Fensterreinigungsroboter=window cleaning robot
Fernbedienung= teleoperation, remote operation
ferngesteuert=remote controlled
Fertigungsroboter=manufacturing robot
Fingerwechsel=change of finger
Flachenfräsen= face milling
fluchtend=in line
Folgeprogrammierung=teach-in process during online programming
Fortbewegungsmittel=traversing agent
Freiheitsgrad=degree of freedom
Führung, schrittweise=lead-by-the-nose guidance
Gangart, verstellbare=modifiable gait
Gelenkarmroboter=articulated arm robot
Gelenkbauform=joint design
Gelenkregulierung=joint regulation
Gelenkvektoren=joint vectors
Gelenkverschiebung=joint displacement

IOP Conf. Series: Materials Science and Engineering **65** (2014) 012036 doi:10.1088/1757-899X/65/1/012036

Gelenkwinkel=joint angle
Genauigkeitsanspruch=accuracy requirement
Genkotsu-Roboter=high speed pick-and-place robots
Geschwindigkeitsphase, konstante = constant velocity phase
Gestell, mechanisches=mechanical frame
Getriebe=drives, gears
Getriebefreiheitsgrad=degree of freedom of drives
Getriebespiel=drive tolerance
gewichtsoptimiert=weight-optimized
Gieren= yaw
Graph der Operationsfolge=Operation Precedence Graph
Greifbefehl=grip command
Greifer=gripper
Greiferwechsel=change of gripper
Greifkraft=gripping force
Greifkrafterhaltung=retention of gripping force
Greifradius=gripping radius
Greifsystem=gripping system
Greifzeit=gripping time
Groβhubgreifer=large stroke gripper
Grundstellung=base position
Gummilochgreifer=rubber hole gripper
Gummimetallunterlage=shear pad
Haltekraft=holding force
Handachse=hand axis
Handhabungsaufgabe=job for execution
Handhabungsgerät mit konstantem Bewegungsmuster=fixed sequence robot
Handhabungsgerät mit programmierbarem Bewegungsablauf=playback robot
Handhabungsgerät mit variablem Bewegungsablauf=variable sequence robot
Handhabungsgerät, bedienergeführtes=manual manipulator
Handhabungsgerät=manipulator
Handhabungsroboter=service robot
Hauptachse=main axis
Hautsensoren, multimodale= multi-modal skin sensors
Hexapodroboter=hexapod robot
Hindernisse, unvorhersehbare=unpredictable obstacles
Hinterfräsen= relief milling
Höhengleichheit=uniformity in height
Höhenstabilisation= height stabilization
Industrieroboter=industrial robot
Intelligenz, künstliche=artificial intelligence
interagieren=interaction
Interner Sensor=internal sensor
Kalibrierung=calibration
Kampfroboter=battlefield robot, robots used in boxing rings
Kardangelenk=universal joint
kartesisch=cartesian
Kette, kinematische=kinematic chain
Kinematik, direkte=direct kinematics

Kinematik, inverse=inverse kinematics
KK-Roboter=Cartesian coordinate robot
Kleinroboter=small robots, tiny robots
Kleinteilgreifer=small parts gripper
Kollisionserkennung=collision sensing
Kollisionsschutz=collision protection
Kollisionssensor=collision sensor
Kollisionsüberprüfung, rapide= fast collision checking
Kollisionsvermeidung=collision avoidance
Koordinate, homogene =homogeneous coordinates
Koordinatensystem (KOS)=coordinate system
Koordinatensystem, absolute= absolute coordinate system
Kopfachse=head axis
Kraft und Drehmoment=force and torque
Kraftvektor=force vector
Kriechkontrolle=creep control
Kugelgelenk=spherical joint
Kugelumlaufspindel=ball screw
Kurzschluss=short-circuit
Kybernetik=cybernetics
Last, multiaxiale=compound load
Laufkontrolle, zuverlässige=robust walking control
Laufroboter=walking robot
Laufstabilisation= walking stabilization
Laufzeit-Architektur=runtime architecture
Leerlaufzeit=idle time
Leichtbauroboter=lightweight robot
Leichtgewicht=lightweight
lernfähig=capable of learning
Lichtbogenschweißroboter=arc welding robot
Linearachse=linear axis
Linearführung=linear guide
Lösungsansatz=set of solutions
Medizinroboter=medical robot
Messfahrt=measurement run
Messwertgeber=transducer
Metastabilität=metastability
Meβunsicherheit=measurement uncertainty
Mikroprozessor, eingebetteter=embedded microprocessor
Militärroboter=military robot
Mittelwaren= middleware
Mustergenerierung, wiederholende=repetitive pattern generation
Nachteachen=re-teach
Näherungsschalter=proximity switch
Nanoroboter=nano robot
Nebenachse=side axis
Nicken= pitch
Not-Aus=emergency off
Nutfräsen== slot milling

Nutzlast=useful load
Offline=Programmierung=programming of a robot in the shut-down condition
Online-Schrittplanung= online footstep planning
Operationsfolge= operation sequence
Orientierung=orientation
ortsbeweglich=moveable
ortsfest=fixed
Ovalkolbenantrieb=oval piston drive
Palettenbeladung=loading of palette
Palletieraufgabe=palletizing job
Parallelität=parallelity
Peripherie=periphery
Planfräsen= face milling
Portalroboter= gantry robot
Pose= position and orientation
Position, befohlene=commanded position
Position, eingegebene=taught position
Programmverifizierung, mit Überwachung=attended program verification
Programverifizierung, bemannter=attended program verification
Prozessoren, eingebettete= embedded processors
PUMA (Programmierbares Universales Montagegerät)=Programmable Universal Machine for Assembly
Quelle=point of origin
Rasenmähroboter=lawn mower robot
Raumkoordinate=spatial coordinate
Reinraumroboter=clean room robot
rekursiv=recursive
Rettungsrobotik= rescue robotics
Roboter mit Erkenntnisvermögen=cognitive robot
Roboter, autonomer =autonomous robot
Roboter, chirurgische= surgical robots
Roboter, ferngesteuerter= guided robot, remote-controlled robot
Roboter, festgeschraubter=fixed robot
Roboter, fliegende= flying robots
Roboter, fliegender=aerobot
Roboter, kugelförmiger=spherical robot
Roboter, medizinische =medical robots
Roboter, menschenähnlicher=humanoid
Roboter, mikro-fliegende, sichtbar kontrollierte= vision-controlled micro-aerial robots
Roboter, NC-gesteuerter =numerically controlled robot
Roboter=robot,
Roboterbeschickungssystem=robotic loading system
Roboterbewegung=robot movement
Roboterchirurgie=robot-assisted surgery
Roboterflansch=robot flange
Roboterfuss=robot base
roboterhaft=robot-like
Roboterhandgelenk= robot wrist
Roboterkalibrierung=robot calibration

Roboterkinematik=robot kinematics
Roboter-Koordinatensystem=robot coordinate system
Roboterkupplung=robot coupling
Robotermensch=cyborg
Roboterregulierung=robot regulation
Roboterschleifring=robotic rotary joint
Robotersteuerung=robot control
Roboterumgebung=robot ambience
Roboterwechselsystem=robot interchange system
Roboterzelle=robot cell
Roboterzubehör=robot accessories
Robotformat=robot format
robotisieren=robotize
Rotationsgelenk=rotational joint
Rückmeldung, sensorbasierte=sensory feedback
Rückschalthysteresis=resetting hysteresis
SCARA (Montageroboter mit selektiv elastischem Arm) = Selective Compliance Assembly Robot Arm
Schaltpunkt=switching point
Schlangenroboter=snake robots
Schlichtfräsen= finish milling
Schnellwechselfinger=quick change fingers
Schnipsen= wrist flip
Schnittstelle=interface
Schnittstellenstromversorgungskasten=Interface/power supply box (IFPS)
Schöpffräsen= scoop milling
Schrittmotor=stepper motor
Schruppfräsen== rough milling
Schubachse=displacement axis
Schubgelenk=displacement joint
Schweißzange=weld grippers
Schweiβroboter=welding robot
schwenkbar=swivel capability
Schwerkraft=gravity
Schwerlastgreifer=heavy load gripper
Schwerpunkt=center of gravity
Sehkraft, maschinelle=machine vision
Sensor, interner=control sensor
Sensoren, verkleinerte= miniaturized sensors
sensorgeführt=sensor guided
Serviceroboter=service robot
Servoantrieb=servo drive
Servoregelung=servo regulation
Servosteuerung=servo control
sich wiederholend= iterative
Sicherheitsbewegung=safe motion
skalierbar=scalable
Spannkraftverlust=loss of clamping force
Spanntechnik=clamping technology

IOP Conf. Series: Materials Science and Engineering **65** (2014) 012036 doi:10.1088/1757-899X/65/1/012036

Spielzeugroboter=toy robot
sprengen=explode
Spritzpistole=spray gun
Sprühwasserschutz65=Ingress Protection Rating 65 (IP65)
Stampfen= pitch
Staubsaugerroboter=vacuum cleaning robot
Staubschutz60=Ingress Protection Rating 60 (IP60)
Steifigkeit=rigidity
Stellantrieb, hydraulischer=hydraulic actuator
Stellantrieb= actuator
Stellungsbezeichnung=positional data
Stellungskommando=positional command
Steuerung=control
Steuerungsaufgabe=control task
Stirnfräsen= face milling
Störkantenradius=interference radius
Störkonturoptimierung=optimization of contour overshoot
Stoßsensor=collision sensor
Strahlungstransportation=beam transportation
Streuung=scatter
Tauchroboter=submersible robot
Tauchsicherheit68=Ingress Protection Rating 68 (IP68) up to 10 m depth in fresh water
Tischroboter=bench-top robot
Tool-KOS=tool coordinate system
Trägheitskraft=inertia
Traglast=payload
Transformation, kartesische= Cartesian transformation
Translationsgelenk=translational joint
translatorisch=translational
Transportroboter=transport robot
Überlast, maximale einzelachsige=maximum single-axis overload
Überlastschutz=overload protection
Überlastsensor=overload sensor
Übersetzungsbewegung=translational movement
Umrissfräsen= contour milling
Umwelterfassung=ambience recognition
Umweltsvertretung, wirksame= efficient environment representation
Verbindungswinkel=joint angle
Verdrehung=distortiion
Verfahrachse=traversing axis
Verfahrweg= traversing path
Vertikalhub=vertical stroke
Wahrnehmung, dreidimensionale=3-D perception
Weltkoordinaten=base coordinates
Werkstückgewicht=weight of the work-piece
Werkstück-KOS=work-piece coordinate system
Werkzeugarbeitspunkt= tool center point (TCP)
Werkzeugbereitstellung=tooling
Werkzeug-Koordinatensystem=tool coordinate system

Werkzeugwechsler=robotic tool changer
Wiederholgenauigkeit=repeat accuracy
Winkelgreifer=angular gripper
Wirkobjekt=active object
Zahlenwert, dimensionsloser=dimensionless numerical value
Zentrischgreifer=centric gripper
Zielframe=final frame of a movement in position and orientation
Zugänglichkeit=accessibility
Zuladung=loading
Zuordnung, räumliche=spatial allocation
Zustimmung, wahlbasierte= selective compliance
Zwangskontrolle=force control
Zwillingsroboter=twin robot
Zykluszeit=cycle time

Badrih Al - Faifi
Maryam AL-Shammary
Zinab AL-Shehry

6-Axis Force/Moment Sensor In Humanoid Robot Foot

Al - Imam Muhammad Ibn Saud Islamic University

ABSTRACT

The foot is the most Important part of the humanoid .Thought the sensor of the robot can understand the environment In which they live , it is important to developed an intelligent foot . In order to walk on uneven terrain safely this poster describes an intelligent foot with 6- axis force/moment sensors for humanoid robot that is one of the solution that can help the robot to walk in uneven terrain safely .

Introduction

Scientist and engineers have been researching to develop a humanoid robot with the function of human being . It is important to provide an intelligent foot in a humanoid robot . To obtain this goal . The foot should measure three forces , three moments and control itself using their forces and moments . The 6-axis force/moment sensor is composed of three force sensors ((Fx , Fy , Fz) sensor) and three moment sensors ((Mx , My , Mz) sensor) in a body to detect the forces and the moment . The 6-axis force/moment sensor should be designed and fabricated by considering robot's weight so the sensors could accurately measure three forces and three moments or the humanoid robot will not be able to walk on uneven terrain safely . The 6-axis force/moment sensor for the intelligent foot of a humanoid robot was developed . The structure of the sensor was modeled using many parallel plate beams (PPBs) the sensing element was designed using finite element method (FEM) and the sensor was manufactured by attaching a lot of strain gages .

6-axis force/moment sensor

1- Design of 6-axis force/moment sensor

the structure of sensing element for 6-axis force/moment sensor. The sensor is composed of the force/moment transmitting block, the fixing block F1–4, the moving block M1–4, the parallel plate-beam PPB1–16.The sensing elements for perceiving three forces and three moments are the parallel plate-beams, PPB1–4 and PPB13–16. PPB1 and PPB2 are for perceiving moment *My* PPB3 and PPB4 are for force *Fz* and moment *Mx*, PPB13 and PPB14 are for force *Fy* and moment *Mz*, PPB15 and PPB16 are for force *Fx*. The size of the plate-beams composed of PPB1–4 is width $b1$, thickness $t1$, length $l1$, that for PPB5–12 is width $b2$, thickness $t2$, length $l2$, that for PPB13–16 is width $b3$, thickness $t3$, length $l3$. The part variables for designing the sensor are width $b1$, $b2$, $b3$, thickness $t1$, $t2$, $t3$, length $l1$, $l2$, $l3$ of plate-beams. The finite element method (FEM) is used to determine the size of the variables . Fig.2

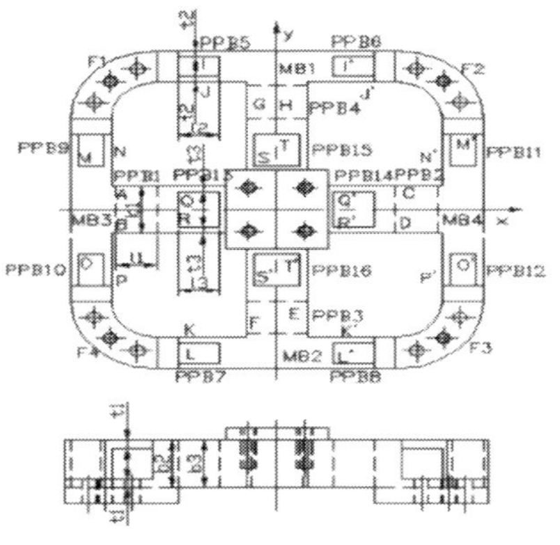

2- Variables for designing the sensor

The design variables of the modeled 6-axis force/moment sensor are the size of the body Wheastone bridges are made for each sensor in 6-axis force/moment sensor [5–10]. Wheastone bridge can be calculated as following equation [5–10]: $\varepsilon = \varepsilon T1 - \varepsilon C1 + \varepsilon T2 - \varepsilon C2$ Eq(1).
where ε is the total strain from Wheastone bridge , $\varepsilon T1$ the strain of a tension strain-gage T1 , $\varepsilon C1$ the strain of a compression strain-gage C1 ,$\varepsilon T2$ the strain of a tension strain-gage T2 *and* $\varepsilon C2$ is the strain of a compression strain-gage C2 . The rated output can be calculated as the equation [5–10]: $Eo / Ei = (1\backslash 4)\ k\,\varepsilon$ Eq(2).
where Ei is the input voltage of Wheastone bridge, Eo the output voltage of Wheastone bridge,
K the factor of strain-gage (about 2.0), ε is the total strain gotten from Eq. (2).

3 -Sensor design using FEM

Table 1 shows the sizes of plate-beams (sensing elements) of two 6-axis force/moment sensor determined by FEM analysis. Table 1

Sensor	Length (mm)			Thickness (mm)			Width (mm)		
	l_1	l_2	l_3	t_1	t_2	t_3	b_1	b_2	b_3
1	8	8	8	1.3	1	1.3	12	12	12
2	8	8	8	2.8	1.5	1.8	12	12	12

Manufacture of 6-axis force/moment sensor

6-axis force/moment sensor with different rated forces *Fx, Fy, Fz* and rated moments 6-axis force/moment sensor with different rated forces *Fx, Fy, Fz* and rated moments *Mx, My, Mz* respectively, is fabricated. The attaching locations for each sensor in 6-axis force/moment sensor were determined in consideration of the results of FEM analysis, and those are shown in Fig.2were determined in consideration of the interference error0%calculated by Eq. (1). The attaching locations of strain-gages for *Fx* sensor are S1–S4, those for *Fy* sensor are S5–S8, those for *Fz* sensor are S9–S12, those for *Mx* sensor are S13–S16, those for *My* sensor are S17–S20, and those for *Mz* sensor are S21–S24. The attaching location of each strain-gage is 1mm from the end of plate beam in the length-direction, and that in the width-direction is the center line of plate-beam in consideration of the grid size of strain-gage (1.2mm × 1.6 mm). These locations were determined in consideration of the interference error0%calculated by Eq. (1). Fig.2

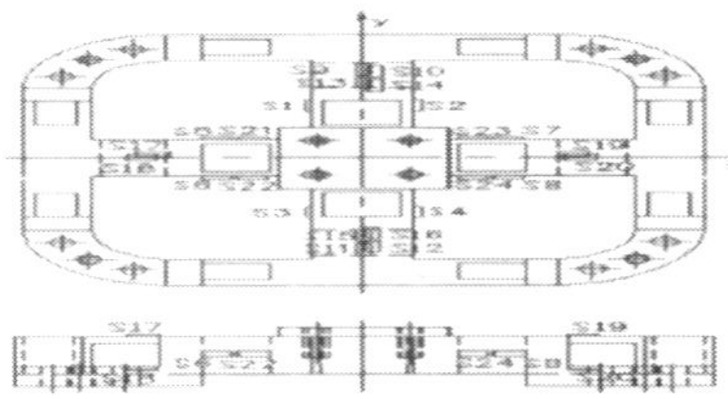

Table 2 shows rated strains from FEM analysis at each attachment location of strain-gages of each sensor in 6-axis force/moment sensor 1 Table 2

Sensor	FEM analysis strain (μm/m)					Reference (μm/m)	Error (%)
	T1	C1	T2	C2	ε		
F_x	236	−236	236	−236	944	1000	−5.6
F_y	225	−225	225	−225	944	1000	−5.6
F_z	277	−277	277	−277	1108	1000	10.8
M_x	274	274	274	274	1096	1000	9.6
M_y	274	−274	274	−274	1096	1000	9.6
M_z	261	261	261	261	1044	1000	4.4

Table 3 shows rated strains from FEM analysis at each attachment location of strain-gages of each sensor in 6 axis force/moment sensor 2 Table 3

Sensor	FEM analysis strain (μm/m)					Reference (μm/m)	Error (%)
	T1	C1	T2	C2	ε		
F_x	236	−236	236	−236	944	1000	−5.6
F_y	225	−225	225	−225	944	1000	−5.6
F_z	277	−277	277	−277	1108	1000	10.8
M_x	274	−274	274	−274	1096	1000	9.6
M_y	274	−274	274	−274	1096	1000	9.6
M_z	261	−261	261	−261	1044	1000	4.4

Conclusions

6-axis force/moment sensor is structure that use (PPBS) to perceive the applied forces and moments. and apply FEM analysis to it . so the robot can have the ability to walk on the uneven terrain safely .

References

1- Gab-Soon Kim\Hyi-Jun Shin^ and Jungwon Yoon^ Development of Six-axis Force/Moment Sensor for an Intelligent Foot of Humanoid Robot
2- Gab-Soon Kim. Development of 6-axis force/moment sensor for a humanoid robot's foot .

AUTHOR INDEX